Heavy Equipment Operations

Level Two

Trainee Guide
Second Edition

PEARSON

Prentice
Hall

Upper Saddle River, New Jersey
Columbus, Ohio

contren®
Learning Series

nccer

National Center for Construction Education and Research

President: Don Whyte
Director of Curriculum Revision and Development: Daniele Stacey
Heavy Equipment Operations Project Manager: Natalie Smith
Production Manager: Jessica Martin
Product Maintenance Supervisor: Debie Ness
Editors: Bethany Harvey and Brendan Coote
Desktop Publishers: Jennifer Jacobs and Jessica Martin

NCCER would like to acknowledge the contract service provider for this curriculum:
Topaz Publications, Liverpool, New York.

This information is general in nature and intended for training purposes only. Actual performance of activities described in this manual requires compliance with all applicable operating, service, maintenance, and safety procedures under the direction of qualified personnel. References in this manual to patented or proprietary devices do not constitute a recommendation of their use.

10 9 8 7
ISBN 0-13-227250-4

PREFACE

TO THE TRAINEE

Being skilled in heavy equipment operations will make you one of the most flexible and versatile workers in several industries. Equipment operators not only work on regular construction building jobs, but also on infrastructure projects (roads, bridges, and ports, otherwise called nonbuilding construction), and in mining and timber operations. As a trained and experienced equipment operator, you will be providing your skills to any project that requires moving and transporting heavy materials, or that demands any kind of earthmoving.

The steady growth in new construction nationwide will provide expanding opportunities for equipment operators through the next few years. There are slightly less than half a million people working as equipment operators today. Yet, as the country's population grows, so will the demands for homes, roads, public buildings, and industrial plants. The need for equipment operators will increase, as experienced operators retire and fewer new operators receive formal training. Those who are skilled in computerized operation of new, more technologically upgraded equipment will be in especially great demand. Highway, street, and bridge construction enjoys the highest wages and is experiencing the fastest growth right now (U.S. Bureau of Labor Statistics).

This is the second level of a three-level curriculum that meets the requirements of an operator apprenticeship program (3 years and 6,000 hours of on-the-job training). You will continue to train in grading work and gain in-depth training on five additional, common pieces of equipment. Modules on blueprint reading and excavation mathematics have also been incorporated into this training to aid in your professional development. This level is another step in mastering ten different pieces of equipment, practicing the principles of leadership, and gaining scientific knowledge of soils and topography. Good luck as you progress further in your construction career.

We invite you to visit the NCCER website at www.nccer.org for the latest releases, training information, newsletter, and much more. You can also reference the Contren® product catalog online at www.crafttraining.com. Your feedback is welcome. You may email your comments to curriculum@nccer.org or send general comments and inquiries to info@nccer.org.

CONTREN® LEARNING SERIES

The National Center for Construction Education and Research (NCCER) is a not-for-profit 501(c)(3) education foundation established in 1995 by the world's largest and most progressive construction companies and national construction associations. It was founded to address the severe workforce shortage facing the industry and to develop a standardized training process and curricula. Today, NCCER is supported by hundreds of leading construction and maintenance companies, manufacturers, and national associations. The Contren® Learning Series was developed by NCCER in partnership with Prentice Hall, the world's largest educational publisher.

Some features of NCCER's Contren® Learning Series are as follows:

- An industry-proven record of success
- Curricula developed by the industry for the industry
- National standardization, providing portability of learned job skills and educational credits
- Compliance with the Office of Apprenticeship requirements for related classroom training (*CFR 29:29*)
- Well-illustrated, up-to-date, and practical information

NCCER also maintains a National Registry that provides transcripts, certificates, and wallet cards to individuals who have successfully completed modules of NCCER's Contren® Learning Series. *Training programs must be delivered by an NCCER Accredited Training Sponsor in order to receive these credentials.*

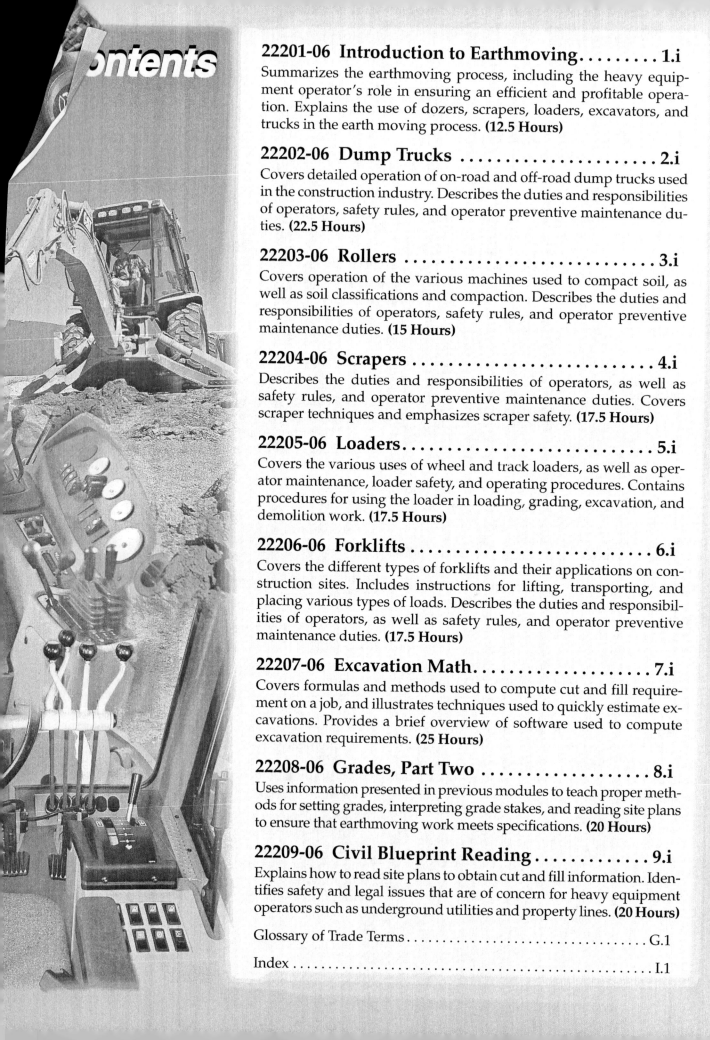

ontents

Contren® Curricula

NCCER's training programs comprise more than 40 construction, maintenance, and pipeline areas and include skills assessments, safety training, and management education.

Boilermaking
Carpentry
Carpentry, Residential
Cabinetmaking
Concrete Finishing
Construction Craft Laborer
Construction Technology
Core Curriculum: Introductory
    Craft Skills
Currículum Básico
Electrical
Electrical, Residential
Electrical Topics, Advanced
Electronic Systems Technician
Exploring Careers in Construction
Fundamentals of Mechanical and
    Electrical Mathematics
Heating, Ventilating, and Air
    Conditioning
Heavy Equipment Operations
Highway/Heavy Construction
Instrumentation
Insulating
Ironworking
Maintenance, Industrial
Masonry
Millwright
Mobile Crane Operations
Painting
Painting, Industrial
Pipefitting
Pipelayer
Plumbing
Reinforcing Ironwork
Rigging
Scaffolding
Sheet Metal
Site Layout
Sprinkler Fitting
Welding

Pipeline

Control Center Operations, Liquid
Corrosion Control
Electrical and Instrumentation
Field Operations, Liquid
Field Operations, Gas
Maintenance
Mechanical

Safety

Field Safety
Orientación de Seguridad
Safety Orientation
Safety Technology

Management

Introductory Skills for the Crew
    Leader
Project Management
Project Supervision

Acknowledgments

This curriculum was revised as a result of the farsightedness and leadership of the following sponsors:

Associated Training Services (WI)
Bay Ltd., A Berry Company
Bilingual Training Works
Ohio Valley Construction Education Foundation
Saiia Construction
Wilder Construction Company
Zachry Construction Corporation

This curriculum would not exist were it not for the dedication and unselfish energy of those volunteers who served on the Authoring Team. A sincere thanks is extended to the following:

Dan Barrow
Michael J. Eggenberger
Ron Kauffman
Erasmo Lopez
Rod Majors
Lynn McKinnie
Tom Piper
Marty Stevens
Jim Row

NCCER PARTNERING ASSOCIATIONS

American Fire Sprinkler Association
American Petroleum Institute
American Society for Training & Development
Associated Builders & Contractors, Inc.
Associated General Contractors of America
Association for Career and Technical Education
Carolinas AGC, Inc.
Carolinas Electrical Contractors Association
Construction Industry Institute
Construction Users Roundtable
Design-Build Institute of America
Electronic Systems Industry Consortium
Merit Contractors Association of Canada
Metal Building Manufacturers Association
National Association of Minority Contractors
National Association of State Supervisors for
   Trade and Industrial Education

National Association of Women in Construction
National Insulation Association
National Ready Mixed Concrete Association
National Systems Contractors Association
National Technical Honor Society
National Utility Contractors Association
North American Crane Bureau
North American Technician Excellence
Painting & Decorating Contractors of America
Portland Cement Association
SkillsUSA
Steel Erectors Association of America
Texas Gulf Coast Chapter ABC
U.S. Army Corps of Engineers
University of Florida
Women Construction Owners & Executives, USA

Heavy Equipment Operations Level Two

22201-06

Introduction to Earthmoving

22201-06
Introduction to Earthmoving

Topics to be presented in this module include:

Overview

Being a heavy equipment operator is more than just being able to safely operate various types of equipment. Collectively, digging, loading, hauling, and dumping soils and other materials is known as earthmoving. It is an important part of the construction process.

This module provides an overview of the earthmoving process, beginning with the planning stage. Knowledge of the overall process helps the operator become a skilled member of the team. An operator must know the terms and equipment used for various earthmoving tasks to be able to communicate with other members of the team.

Different types of equipment can be used to perform a specific task. A skilled operator can select the equipment best suited for the job conditions. Knowledge of the correct order of operations makes an operator a more productive member of the team.

Objectives

When you have completed this module, you will be able to do the following:

1. Identify and explain earthmoving terms and methods.
2. Describe how to safely set up and coordinate earthmoving operations.
3. Identify and explain earthmoving operations.
4. Identify and explain soil stabilization methods.
5. Identify the best equipment for performing a given earthmoving operation.
6. List in the correct order the steps involved in an earthmoving operation.

Trade Terms

Backhaul	Inorganic
Bedrock	Pay item
Center line	Pit
Cohesive	Organic
Consolidation	Riprap
Core sample	Select material
Cycle time	Shoring
Dewatering	Soil test
Discing	Spoils
Dragline	Stormwater
Embankment	Test boring
Expansive soil	Test pit
Gradation	Trench
Groundwater	Water table
Impervious	Windrow

Required Trainee Materials

1. Pencil and paper
2. Appropriate personal protective equipment

Prerequisites

Before you begin this module, it is recommended that you successfully complete *Core Curriculum*; and *Heavy Equipment Operations Level One*.

This course map shows all of the modules in the second level of the *Heavy Equipment Operations* curriculum. The suggested training order begins at the bottom and proceeds up. Skill levels increase as you advance on the course map. The local Training Program Sponsor may adjust the training order.

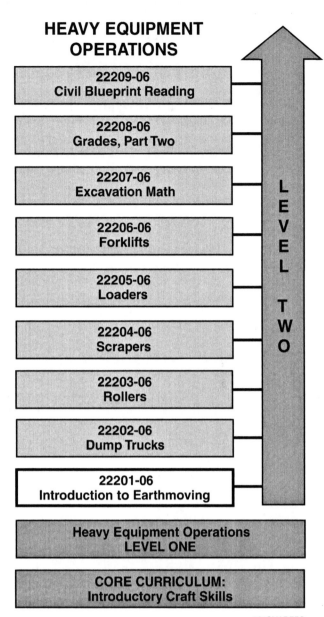

HEAVY EQUIPMENT OPERATIONS

22209-06
Civil Blueprint Reading

22208-06
Grades, Part Two

22207-06
Excavation Math

22206-06
Forklifts

22205-06
Loaders

22204-06
Scrapers

22203-06
Rollers

22202-06
Dump Trucks

22201-06
Introduction to Earthmoving

Heavy Equipment Operations
LEVEL ONE

CORE CURRICULUM:
Introductory Craft Skills

LEVEL TWO

201CMAP.EPS

1.0.0 ◆ INTRODUCTION

Earthmoving is the process of digging, loading, hauling, and dumping any material that is needed for construction or is in the way of construction. The process may also include digging and removing material from a **pit** or mine and processing the material. The primary process operations are soil or rock excavation, backfill, and **embankment** construction. Earthmoving may also include the transportation of materials, stabilization of embankments, and control of **groundwater**.

The word earth refers to any material in the earth's crust that is above **bedrock**. Bedrock is the earth's hard, **impervious** foundation, and includes soil and rock. Rock is a natural aggregate of minerals that are joined by strong and permanent **cohesive** forces. Soil includes **organic** and **inorganic** material, as well as large and small pieces of rock.

The methods of excavating and removing rock differ from those required for removing soil. They are also more expensive. Because of this cost, contractors must first define what makes up rock excavation, and then consider everything else to be soil excavation. The common practice is to call all excavation unclassified until the earthmoving contractor defines how much of the excavation is rock and how much is soil. This allows the contractor to determine what equipment is needed and thus the cost of the excavation.

2.0.0 ◆ MINING

Mining is a type of earthmoving work and is a form of excavation. Mining requires special equipment and methods, so it is described separately.

The term pit is used to describe any open excavation that is made to obtain material of value such as coal, mineral ore, select fill material, gravel, or quarry rock. See *Figure 1*. Usually, any pit operation involves not only the use of heavy equipment, but also blasting, processing, and drainage. The size of pits will vary from very small project pits, where **select material** is quarried for fill, to large open-pit mining operations such as those used for mining iron ore or bauxite. Quarries that yield construction materials such as sand, limestone, and granite are located all over the United States. Some are set up temporarily for the purpose of supplying material to a specific project, whereas others are set up as permanent business ventures.

The type of heavy equipment used for a pit operation will depend on the size of the pit and the material being excavated. For small pits, bulldozers, front-end loaders, and dump trucks are sufficient.

For large or more permanent operations, **draglines** like the one shown in *Figure 2* may be more economical. Even in the larger pit operations, there will still be a need for other larger specialized equipment to perform jobs such as heavy-duty hauling (*Figure 3*) and the construction of haul roads, drainage work, and stockpiling.

201F01.EPS

Figure 1 ◆ Pit mining.

201F02.EPS

Figure 2 ◆ Dragline.

201F03.EPS

Figure 3 ◆ Off-road dump truck.

3.0.0 ◆ EARTHMOVING OPERATIONS

Earthmoving consists of moving material, usually soil or rock, from one place to another. This can be a very short distance, as in the case of excavating a **trench**, or a longer distance, such as a cut and fill operation for construction of a highway.

3.1.0 Preliminary Activities

Several activities are required before any excavation and embankment work can be started. The type, quantity, and location of material on the site will determine what kind and how many pieces of equipment are needed to do the job. Each project has a set of grading plans that show what the site is to look like after the earthwork is completed. Survey crews will transfer the information on the plans to the actual building site using grade stakes so that workers will know what areas require excavation and fill.

Projects adjacent to public highways or other areas where people pass by require traffic and pedestrian control. The control must be put in place before construction begins. The project drawings and specifications usually show the required signs and layout.

3.1.1 Soils

The success and stability of any construction project depend upon the ground on which it is built. Most people give little thought to soil, but there is nothing more important in determining the suitability of a site for building. Site selection involves a series of complicated and rigorous tests performed by specially trained workers to determine the site's soil composition and properties.

There are many different types of soils, and each has its own unique characteristics. Soil composition varies by region, with most areas having a combination of two or more types. It is very rare to find a site that is composed of a single soil type. There are several classification systems for soil, which you will learn about in the *Soils* module in *Heavy Equipment Operations Level Three*. A general description of the different types of soils follows:

- *Gravel* – Any rock-like material greater than 0.125 inches (⅛ inch) in diameter. Larger particles are called cobbles or stones, and those larger than 10 inches are called boulders. Gravel occurs naturally or it can be made by crushing rock. Natural gravel is usually rounded from the effects of water, while crushed rock is usually angular.
- *Sand (coarse and fine)* – Mineral grains measuring 0.002 to 0.125 inches. Sand is made from grinding or decaying rock. It usually contains a high amount of quartz. It is called granular

material because it separates easily, giving it almost no cohesive strength. Coarse sand is frequently rounded like gravel and is often found mixed with gravel, but fine sand is usually more angular.

- *Inorganic silt* – Very fine sand with particles that are 0.002 inches or less. Silt is sand that has been ground very fine. It has a dusty appearance and powdery texture when dry, but sticks together when wet. Silt has almost no cohesive strength, so dried lumps are easily crushed. It has a tendency to absorb moisture by capillary action, which means that moisture wicks up through the soil, making it problematic in areas where the water table is shallow.
- *Clay* – The finest size of soil particles. Clay is very cohesive. When wet, clayey soils feel like putty and can be easily molded and rolled into ribbons. When dry, clay is very strong and clumps are difficult to crush. A small amount of clay in the soil at a building site makes it ideal for building, but clay is an **expansive soil**; that is, it swells and shrinks with moisture changes, so pure clay is not suitable for building.
- *Organic matter and colloids* – Partly decomposed vegetable and animal material. Organic matter is usually soft and fibrous and is odorous when warm. This material is not suitable for building or as fill because it decays and it loses volume, which causes air pockets, making the ground unstable. Colloidal clays are very fine particles that can be suspended in water and do not settle quickly. Individual particles cannot be seen with the naked eye. These materials are very susceptible to swelling and shrinking, so they are unacceptable for building.

With the exception of some small foundation excavation, small trenching, utility work, and landscaping, there will be a need for **soil testing** at a job site. For large projects, testing will be performed before any earthmoving work begins. From this information, the earthmoving contractor will have a good idea of what kinds of equipment are needed and how to plan the work.

On some jobs, it may be necessary to examine the soils or determine rock surfaces by digging a **test pit**. Although the pit will furnish the best information on the different layers of soil and rock, digging pits is slow and costly. They are usually dug only where special foundation conditions need investigation. **Test boring** is the preferred method where any depth is required. **Core samples** taken from the borings provide the needed information about the various soil types.

Site plans are developed based in part on the information gathered during soil testing, so it is important for equipment operators to understand

and accurately follow the project plans and specifications. When in doubt, you should consult your supervisor for guidance.

3.1.2 Review Plans and Specifications

Construction personnel need to understand what is included in project plans and specifications and how that information is transformed into directions for doing the work.

Most new construction work will require a single set of plans. Other work, such as major reconstruction or rehabilitation of roads and vertical structures, will have two sets of plans: the as-built plans documenting the existing roads or structures, and a set of plans for the new work. As-built plans can be very helpful in determining what kind of problems the operator might have in excavating, **dewatering**, or demolishing the existing site.

Project specifications are important documents that provide directions, provisions, and requirements for performing the work illustrated and described in the plans. The items in the specifications cover the methods of performing the work or describe the quality and quantity of materials and labor to be furnished under the contract. Addenda are issued as approved additions and revisions to the original specifications.

A number of other documents are normally incorporated by name into the specifications, including standards for material specifications, test procedures, safety regulations, and traffic control devices.

Plans and specifications should be accurate and complete and leave little room for assumptions or interpretation. They should also define specific responsibilities under the contract. If there is a discrepancy, the authority hierarchy is usually established in the contract documents. When in doubt, consult your supervisor.

A typical set of plans is made up of pages of drawings and notes that provide information about the dimensions, materials, and construction sequence of the project. Information in plans will vary depending on the type of project being built. Plans for horizontal construction projects, such as roads, canals, and airports, normally contain many drawings of cross section details and plan and profile sheets. For vertical construction such as warehouses, office buildings, and houses, the plan set will have foundation drawings and any special **shoring** details for excavating the foundation, as well as the normal building design drawings.

Figure 4 shows a typical cross section for a road construction job. This plan shows how the highway would look if it were cut in two from side to side. The distances shown are from the **center line** to the various outer points. If the grade is the same on both sides of the road, only one side of the cross section, with the distances shown, is required. If the grade is different throughout the section, then a total layout of the cross section should be shown from curb to curb or right-of-way limits. This occurs when there are many cuts and fills in rough terrain.

The plan sheet is usually drawn on the top half of the same sheet as the profile. It is an aerial view of the project. The stationing moves from left to right and should correspond to the profile stationing. The center line is shown as a solid line in the middle of the road and can be marked at the construction site with a center line stake. For a highway project, the plan sheet typically shows the following:

- The center line
- The edge of pavement
- The ditch line
- The right-of-way limits
- Other details, such as drainage improvements and structures

Figure 5 shows a typical plan sheet for highway construction.

Profiles are printed on grids to allow for quick reference to the elevation information. The profile sheet shows the elevation of the natural terrain and designated station points, as well as the proposed finish grade at the same points. The finish grade usually is the elevation at the center line for the stations shown at the bottom of the sheet. This value corresponds to the value on the cross section for that particular station.

Figure 6 shows a typical profile sheet for highway construction.

Survey control must be verified before the surveying and construction stakes can be placed. Once the control stakes and reference marks have been placed, the following layout stakes are placed:

- Right-of-way or limits of construction
- Reference stakes
- Slope stakes
- Center line stakes
- Ditch stakes

Figure 7 is an example of a site plan for part of a subdivision. It shows property line boundaries, existing roadways, and building foundations. In addition, it shows the existing contours of the ground in dashed lines, and the planned grades in solid lines.

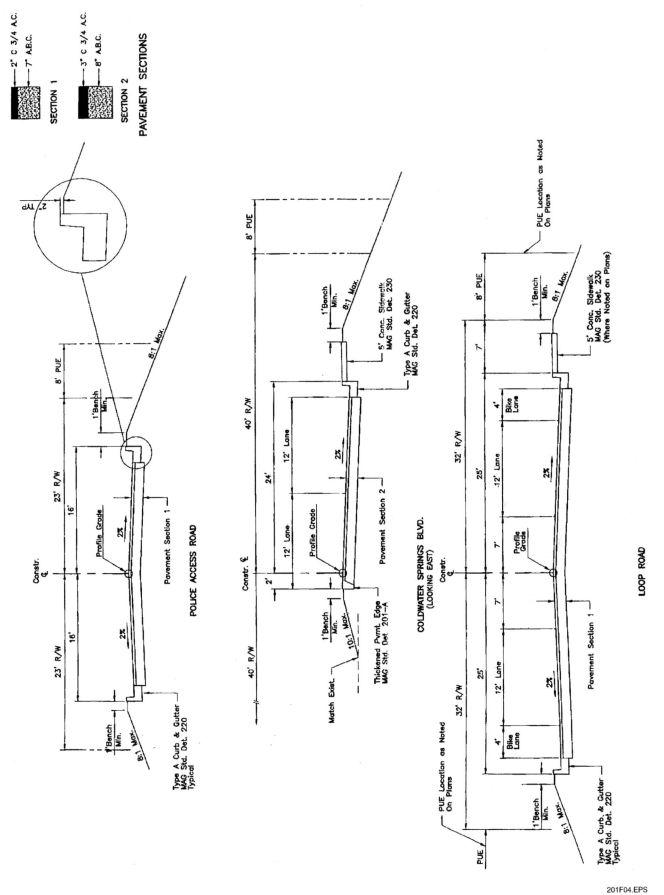

Figure 4 ◆ Typical cross section for a road construction job.

201F04.EPS

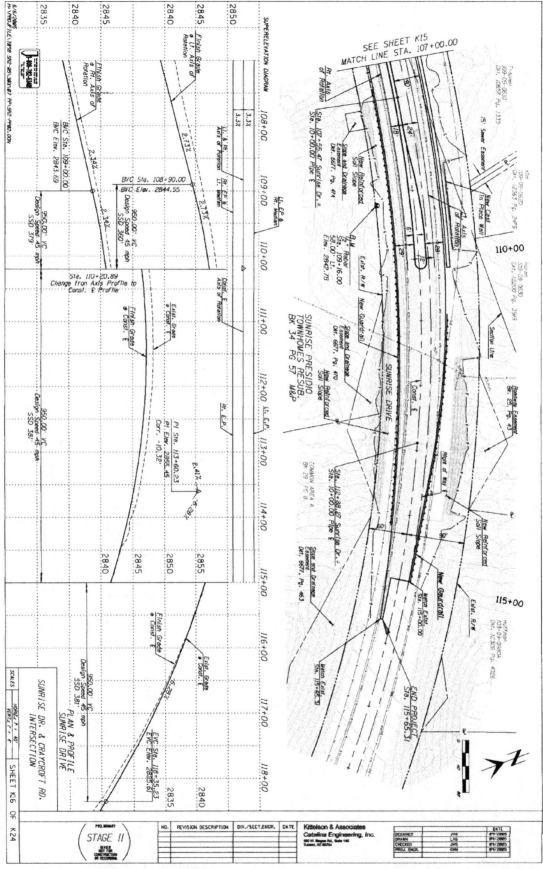

Figure 5 ◆ Typical plan sheet for highway construction.

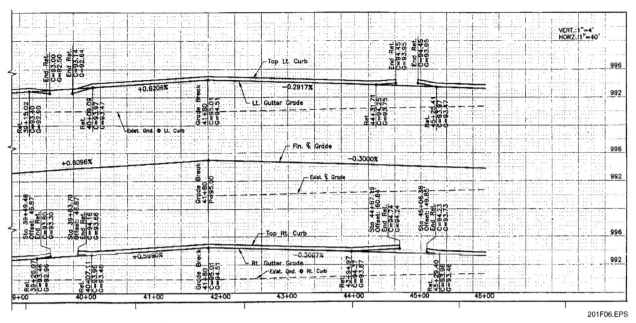

Figure 6 ◆ Typical profile sheet for highway construction.

3.2.0 Laying Out Slopes and Grades

Layout of construction sites will vary according to the kind of construction required, the terrain, and the location. This is normally done using wooden and metal stakes placed in the ground or spray paint and flagging attached to some solid object in the construction area.

Right-of-way stakes determine the limits of highway and road projects.

> **NOTE**
>
> The area beyond the stakes may be private property and should not be entered unless noted on the plans.

As you learned in the *Grades* module of *Heavy Equipment Operations Level One*, grade requirements for the site are written on the stakes set by the survey crew. These stakes tell workers about various cut and fill requirements. It is essential for all workers to understand the importance of stakes and to avoid damaging or removing stakes as work is performed. Resetting stakes is a costly effort that can delay the completion of a job and cause cost overruns.

3.2.1 Layout Control

Layout controls are established by accurately setting specific stakes to use as permanent reference points. These are called bench marks. In turn, the bench marks are used for placing other stakes, such as slope and grade control stakes. All of these points are shown on the site construction plans.

3.2.2 Slope Control

During excavation and fill activities, slope control is handled with slope stakes that are placed at the edge of the construction limits by the survey crew. These stakes are the principal guides for grading operations. It is important that they not be disturbed by construction equipment.

After rough grading work, slopes must be checked against the requirements written on the slope stakes. Although this is usually the responsibility of the contractor, the owner's representative will also check them to be sure the slope is built properly. Slopes should be checked when:

- The equipment operation is complete but before it is moved out of the area.
- There is an indication that the slope is flatter or steeper than indicated on the plans, even though the slope agrees with the stakes. The possibility of human error should be considered when checking the slopes.
- The equipment is finishing the backslope or inslope. These are checked as they are being built to prevent extra work later.

3.2.3 Grade Control

The grade is the rise and fall of the constructed surface along the horizontal plane. The basic grade requirements are shown on the profile sheets in the project plans. Slope stakes provide the primary information needed to control longitudinal grades for roadway projects. These normally show the vertical distance from the stakes to the subgrade shoulder points, along with cut and

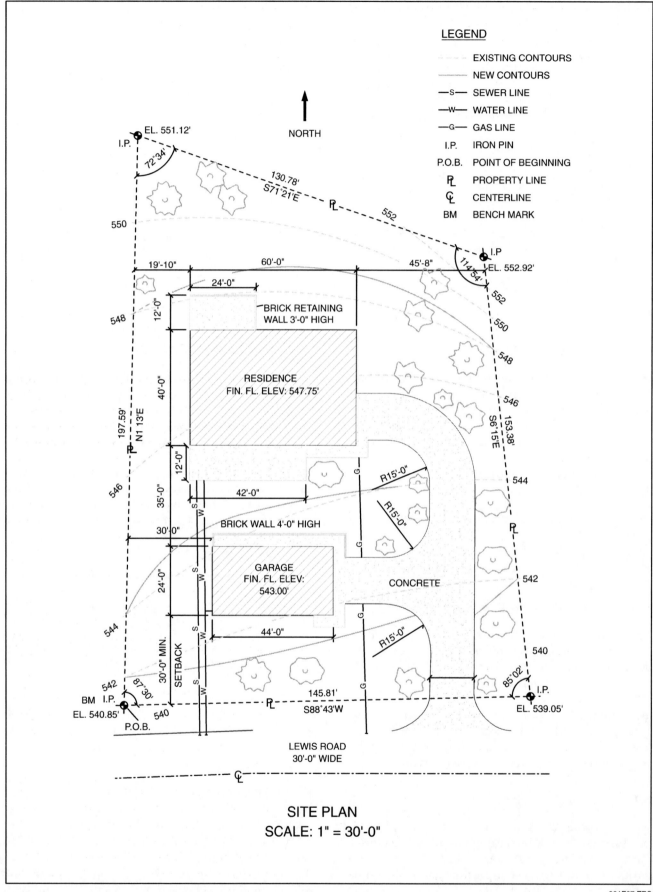

Figure 7 ◆ Site layout for part of a subdivision.

fill information. This information is used to maintain grade control during rough grading.

Finish stakes, called blue tops, are placed by the survey crew to guide finishing work and trimming of the subgrade. These blue tops are driven so that their tops are flush with the proposed grade lines along each subgrade shoulder and the center line.

Heavy equipment operators are responsible for cutting or filling areas to the prescribed grade as stated on the grade stake.

3.3.0 Setting Up and Coordinating Operations

Before beginning any earthwork, the contractor must set up the job. The setup depends on the type of project. A highway project is characterized by a long, narrow area where access may only be available from one or both ends. To get to intermediate points along the route, access roads may have to be cut so that equipment can be carried to the job site. Setting up for a commercial or industrial site is a little different from setting up for highway work. Because these sites can be accessed from existing roads, equipment may be easily transported to the site and unloaded. These areas are often located in populated or highly trafficked areas, so security and site access may be a more significant issue than for highway work. In addition, these sites may have added complications of excavating for parking areas, landscape areas, and drainage. Also, there may be different requirements for excavating and hauling that will affect the type of equipment to use. For example:

- Are there requirements to fill planters and other landscape areas with topsoil?
- How much contour grading is required?
- Is there any topsoil that must be stripped and saved, or does it have to be hauled away? If there is any vertical excavation, how much shoring will be required?
- Have all underground utilities been accounted for?

Regardless of the type of project, preparations for beginning the construction usually include a pre-construction conference that is held before any work begins. At the pre-construction conference, the owner's representative, the designer, the construction manager, and the contractor discuss each item of work. The main topics may include the following:

- Foreseeable problems and unusual conditions
- Structures designated for removal
- Erosion, sediment, and **stormwater** runoff control
- Traffic control
- Utility locations

- Debris disposal
- Signing requirements
- The contractor's work plan and schedule of operation
- Project requirements and specifications
- Equal Employment Opportunity Commission (EEOC) regulations
- OSHA regulations and Minr Safety and Health Administration (MSHA) regulations
- Any applicable local, state, and federal rules or regulations

The purpose of the meeting is to bring the contractor and the owner to the same understanding of the project requirements.

3.3.1 Staging Areas, Equipment Parking Areas, and Site Access

During the pre-construction conference, requirements for equipment staging and parking areas for servicing may be reviewed. On smaller jobs, this may not be a problem because there may be some space that is not being excavated or filled. On highway projects where larger equipment, such as scrapers or bulldozers, are used, there may be a problem with keeping idle equipment out of the way. Most contractors are very concerned about the condition and security of their equipment. They prefer to move it to an area around the work site or to a fenced location when it is not in use. This may not be practical for highway jobs because of the distance a machine must be roaded to get it to a secured area. For highway work, the equipment is often shut down and parked on some level spot close to where the machine is working.

Site access is a concern for many contractors because of the liability issue from accidents. Security of the equipment must be considered because of vandalism and theft. For highway projects where the work is strung out over several miles, it is difficult to secure the area with any kind of fencing or enclosures. Depending on local conditions, the contractor may decide to park equipment in only one or two locations that can be guarded. For commercial and industrial construction sites, the area is usually smaller and the boundaries are well defined.

Equipment operators should know the policy and procedures for securing their equipment during non-working periods. This includes what can be left in and on the equipment after shutdown, as well as where the equipment is to be parked.

Access to the site should be limited to authorized persons at all times, so operators should be alert to any unauthorized persons on the job site. This includes people who wander onto the site by accident, people who are trying to take shortcuts, and children who are curious about the equipment. Normally, requirements for being on the job site include the use of hard hats and safety vests, so it is relatively easy to spot someone who does not belong there. If you see anybody who looks like they do not belong on the site, call it to the attention of your supervisor.

3.3.2 Construction Signs

In some cases, it may be necessary to post construction signs to warn the public of the construction activity. This may include flag persons or other forms of traffic control, especially where construction equipment will be crossing the road or when working very close to a public area such as a commercial building site in an urban area. No work may begin until all signs and warning devices are installed according to the project plans. You may wish to refer to the *Safety* module in *Heavy Equipment Operations Level One* for more information about the various types of signs used on a construction site.

3.3.3 Clearing and Grubbing

The work site must first be stripped of all vegetation, obstructions, and unwanted structures before excavation or other construction activity. The removal of this material is called clearing and grubbing, or right-of-way preparation. It also includes the preservation of trees and other objects that are to be saved. Generally, the distinction between the two operations is as follows:

- Clearing is the removal of objects above the original ground.
- Grubbing is the removal of objects below the original ground.

The contractor is generally required to clear and grub all areas of the site. In the cross section shown in *Figure 8*, the area within the property lines is to be cleared and grubbed of all trees, roots, stumps, and other protruding objects not designated to remain. The depth of cover for objects usually determines whether they need to be removed.

Clearing and grubbing may include the removal of structures and obstructions. This includes buildings, fences, old pavements, and abandoned pipe or culverts. After the area has been cleared and grubbed, it is scalped of brush, roots, and grass. The top soil is generally removed and stockpiled during this phase.

If the area is undeveloped, there is probably some vegetation that will need to be cleared. Usually the vegetation is removed by scraping if off with earthmoving equipment and placing it in a designated area. If there is very little significant vegetation, it can be **disced** under. When clearing areas, remember the following points:

- When using a scraper to remove vegetation material, remove shanks and teeth from the scraper's cutting edge to reduce the amount of soil taken with the vegetation.
- Do not leave any vegetation in the building or construction area. Anything that grows is unsuitable for compaction because as it decays, it will lose volume and cause a depression to form on the excavation surface.
- Use the proper equipment for grubbing. There may be large trees or boulders that need to be removed before a scraper or grader can start pushing the soil around. A bulldozer or track loader can be used to remove these obstructions. The bulldozer can move large boulders out of the way and can knock down large trees.

There are several ways to dispose of debris. It may be hauled away from the construction area, burned (if allowed), buried, or mulched. High-quality top soil may be stockpiled to be used during landscaping at the end of the project, or it may be sold for profit. Heavy equipment operators should be briefed on the desired disposal method to avoid costly errors.

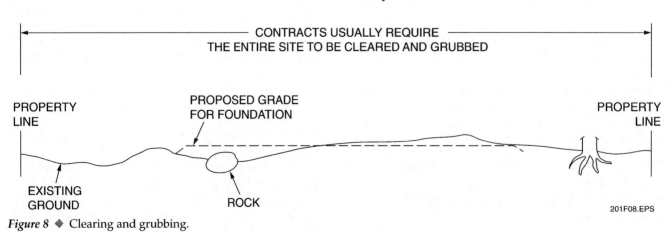

Figure 8 ◆ Clearing and grubbing.

3.4.0 Production Measurement

Although equipment operators usually do not get involved in the details of planning the overall project work, they need to understand how it is done. At some point, they should also provide suggestions or input based on their experience with different types of soils and the handling of different pieces of equipment under various conditions.

3.4.1 Mass Haul Diagram

The mass haul diagram is a graphic method of analyzing the movement of soil from one loading area to the dumping site. It is mainly used in highway, canal, and airfield projects where the cut and fill work is spread out in a long narrow band. The mass diagram relates the proposed grade to the existing grade and is useful in determining the location and volume of excavations and the haul distances to points in the fill.

This analysis identifies any areas where it would be better to obtain material from a place outside the project limits rather than haul material a very long distance. Also, it shows where to dump surplus excavated material in a **spoils** area rather than haul it to a fill that is farther away. These decisions depend on the cost of the equipment and materials being used.

3.4.2 Cycle Times

Cycle time is the measurement of how long it takes to perform a specific operation, such as the time for a piece of equipment to load, haul, dump, and return to the loading point to start the next cycle. This allows the contractor to estimate how long it will take to complete a specific activity, like rough grading or excavating a ditch.

Cycle times are computed differently for different types of equipment. For scrapers that haul and dump as well as load, the cycle time includes picking up the material from the ground, hauling a full bowl to the dump site, spreading the material, and then driving back to the excavation area to scrape another load. In the case of a front-end loader, the cycle time is measured from the time the loader begins digging with the bucket, through maneuvering to load the truck or stockpile, loading (or dumping the bucket), and then returning to the point of digging.

Most equipment manufacturers provide information about average cycle times, which can be used to estimate the time and cost of doing the work. Contractors also use cycle time measurements to determine the operator's efficiency.

3.5.0 Maintaining Haul Roads

Well-maintained haul roads are important for efficient operation. On long haul roads, a grader and water truck should make a pass over the road periodically to keep it in good condition. This reduces the wear and tear on equipment as well as cycle time.

On small jobs, a grader may not be available for use on the haul road. In this case, the scraper operators can maintain the road. If the road is being watered to keep the dust down, the scraper operators must be careful. Wet soil can cause a scraper to skid out of control.

3.6.0 Drainage Requirements

Excavation descriptions in this module are based on the movement of dry material. However, most materials are not completely dry in any natural environment. Even in a desert, the first couple of feet of soil are dry, but as you dig farther, the earth feels cool and moist to the touch. Millions of yards of material are moved every year without major problems with water. Rain may slow or temporarily halt operations because of the effect on the surface, but usually this will not produce impossible excavating conditions. Many construction sites will have plans to control the flow of stormwater over the site to prevent erosion and water pollution. The federal government requires these plans, so it is important that you understand and follow them. Other drainage problems on construction sites that are related to earthmoving are the control of groundwater and the allowance for drainage from embankment materials.

3.6.1 Control of Stormwater

The Environmental Protection Agency (EPA) requires that contractors performing any activity that disturbs one acre or more of land obtain a National Pollutant Discharge Elimination System (NPDES) stormwater permit before work is started at the site. The purpose of the permit is to ensure that the contractor has a workable plan to prevent pollution and control stormwater runoff from the site, as well as to prevent erosion.

NOTE

Construction sites that are smaller than one acre, but are part of a larger development, need to have this permit, too.

Contractors need to use adequate measures to prevent erosion and sedimentation and to control stormwater runoff. Erosion is the eating away of soil by water or wind. Sediment refers to soil that has been moved from its original place by wind, water, or other means. Stormwater runoff is water from rain and snow that is shed from the ground rather than being absorbed. As a heavy equipment operator, you need to know and understand the site's erosion, sediment, and runoff prevention plans.

Every site is different, so each will have its own erosion and sediment control plan. The best way to prevent erosion is to avoid disturbing the surface of the soil, so it is important to follow plans closely and disturb only the earth that needs to be graded. Other measures include reducing the steepness of slopes, covering disturbed soil with mulch, erosion cloth, or **riprap**, and contouring the grade so that water is absorbed rather than shed. A slope that is graded with a rough surface will slow the flow of water and promote absorption of water rather than runoff, thus helping to prevent erosion.

Because sediment is soil that has been moved from its original place, the best way to avoid sediment is to prevent erosion. In earthmoving, some erosion is unavoidable, so most sites will use various devices, such as silt fences, to trap displaced soil. You can help to prevent sedimentation by avoiding walking or driving across disturbed soil, washing soil off truck tires before leaving the site, and limiting vehicle operation to approved haul roads.

As rain washes over a construction site, it picks up loose soil, debris, gasoline, oil, and other chemicals from soil or paved areas that can pollute water and destroy wildlife habitat. The first step in preventing such pollution is to prevent the release of pollutants into the environment. You can do this by being very careful not to spill solvents, chemicals, oil, grease, paints, and gasoline onto the ground. Your employer will probably have designated areas for equipment maintenance and wash downs and a protected storage facility for chemicals, paints, solvents, and other potentially toxic materials. You can also help to prevent pollution by placing trash in the proper receptacle and using the sanitary facilities provided on the site.

3.6.2 Control of Groundwater

Water existing below the ground surface is typically divided into two zones. The zone of aeration contains varying amounts of moisture and air. Beneath it, water is continuously present in the zone of saturation. The water in this lower area is called groundwater. The top of this layer is called the **water table**. *Figure 9* shows a cross section of the zones of aeration and saturation.

The depth of the water table varies widely. It depends on the amount of annual precipitation, the terrain, and the amount of extraction due to irrigation and other uses of the water.

The depth of the water table and the amount of groundwater may be determined by soil borings. Borings are made at specific locations over the job site to plot the profile of the water table. From this profile, the need for any type of special dewatering activity, such as sumps or various types of drains, can be established.

Sumps are open pits or holes built to collect groundwater. They are then pumped out mechanically. The requirements for sumps vary with the depth and amount of flow rate of the groundwater.

Drains are various types and sizes of ditches that collect and channel the flow of water away from the excavation area. They should be limited in width and depth to the minimum requirements and should intercept the groundwater as close to the bottom of the excavation as possible. The layout of these temporary drains is determined by soil conditions and the location, direction, and quantity of flow.

3.6.3 Drainage of Stockpiles

A typical problem with stockpiles is the storage of wet materials. Wet material does not stack well and tends to spread out or flow. The ease of outflow of water from the stockpile depends on the type of material.

Water that is drained off the stockpile must be controlled. When building a large stockpile of select material, a shallow channel should be cut on the stockpile to collect the runoff and drain it away from the storage area to a holding area. Silt fences must be used to control water flow and discharge of sediment.

3.7.0 Site Excavation

In excavation work, planning is important. The equipment should never stand idle while grades are being checked or while a decision is being made on where to start the next cut or fill. The operator should study the plans and stakes carefully before starting to work, taking time to analyze the type of equipment needed. For example, is a scraper sufficient to do the cutting of the soil, or is the soil hard enough that some type of ripping is needed to loosen it before using the scraper?

Before breaking ground, it is important to determine if there are any utilities or other structures buried on the site. Today, utility companies go to great lengths to inform the public and contractors where their lines are to prevent damage to them. The local utility company will set stakes to mark underground utilities. Survey crews lay out locations of

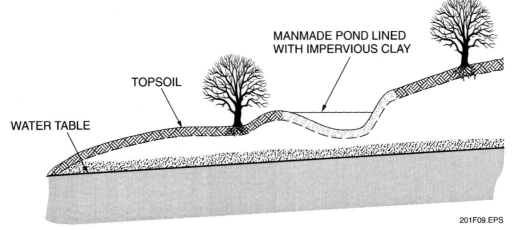

Figure 9 ◆ Cross section of a water table.

new utilities. These markers are color-coded using the standard underground color code found in *Table 1*. When you see these markers, act accordingly. If you do not know what to do, ask your supervisor. This subject is covered in more detail later in this module.

There are many different types of equipment to use for excavating. The machine that is chosen for a particular job must fit the requirements of the work. Typically, the equipment must match the type of material that will be excavated and be able to work in the area without being unstable or inefficient. In general, the following types of heavy equipment are used for excavation:

- Scrapers (*Figure 10*)
- Excavators (*Figure 11*)
- Motor graders (*Figure 12*)
- Bulldozers (*Figure 13*)
- Wheeled and tracked loaders (*Figure 14*)
- Backhoe loaders (*Figure 15*)

There are many different types of excavations, each presenting its own challenges. On some jobs you will be working at several different excavations sites, including the following:

- Highway/roadway excavation
- Several types of bulk excavations
- Limited-area vertical excavation
- Trenching

3.7.1 Roadway Excavation

Roadway excavation is a leveling process used in highway construction and airfield grading. Ideally, it involves the movement of material from high ground to low areas. This is called a balanced job because the soil needed for fill areas is obtained from cut areas and there is no need to haul material in from outside the project (*Figure 16*). In other

Table 1 APWA Underground Color Codes

Color	Meaning
Red	Electric power lines, cables, conduit and lighting cables
Yellow	Gas, oil, steam, petroleum, or gaseous material
Orange	Communication, alarm or signal lines, cables or conduits
Blue	Potable water
Green	Sewers and drain lines
Purple	Reclaimed water, irrigation and slurry lines
White	Proposed excavation
Pink	Temporary survey markings

cases, cut material may be wasted because it is not needed or is unsuitable for fill. Material moved from a borrow area to fill low spots for embankment construction may also be included.

Roadway excavation is normally shallow and the site is easily accessible. *Figure 17* shows a typical layout for roadway excavation. Equipment used includes scrapers, bulldozers, and possibly front-end loaders. Selection of the specific pieces of equipment will depend on the size of the job, characteristics of the material being moved, and haul distance.

3.7.2 Bulk-Pit Excavation

Bulk-pit excavation is, as the name implies, the digging of a pit. It involves digging and loading material of great depth and volume and then hauling the excavated material to another site. The pit may be at the project site, such as when a building with a basement is being built in a urban area, or it may be outside the work site, such as at a borrow pit.

When at the project site, a pit is usually required because of some site limitation such as a

Figure 10 ◆ Scraper.

201F10.EPS

Figure 11 ◆ Excavator loading a dump truck.

201F11.EPS

Figure 12 ◆ Motor grader.

201F12.EPS

Figure 13 ◆ Bulldozer.

201F13.EPS

Figure 14 ◆ Wheeled loader.

201F14.EPS

Figure 15 ◆ Backhoe loader.

201F15.EPS

neighboring building or street, which limits the access to the site. The resulting pit has vertical or nearly vertical walls and requires that the equipment operate against the face of a bank. The type of equipment used at these locations will depend on the size and accessibility of the site. Equipment with a large payload is often more efficient and economical to use, but when the site area is limited, smaller equipment that is more maneuverable may be used despite higher costs.

As you learned previously, the cycle time of any equipment is the measurement of how long it takes to perform a specific function. When excavated material is loaded onto a truck for removal from the site, it is important to minimize the movement of the excavator or loader for the shortest cycle time. *Figure 18* shows two methods of loading trucks. In Method 1 the truck is positioned so that the excavator boom swings between the excavation face and the truck bed. In Method 2, the loader must

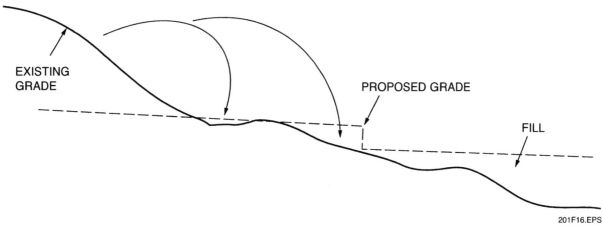

Figure 16 ◆ Balanced excavation.

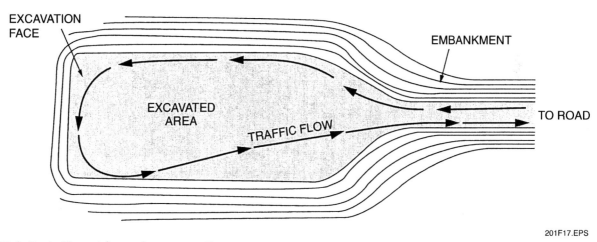

Figure 17 ◆ Typical layout for roadway excavation.

move to the excavation face for loading, than back up and move into position to dump its load into the truck bed. Method 2 requires excessive movement of the loader, thus increasing the cycle time and costs. Method 2 should be used only when access to the excavation site is too narrow to permit both the truck and loader access at the same time.

NOTE

During loading, the truck must be positioned as close to the point of excavation as possible to minimize the movement of loading equipment.

There is a third method of moving excavations away from the excavation site, and it is used in special conditions where it is not necessary to do more than push the excavated material aside. In this method, a crawler-mounted loader or bulldozer uses its bucket or blade to dig and collect material

on the front, and then pushes beyond the access ramp to the excavation site.

Excavated material that has been loaded onto a truck is removed from the site in a path similar to the one shown in *Figure 19*. In the scenario in *Figure 19*, using the paved road above the site for entry and limiting the one below the site to exiting vehicles would permit more efficient traffic flow. Although you may be fortunate enough to experience this on large jobs, local permitting may limit the construction traffic flow to a single point. In addition, the construction of two access roads would increase the cost of the job.

3.7.3 Bulk Wide-Area Excavation

Bulk wide-area excavation is similar to bulk-pit excavation, but this type of excavation permits access to the area from many directions, making it easier to access and leave the excavation site. Bulk wide-area excavation is usually shallower in depth but larger in area than bulk-pit excavation.

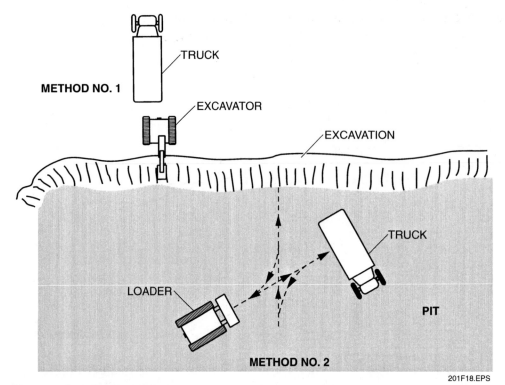

METHOD NO. 1

TRUCK

EXCAVATOR

EXCAVATION

TRUCK

LOADER

PIT

METHOD NO. 2

201F18.EPS

Figure 18 ◆ Loading a truck at the digging face.

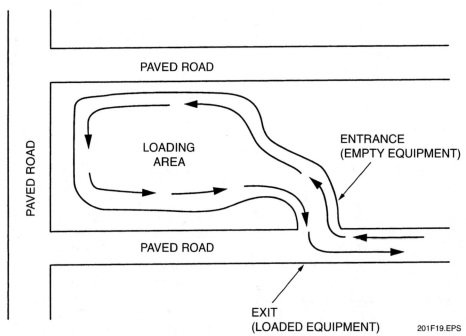

PAVED ROAD

PAVED ROAD

LOADING AREA

ENTRANCE (EMPTY EQUIPMENT)

PAVED ROAD

EXIT (LOADED EQUIPMENT)

201F19.EPS

Figure 19 ◆ Bulk-pit excavation.

Bulk wide-area excavation is used in highway construction, airfield grading, the removal of top layers of soil (as in quarrying or strip mining), and the building of earth dams. In some cases, it only involves moving material from high ground to low areas. Other times, it involves wasting material that is not needed or that is unsuitable for fill. Material moved from a borrow area to make up deficiencies in embankment construction may also be included.

Equipment for this type of operation usually involves scrapers, bulldozers, and possibly front-end loaders. Selection of the specific pieces of equipment will depend on the size of the job and the characteristics of the soils being moved. Scrapers have difficulty in loading loose, dry sands and rock, even if it has been crushed. They also have trouble unloading wet sticky clays. The **gradation** of the material is also an important consideration when choosing the type of equipment.

In this type of excavation, the ideal location ensures short haul distances. For short distances, bulldozers and graders can be used to scrape the material, push it the required distance, and then level it as required. If haul distances are longer, scrapers should be used when the material can be loaded efficiently. They can cut a layer of material while moving forward, and they can travel at speeds up to about 30 miles per hour (mph) when fully loaded. At the dump area, they can slow down and dump the material by spreading it in layers at the specified thickness.

When this type of excavation is located far from the dump site, material is loaded onto trucks or other carriers and carried to the dump site. If this is the case, the loading area is set up to permit the smooth flow of traffic and quick loading of material.

3.7.4 Channel Excavation

Channel excavation is the excavation of loose, unconsolidated materials, of material lying under water (such as a drainage channel), or of saturated soils that prevent the equipment from traveling over the excavated area. This type of excavation is usually performed by equipment standing on solid ground at the same level or higher than the material being worked. See *Figure 20*.

3.7.5 Limited-Area Vertical Excavation

Some excavation must be done by lifting the material vertically out of the pit because the sides of the excavation require support with some type of shoring material. *Figure 21* shows a limited-area vertical excavation.

The piece of equipment for this type of job is often the clamshell. The clamshell is a bucket-type attachment that is used on a front-end loader or a backhoe. The clamshell is versatile but requires close attention by the operator and is not very efficient in its digging action. This is because the bucket is on the end of a line and cannot be pushed into the soil to dig with any force; it can only pick up loose material from the bottom of the pit. *Figure 22* shows an example of a clamshell.

3.7.6 Trench Excavation

A trench is a temporary opening in which something, such as a pipe or box culvert, is placed and covered. The trench width, and usually the depth, are limited to 15 feet. A large-scale type of trench

201F20.EPS

Figure 20 ◆ Excavating a drainage channel.

www.SHORING.com
201F21.EPS

Figure 21 ◆ Limited-area vertical excavation.

201F22.EPS

Figure 22 ◆ Clamshell.

excavation is known as cut and cover. This includes excavating a large and sometimes deep trench, laying the specified pipe or culvert, and then covering everything over or building on top of it. This option is considered for highway and transit system tunneling when the excavation of the trench would cost less than boring a tunnel.

> **NOTE**
>
> Do not confuse a trench with a ditch. A ditch is a narrow slot cut in the ground and left open. A trench is an opening in which something is placed and then covered.

Trench excavation is different from pit or wide-area excavation. Pit or wide-area excavation focuses on moving large quantities of material. With trench excavation, the focus is on the rate at which the pipe or culvert can be placed. This factor affects the type and size of equipment selected for the trenching job. The trench should not be excavated any wider than is necessary for bedding and setting the structure. Various shields, such as the one shown in *Figure 21*, can be used to shore the sides of the trench.

Three main classes of equipment are used in trenching: backhoes, excavators, and trenching machines. The trenching machine is specifically designed for trench excavation. *Figure 23* shows a trencher.

3.8.0 Loading

Regardless of the type of project, if the material is to be moved from one area to another, it must be loaded, hauled, and dumped. The equipment that transports the earth must return to the loading point for another load. This is called **backhauling**. Selection of the right type and size of equipment to do these basic operations is important for efficiency and cost. Usually, there is more than one way to perform the work, but the most efficient and cost-effective method should be selected.

Loading is the first step in the earthmoving cycle. Loading is performed as a basic function, such as loading material from a stockpile or spoils, or as a secondary function when some other work is being performed. This could be some type of excavation, such as trenching or cutting with a scraper for cut and fill operations.

The specific job and the type of material encountered will have a great influence on both the selection of equipment and the method used.

3.8.1 Selection of Loading Equipment

Selection of loading equipment depends on factors such as cycle time, type of material, and equipment efficiency. For operations that move large amounts of material, the type of haul equipment must also be considered when choosing the loading unit. Another important factor is the overall cost of the operation, which includes the hourly cost of using a certain type of equipment.

The options available for mobile loading equipment include the following:

- Front-end loaders (and variations)
- Excavators
- Scrapers
- Shovels

Front-end loaders can be used for loading from stockpiles, **windrows**, and spoils piles. They can also be used to excavate and load in the same operation. Configurations include wheel and trench loaders. Their size range and tight turning radii make them favorite candidates for loading operations.

Excavators come in many different sizes and configurations. The term excavator includes backhoes and telescoping excavators. Although they are primarily used for excavation of material below the grade or into banks, they can also load material into dump trucks and other haul units. When the job requires only one operation, such as clearing or excavating ditches, an excavator can be used to do the excavation and load the truck in one operation.

Scrapers are special pieces of equipment that typically are used where large quantities of earth need to be moved. They are configured to pick up the material through a scraping action, load it into the bowl, and transport it to the dump site. Scrapers also can be used as haul units with other pieces of equipment loading into the scraper bowl.

Shovels are mainly used to excavate material in pits and quarries. Because they are track mounted, they do not move very quickly. However, shovels can be manufactured in large sizes that can dig and load large quantities of material in one cycle.

Other types of loading equipment include conveyors, overhead buckets, and cribs.

201F23.EPS

Figure 23 ◆ Trencher.

3.8.2 Methods of Loading

Equipment unit size and maneuverability establish the method of loading. As described earlier, some equipment can load as well as excavate. Loading consists of two basic methods: loading from piled material and loading from an excavation.

When loading from stockpiles and spoils, the front-end loader is the primary piece of equipment used. It is able to maneuver well and work with many different configurations of haul units. Loaders usually follow an I or Y pattern when loading from a bank of material. The configuration used will depend on the amount of available work area. *Figure 24* shows a front-end loader working against a bank while loading trucks using the Y pattern.

Because of their configuration, most excavators must remain stationary while they work. Their method of loading is limited to loading directly from the excavation or first dumping the material in a pile next to their position and then loading from the pile into the haul unit. This method is not suitable if the area is too narrow to permit the excavator and haul unit access to the site at the same time.

3.9.0 Hauling

Hauling is the transport of material from one point to another. This may be a very short distance within the construction site or a long haul to dispose of material, such as that from clearing and grubbing operations. Hauling is normally broken into two types:

- Over-the-road hauling is hauling over public highways using a haul unit, such as a dump truck.
- Off-road hauling is any hauling that is done within a job site and not on public highways. Equipment used for off-road hauling includes large trucks that are not allowed on public highways, scrapers, front-end loaders, and regular trucks.

Figure 25 shows effective haul distances for various types of hauling equipment.

3.9.1 Selection of Hauling Equipment

Equipment used for the two types of hauling is very different in design and construction. The specific operation determines which type to use.

Over-the-road haul units have very distinct limitations because highways are designed for specific load conditions, and truck load limits are established according to their loadbearing capacity. These units are also restricted in width and height. These limitations are enforced by state or federal law.

Off-road hauling units may include scrapers, front-end loaders, or trucks of various types. Off-road equipment is usually wider and higher than over-the-road dump trucks. This allows for an overall increase in the size and capacity of the equipment. For example, a tandem-axle over-the-road

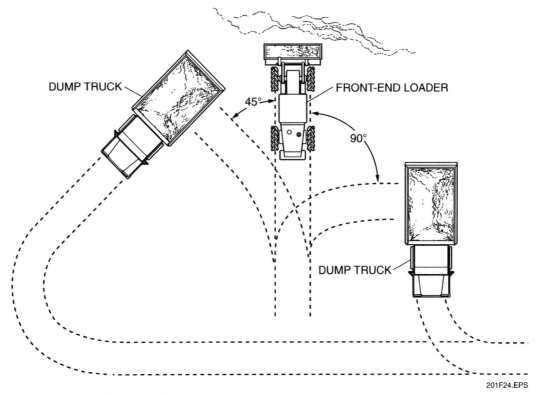

201F24.EPS

Figure 24 ◆ Loading dump trucks using the Y pattern.

dump truck can carry a load of about 12 cubic yards of material; a large scraper can carry up to 44 cubic yards.

There are several considerations in the selection of the most appropriate haul unit. The first is whether any portion of the haul will be over public highways. If so, use restrictions must be checked. The person operating the equipment is responsible, and if there is a violation, the operator will get the citation. Also, the haul route needs to be checked for problems such as bridge height, width, and weight limitations. Hauling over narrow, curving roads with high crowns can be hazardous for larger trucks. High road crowns that can tilt the truck may limit the allowable heaping when the unit is loaded.

Individual job conditions dictate off-the-road types of hauling equipment. There are basically three configurations to be considered.

- *End-dump (rear dump)* – These types of trucks should be used if material consists of rock fragments or masses of shale that are too large for other units to unload, or if dumping is to be done over the edge of a waste bank or fill. Rear dumps provide maximum flexibility for a variety of job conditions. They can be straight body or articulated.
- *Bottom-dump* – These units are good for transporting free-flowing materials over reasonably level haul routes that permit a high travel speed. They are best for dumping in windrows over a wide area.
- *Scrapers* – Scrapers can haul a large amount of material at relatively high speeds for short distances. They have an added advantage of being

able to load the material directly by cutting or scraping a thin layer of soil while moving forward at a slow speed. When the bowl is full, the scraper can travel at a higher speed to the dump site and spread the material in a thin layer. Because of their large size, scrapers are not normally used in small or confined areas. They are good for cut and fill work on highways and airports, as well as general contouring of the ground.

Another type of truck that may be used under special circumstances is a side dump truck. The trailer on this type of truck is built to pivot to the right or left and quickly dump its load. This truck has limited use because the dump site must of sufficient length and be configured so that the entire load can be dumped at once.

3.9.2 Hauling Methods

Hauling is the largest single factor in cycle times for earthmoving operations. Assuming that the material is transported by some type of haul unit instead of a cable or conveyor system, the synchronization of the hauling operation is the main concern. This includes the timing of the hauling and the traffic pattern. If the job requires a considerable number of haul units, a lot of time can be lost if work operations are not timed properly. Also, for jobs requiring the placement of large amounts of cut and fill or the removal of excavated material, it may be necessary to set up special haul roads and to lay out traffic patterns.

A traffic control supervisor may be required for operations employing a substantial number of haul

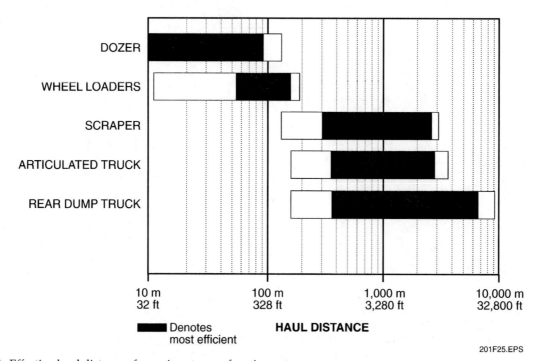

Figure 25 ◆ Effective haul distances for various types of equipment.

201F25.EPS

units. The supervisor controls operating conditions at both the loading and dumping sites and keeps track of the haul units in transit.

3.10.0 Dumping

Material may be dumped either to build up an area to make an embankment or to dispose of spoils. The dumping must be more closely controlled when building up an embankment than when dumping at a spoils site.

3.10.1 Dump Sites

Spoils areas are used for the disposal of surplus excavation materials or materials that are unsuitable for fills. The location of the spoils area may depend on the type of material and the haul distance involved. Soft materials going to a spoil area may require mixing with firmer soils. Organic debris may be burned or spread out in thin layers. Rocks and boulders may require placement in a specific way to reduce environmental concerns, such as erosion. Dumping of spoils outside the project site may be more convenient, but the cost of each option must be considered.

If the material to be dumped is for filling, backfilling, or making an embankment, these areas are shown on the plans and indicated with the survey and slope stakes. Material is sometimes dumped alongside the area to be filled and then moved into place with bulldozers or graders. The project engineer or activity supervisor marks these areas with stakes, paint, or lime so the equipment operator knows where to dump each load.

Another type of dump site is a stockpile. Stockpiles are used to store material that will be used later in the project. A good example of this is the topsoil that is removed at the beginning of a project. Because good topsoil is an expensive **pay item**, it is saved for landscaping and contouring at the end of the project.

3.10.2 Dump Patterns

Dump patterns depend on the material being dumped and the type of haul equipment being used. For spoils dumping, the material may be dumped into another excavation, such as an abandoned pit or over an embankment, or it may be dumped in a level, open area. Often, spoils are hauled in a rear dump truck. Depending on the amount of spoils, dozers or front-end loaders may be needed at the dump site to move the material around and blend it in with the surrounding terrain.

The haul equipment for fill material usually includes bottom or rear dump trucks or scrapers. When backfilling, the material is dumped directly into the excavation or beside the excavation, and then another piece of equipment, such as a loader or dozer, is used to place the material.

Embankment construction fill material can be placed by either trucks or scrapers. *Figure 26* shows several different dumping patterns.

A dumped fill pattern for level ground is indicated in *Figure 26A*. The material is placed in many individual piles so it is easier to spread. This method reduces the amount of work required to move material over longer distances.

Figures 26B and *26C* show dumping alternating coarse and fine material along an embankment in order to provide for thorough mixing of the material. Once the loads have been dumped, a motor grader is used to blend the coarse and fine material together. Another approach to placing two different types of material prior to blending is to completely spread one type of material before the second type is dumped and spread on top. A scarifier is then used to mix the coarse and fine material before compacting.

In highway construction, material to construct an embankment is usually dumped in layers. This can be accomplished with scrapers, which can spread the material as they dump, or with trucks, which dump in piles or windrows. These piles and windrows are then leveled out with a dozer or motor grader.

3.10.3 Backhaul

Backhaul is not strictly defined as part of the dumping process. Backhaul is the return trip that the operator makes after dumping a load. Backhaul is part of the cycle time, and because the truck is empty during this period it is considered unproductive. The success or failure of an earthmoving operation is based in part on how quickly a truck can safely return to the excavation site for another load. Equipment operators can help make the operation a success by exhibiting the following behaviors:

- Drive cautiously at all times.
- On the return trip, use the prescribed return route from the dump site. It is usually set based on the distance between the excavation area and the dump site.
- If haul roads are used, the pattern for the return trip is set and must be followed by all operators to avoid disrupting traffic flow.
- If there are no haul roads, then the truck must drive through the construction site. In this case, the project engineer should establish a pattern for the hauling and backhauling. Operators follow markers laid down for this purpose, and ground personnel guide the equipment out of the area once the dump is made.

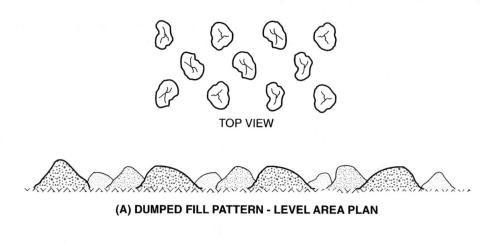

TOP VIEW

(A) DUMPED FILL PATTERN - LEVEL AREA PLAN

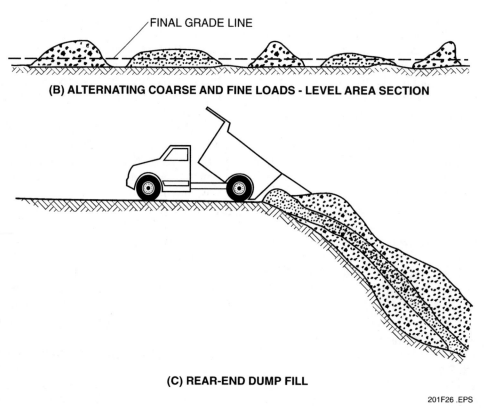

FINAL GRADE LINE

(B) ALTERNATING COARSE AND FINE LOADS - LEVEL AREA SECTION

(C) REAR-END DUMP FILL

201F26 .EPS

Figure 26 ◆ Dumping patterns.

- Stay in line and follow the directions of the traffic supervisor.
- Think ahead as to how to maneuver the machine when approaching the loading site for the next cycle.

For scraper operations, the scraper may need to be positioned where a bulldozer can come up behind and start the pushing operation. A good scraper operator, working in synchronization with the dozer operator, can perform this operation smoothly and quickly. Otherwise, much time is lost in starting, stopping, and lining up the two machines.

Dump truck operators can help loader operators by getting into position quickly and keeping an eye on the loader. If a spotting marker has been placed, the operator should know where it is and how to position to it.

3.11.0 Fill and Embankment

Excavated materials can often be used as fill, backfill, or embankment, but it may be subject to one or more soil tests before it can be used as backfill or embankment. The number of tests required depends on the intended use of the material. Before filling, backfilling, or creating an embankment, always check federal and local regulations for the type and content of the materials allowed.

3.11.1 Fill Material

Some fill will be material that has been stockpiled from excavations within the site. Other fill will be brought into the site from another area as needed. Fill material can be classified as three types: ordinary fill, waste fill, and select fill. In general, all fill should be kept free of organic matter if any future use of the filled area is planned.

Ordinary fill meets a pre-established project requirement but is not selected based on any specifications. It is usually brought into the construction site and dumped in piles. It is then spread uniformly over an area and compacted according to the job specifications.

Ordinary fill can be used to build up an area where there is little or no compaction required. The fill is simply dumped into the area such as an embankment, scraped area, or around some types of buildings. When fill is dumped into a deep fill area and allowed to roll down the slope, it tends to separate because the larger fragments roll to the bottom and the fine material stays close to the top. Dumped fills should be kept free of tree stumps, organic matter, trash, and sod if any future use of the filled area is planned. An exception to this process is the special case of sanitary landfills, where garbage and trash are spread several feet thick over the ground surface, covered with additional clean earth, and left to compost.

> **NOTE**
> Any fill that contains a great deal of rock must be spread over the area rather than dumped to ensure a uniform thickness.

Waste fill is recycled, treated material approved for use in certain landfill areas. It is used to build up areas where no compaction is necessary. This could include embankment, landscaped areas, or backfill around some types of structures. Waste fill normally has little or no compaction requirement. When rock is used, it should be spread over the area rather than piled at a single point.

Select fill is material with specific properties that can be obtained only by selective excavation or manufacture. This category includes selected impervious clay materials; gravel or sand; selected rock used for stabilizing slopes; and dumped riprap. Again, the equipment used to place fill material includes the dump truck, scraper, dozer, and sometimes the front-end loader. Spreading of the material is accomplished by the scraper, dozer, and motor grader. Compaction can be accomplished by wheel or tractor compaction or by a mechanical roller of some

type. Select fill is usually more expensive than other types of fill and must be used, placed, and compacted as specified by the project plans.

3.11.2 Backfill

Backfill is the returning of excavated material to an excavation site. Before the excavations, the ground was fairly stable because the soil had settled—this is called **consolidation**. It is important to backfill and compact excavations so that a minimal amount of future settling occurs at the site. This may seem simple, but because of compaction issues, some backfills are poorly made and cause damage to the buried structure and the surrounding area.

The main concern with backfilling is choosing the best method for compacting the fill material. Although the specifications for some work require that all excavated material be disposed of and specifically selected materials be hauled in for backfilling, most contracts allow the use of excavated or dumped fill for backfill. If dumped material is used, compaction criteria must be varied to suit the soil encountered. While poorly compacted material will ultimately settle, over-compacted material will not settle at the same rate or under the same loads as adjacent natural soils, and this effect can be just as damaging.

For backfilling around small structures and buried utilities, a backhoe is the ideal piece of equipment. The backhoe operator needs to understand the process of compacting backfill material so that he or she can place material at points where the workers need it to build up the backfill. By compacting too much in one area, pressure will build up against the structure and cause it to bend out of alignment, or even break. See *Figure 27*.

3.11.3 Embankment Construction

Embankments are fills that are placed for specific purposes under carefully controlled conditions. Embankments usually consist of soils of a maximum particle size placed with careful selection, compaction, moisture control, and mixing.

Embankments are used for highways and roads, earth dams, levees, canals, and runways and are classified according to method of compaction:

- *Equipment-compacted fill* – This is generally a fill, often of select material, that is compacted by the wheels of the haul units. Moisture control may be required. This type of embankment construction normally does not have a compaction requirement.
- *Rolled-earth fill* – This type of compaction is used where highly impervious fills are required for earth core dams or canals. Degree of compaction

is normally specified for rolled-earth fills. The design is based on the properties of the soils to be used. *Figure 28* shows a rolled-earth fill.

The embankment may be placed on the natural ground, or the embankment may require excavation to another layer under the ground. Embankment material is spread and leveled in layers or lifts.

The equipment and procedures involved in dumping and spreading the embankment material depend largely on the type of equipment the contractor has on the project and the type of material being used. Embankments are built in lifts, one on top of the other, for the full width unless otherwise specified. The contractor must have enough spreading, mixing, and compacting equipment and use the procedures necessary to meet moisture-density

requirements and obtain uniform, well-mixed layers of embankment.

The scrapers usually follow patterns in picking up and depositing material that does not interfere with other equipment operations. The work should progress at a uniform rate without slowdowns due to improper direction of equipment. It is extremely important that embankments be constructed uniformly. The following work activities will ensure uniformity:

- Fill should be placed evenly from one side of the embankment to the other.
- Each lift should be spread to a uniform thickness that does not exceed the specified thickness before compaction. When excavated material consists predominately of rock fragments that

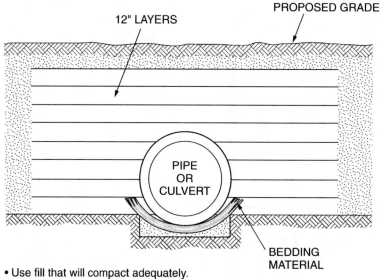

- Use fill that will compact adequately.
- Sift fill into trench in 12" layers to prevent air pockets.
- Bury pipe or culvert at least 2'.
- Compact and fill as necessary to achieve proposed grade.

201F27.EPS

Figure 27 ◆ Backfill.

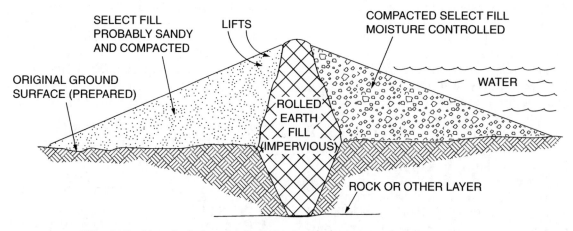

- Fill on both sides of rolled earth will provide it with structural support for added strength.

201F28.EPS

Figure 28 ◆ Rolled embankment.

exceed the specified thickness, the specification may allow the material to be placed in the embankment in layers not exceeding the thickness of the average size of the larger rocks. Each of these rock lifts must be leveled and smoothed with finer material, but the final rock lift must be several feet below the finish grade.

- Each lift must be compacted before the next lift is placed. Double dumping, or placing two lifts without compaction, should not be allowed.
- Where large chunks of material are placed, extra compaction and blading will be needed to spread the material uniformly.

Under embankment areas where fill is to be placed, the original ground must be broken up by scarifying it to a specified depth and then re-compacting it.

3.12.0 Compaction

Compaction increases the density of soil to create a stable ground surface. It involves pressing soil particles together to force out air and water from the spaces between the particles. Compaction reduces the risk of future settling, so it provides a solid surface on which to build. After soil is disturbed, compaction occurs naturally over time as a result of wetting, drying, freezing, thawing, groundwater movement, and the weight of the top layer of soil.

In construction, it is necessary to speed the compaction process with equipment that imitates nature. Soil on construction sites will be compacted with various types of equipment by pressure, kneading, vibration, and impaction, or by using a combination of these methods. The equipment will provide the forces needed to quickly settle soil and allow building to begin.

4.0.0 ◆ STABILIZING SOILS

Soil stabilization is the process of preparing subbase soils to provide a higher loadbearing capacity. The process involves pulverizing and mixing the soil thoroughly, often with binders, so that after compaction and curing, the soil is more dense, more stable, and stronger. The general requirement is to stabilize the area well beyond the points of where heavy loads will occur. For a road subgrade, this means stabilizing past the road boundaries. *Figure 29* shows the subgrade stabilization limits.

Stability and loadbearing capacity depend on two factors, the internal friction and the cohesion of the soil particles. The higher the internal friction of soil, the higher the cohesion and the greater the capacity to bear loads. Chemical binders are used to stabilize existing soils by adding internal friction and cohesion.

The first three steps in stabilization are to determine soil properties, the amount and type of binder required, and the degree of compaction needed. These are all a result of analyzing the soil or material to be stabilized. Analysis includes gradation, moisture content, density, internal friction, and cohesion. Tests and classification of soils were discussed in an earlier section.

There are many advantages to the stabilization of soils. These advantages include the following:

- Allows use of in-place soil
- Eliminates need for excavating
- Eliminates the cost of replacing soil
- Reduces cost by mixing in place
- Reduces construction time

The stabilization process includes the following series of steps:

NOTE: SHAPING THE SUBGRADE IN THIS MANNER IS TERMED BOXING OUT.

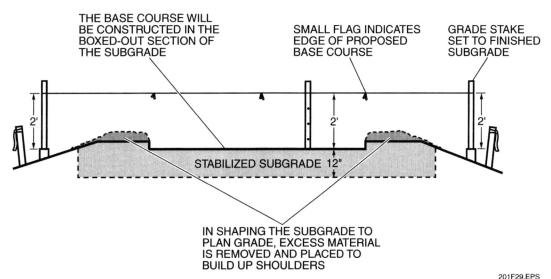

THE BASE COURSE WILL BE CONSTRUCTED IN THE BOXED-OUT SECTION OF THE SUBGRADE

SMALL FLAG INDICATES EDGE OF PROPOSED BASE COURSE

GRADE STAKE SET TO FINISHED SUBGRADE

STABILIZED SUBGRADE 12"

IN SHAPING THE SUBGRADE TO PLAN GRADE, EXCESS MATERIAL IS REMOVED AND PLACED TO BUILD UP SHOULDERS

201F29.EPS

Figure 29 ◆ Subgrade stabilization limits.

- Choosing a binder
- Preparing the subgrade
- Spreading the binder
- Determining the logistics of the stabilization job
- Stabilizing
- Compacting
- Checking quality control

4.1.0 Choosing a Binder

Binders are selected to complement the particular soil being treated. The type and amount are dictated by the soil type, the size or gradation of its particles, its moisture content, and the loadbearing specifications of the project.

Binders may be dry, such as lime or cement, or wet, such as lime slurry. Binders can be spread by dump trucks, spreader trucks, or tanker trucks ahead of a soil stabilizer. A soil stabilizer is a machine with a flat-bottomed mixing chamber attachment. The soil is picked up and can be mixed, blended, or pulverized in the chamber. The soil is then put back down and flattened on the roadbed or surface. A soil stabilizer can stabilize a road base to approximately a 12-inch depth in a single pass. If these types of machines are used, binders can also be added directly to the soil in the stabilizer mixing chamber.

The method of adding binders depends on the type of stabilizer equipment used, the size of the project, and the types of additives. The binder must be thoroughly mixed with the base material and the mixture pulverized for uniform color and size. Common binders for roads include:

- Cement
- Lime
- Fly ash
- Asphaltic binders
- Calcium chloride
- Polymer

4.1.1 Cement

Cement is a fine, powdered hydraulic binder comprised of calcified lime compounds mixed with silica, alumina, and iron oxide. When mixed with water, cement will set and harden in air or under water. Soil cement is made by mixing dry portland cement and water with the soil. The soil is compacted and paved.

Cement binder provides the highest loadbearing capacity of all binders. It can be used in gravel, sand, or reclaimed asphalt. Silt and clay may also be stabilized with cement.

4.1.2 Lime

Lime is especially effective as a binder in clay-bearing soils and aggregates because it reacts both chemically and physically with the soil. Lime used for stabilization includes quicklime (calcium oxide) and hydrated lime (calcium hydroxide). Both are burned forms of limestone.

Lime is applied as a dry powder, as a slurry, or in pellets. When reacting chemically with clay soils, lime produces coarse, friable particles from the clay and produces a cementing action with silica. Lime reacts favorably with soils in the higher plasticity index ranges. Lime treatment is used for sub-base stabilization, base stabilization, and lime modification.

4.1.3 Fly Ash

If lime is used on low plasticity index soil, a second additive may be used to produce the required reaction to the lime. This is called fly ash treatment. Fly ash is a residue recovered from smoke produced by coal-fired burners. Fly ash consists of minute spherical particles of glass, crystalline matter, and carbon. Its chemical analysis resembles portland cement without lime. It reacts chemically with lime to enhance the strength of the soil, giving it cement-like properties. Fly ash combined with lime also produces an excellent road base when mixed with aggregate.

4.1.4 Asphaltic Binders

Asphaltic binders are often used as stabilization binders, particularly when rehabilitating or reclaiming deteriorated surfaces. They can be used in all coarse-grained and mixed-grain soils. Asphaltic binders include:

- Asphalt emulsions
- Cutback asphalt
- Road tars
- Straight asphalt
- Foamed asphalt

Depending on the type, asphaltic binders may be mixed cold or warm with the material being stabilized. Generally, the asphaltic binder content is about 1 to 2 percent by weight of the dry weight of the material.

4.1.5 Calcium Chloride

Calcium chloride is another material used as a binder. It increases soil density like other binders and is particularly valuable in colder climates.

Calcium chloride lowers the freezing point of water. The roads that use it for stabilization of the base material are better protected against frost damage. The use of calcium chloride for stabilization may require less compaction. It can be applied dry or in a solution.

4.1.6 Polymers

Widespread use of polymers as soil stabilizers is a fairly recent occurrence. A simple way to explain polymers is that it and one or more other materials are bonded together by chemical reaction. In soil stabilization, these products bind with soil particles to create a durable surface that permits motorized traffic. Application requirements vary with each product, so it is important to follow the manufacturer's instructions. Most products are applied as a viscous liquid that dries quickly (in a matter of minutes to hours, depending on the weather). One such product was used in Afghanistan by US Marines, who nicknamed it Rhino Snot because of its appearance. Some polymers can be used in a weaker solution to suppress dust for erosion control.

4.2.0 Preparing the Subgrade

To prepare the subgrade, top soil and large stones are removed from the proposed right-of-way, and the subgrade is carefully shaped to grade. The desired site profile and the mix uniformity depend on accurate grading. The grading is done by a motor grader or by a fine grade trimmer. A trimmer controls profile with extreme accuracy and can accomplish the job in one pass.

After grading, scarifying or disc harrowing to the depth of stabilization is sometimes used. However, a stabilizing machine capable of mixing the binder and pulverizing the soil in one operation can eliminate this step.

4.3.0 Spreading the Binder

Dry binders must be evenly distributed over the area to be stabilized to assure that the specified soil-to-binder ratios are met. A box-type spreader attached to a dump truck or a rear-mounted spreader on a pneumatic truck is used for this. The spreader meters out the material as the load is discharged ahead of the soil stabilizer.

A lime slurry or other liquid binder is best distributed by tanker truck, which will spread the product to pre-calculated depths as called for in the specifications.

4.4.0 Determining the Logistics of the Stabilization Job

When planning the stabilization job, logistics play an important part. Planning considerations include the following:

- *Atmospheric conditions* – The temperature should be mild with little or no precipitation expected. The initial mixing process should not be done in heavy rain. After the initial course is compacted and set, moderate rain is no problem.
- *Binder material supply* – Binder material must be available as needed to avoid delays in application.
- *Water supply* – When cement is being used for a binder, the optimum moisture content is critical. Equipment must be available to supply water for the mix as needed. Lime is usually added to soils with high moisture, so water is not an issue for mixing. However, sufficient water should be available to control dust when using lime.
- *Compaction equipment* – Appropriate equipment should be available for compaction to begin immediately after the stabilizer finishes the mixing and grading process.
- *Trimmer equipment* – After compaction, the surface should be trimmed to accurate grade and crown specifications.

4.5.0 Stabilizing

A soil stabilizer machine cuts, mixes, and pulverizes native, in-place soils or select material to modify and stabilize the soil for a strong base. Stabilizing agents may be mixed into the soil by the machine during this process. Stabilizers use blades on rotors or teeth on mandrels for cutting, and many have automatic depth control. Some models have a mixing chamber, which provides for consistent blending of materials. *Figure 30* shows a large stabilizer.

201F30.EPS

Figure 30 ◆ Large soil stabilizer.

To carry out stabilization by mechanical methods, the entire area must be mixed thoroughly to the designated limits. The simplest type of equipment is a traveling mixing plant with a single rotor mixer design. This machine can move rapidly and can mix an area in several passes.

Cutting depth is usually up to 12 to 15 inches, and the cutting width is typically 8 feet. The production rate of the stabilizer can be calculated with these numbers and the travel speed, or it can be looked up on production charts.

Stabilizer features may include the following:

- Mechanical rotor drive
- Microprocessor control of major machine systems
- Four steering modes with automatic rear wheel alignment
- Choice of rotors
- Roll-over protection
- Liquid additive and/or water spray systems

The machine should be able to cut up or down, since each operation will determine which type cutting will result in more complete mixing. The ability of the machine to accomplish the specified mixing depth is critical to permit one-pass operation. Sometimes it is desirable to make an initial pass to establish proper gradation of the soil, followed by application of the binder, and then make a second pass with the soil stabilizer. Some stabilizer models offer the ability to reverse direction without turning around—a time-saving feature.

4.6.0 Compacting

The stabilized area must be compacted after the stabilizer has finished. Only adequate compaction produces the required density. If cement is used, the compaction must be done immediately after mixing because the cement in the mixture will begin to set. A cement mixture also requires that a vibratory roller compaction machine be used to obtain the needed results. *Figure 31* shows a vibratory roller.

When compacting a lime-stabilized area, the compaction must be done when the moisture content has dropped to a point within the specified range. This may be immediately or after a short time.

The number of passes made by the compactor depends on the amount of compaction to be achieved. This will, in turn, depend on the material moisture content, the layer thickness, the compactor type, and the degree of compaction called for in the specifications. With proper compaction, a final curing stage may be eliminated before the next course is applied.

201F31.EPS

Figure 31 ◆ Vibratory roller.

4.7.0 Checking Quality

Each layer of the roadbed must meet design specifications to ensure that the final product will stand up under projected loads and will meet smoothness requirements. Each stabilized layer must be checked for:

- Compaction or density
- Thickness
- Mix uniformity
- Gradation
- Loadbearing capacity
- Moisture content
- Binder content

5.0.0 ◆ SAFETY GUIDELINES

Once you are comfortable reading the construction plans for your project, it is important that you relate them to the physical building site. As you have learned in previous modules, safety is your first priority on the job, so take responsibility for your own safety. Once you are familiar with the site plans, walk the construction area to get a firm idea of the work that you need to perform and to identify potential safety hazards. Report any potential problems you see to your supervisor. Never assume that the project engineer or your supervisor is aware of a potentially unsafe condition.

5.1.0 Personal Safety

Before you start any earthwork, walk the site if possible. Be alert for large depressions or other obstacles that are not shown on the plans. This can be especially important in undeveloped areas where large animal burrows (such as turtle holes) are present. Driving over unexpectedly rough terrain can throw your equipment off balance and place you in danger of tipping over. These areas may need to be corrected during clearing and grubbing.

In the warm weather, look for swarming insects that could mean a hornet's nest or beehive in the area. Many people are highly allergic to insect stings and need to carry special medication to counteract the sting effects. Notify your supervisor of your findings. The nest may not be able to be removed, but other workers should be notified of its existence.

When it is not possible for you to walk the site before starting work, you must be especially careful the first time that you drive over a section of land. Always scan the terrain looking for potential hazards, drive slowly, and be ready to stop.

5.1.1 Site Contamination

Whether you are working at a new construction site or performing demolition on a building, there is always the possibility that you will uncover a contaminated site. While you are walking the site, look for signs of illegal dumping and other contamination. Contaminants can be in the soil, buried underground, or stored in some container such as a barrel. Some warning signs of potential contamination are as follows:

- Puddles of fluid on the ground, especially fluid with an unusual color or odor
- Any fluid seeping out of the ground
- Unusual or foul odors
- Unmarked barrels or tanks—often buried or otherwise camouflaged
- Unexplained and remarkably bare patch of land in an area that is otherwise heavy with vegetation

Identifying contaminants is a complicated and time-consuming procedure that requires advanced education and training. Heavy equipment operators are not expected to identify contaminants, but you are required to report any potential contaminations to your supervisor.

5.1.2 Sewage

Construction workers working at an excavation site may accidentally uncover live sewers or septic tanks. Not only is this unpleasant, but contact with raw sewage can place you at risk for disease. One of the most common diseases is Hepatitis A. Hepatitis is a viral disease that attacks the liver. It is contracted when some of the contaminant is swallowed and can be passed on by person-to-person contact. To prevent infection, wear latex or similar gloves when contacting sewage, wash your hands frequently, avoid touching your face, and do not eat or drink in the area.

WARNING!
Some people are very allergic to latex. Some symptoms can be mild, such as an itchy red rash where the latex contacts the skin. Other symptoms are life threatening, such as anaphylactic shock, which is characterized by swelling of the airway, causing shortness of breath. If you have these symptoms when you wear latex gloves, remove them immediately, wash the affected area, and seek medical assistance if necessary. Then use non-latex gloves.

If you do become ill after working on a job where you were exposed to sewage, tell your doctor. Hepatitis A is not often fatal, but there have been cases where healthy individuals have died from this disease. If you frequently come in contact with sewage at your job, talk to your supervisor about being vaccinated against this disease.

5.1.3 Soil Contaminants

Excavation work generates a great deal of dust from soil. Mixed in with the dust can be any number of disease-causing organisms. Many of these organisms come from animal droppings, especially bird droppings, so keep an eye out for animal nesting areas when you are walking the site. Two diseases that can be caused by bird droppings are histoplasmosis and cryptococcosis. Both of these diseases usually attack the lungs. The symptoms from these diseases are usually mild but can be severe in people with pre-existing conditions. Both of these diseases are especially dangerous to people who have acquired immunodeficiency syndrome (AIDS).

You can help to limit your exposure to soil contaminants by wearing a well-fitted respirator with a HEPA filter, showering and changing into clean clothes at your job site when possible, avoiding eating or drinking in the dusty area, and periodically spraying the area with water.

5.2.0 Safeguarding Property

As an employee, you are responsible to your company for your actions on the job. When you think of safeguarding property, you probably think of the equipment you operate. Use it responsibly, report defects, and secure it when not in use. In addition, you need to treat the building site responsibly because the site owners have contracted with your employers to construct a project to certain specifications that are shown on the plans.

You can treat the job site responsibly by following the plans and being careful not to unduly damage the site. This is especially important on small jobs where a designated equipment staging and maintenance site has not been selected. It may be up to you to select a site, so do so with the construction plans in mind. For example, it would be hard to maintain a flowerbed if oil and hydraulic fluid dripped into the soil, so if your job site has no designated areas for loading, unloading, parking, and maintenance, select a site well away from the building.

Know the general plans well enough to avoid damage that heavy equipment can cause. Locate any existing structures, roads, or walkways to ensure that you can avoid driving over these with any equipment that might damage them. Know the location of underground structures such as culverts, pipes, and septic systems at the site. Driving over these structures with heavy equipment can crush them and cause your equipment to be thrown off balance, causing an accident.

Never cross property lines without permission. Not only is this trespassing, but heavy equipment can also damage the terrain and cause the property owner to have bad feelings towards your employer. Further, there may be underground structures on the adjacent property that your equipment could damage.

When the plans call for certain trees to be preserved, know their locations. Don't guess. Ask your supervisor if you need help. Look on trees and in bushes for survey markers or bench marks (*Figure 32*). Never destroy or move these markers. Notify the project engineer or your supervisor of the locations of any markers not shown on the site plan.

5.3.0 Underground Utilities

The development of site/plot plans requires that the locations of underground utilities and other existing conditions be accurately shown on both the original site/plot plans and the final as-built plan drawing(s). This information is especially critical when the site for a proposed new construction contains previously installed underground facilities.

The locations of existing underground utilities at a particular site can be obtained by contacting the One-Call system in the state where the construction is to be performed. Some states have a state-wide One-Call system. Other have multiple systems, but no area in the state is without coverage. The One-Call system is a communications system established by utilities, governmental agencies, or other operators of underground facilities to provide a single point of contact that contractors can call to give notification of their intent to do excavating at a particular construction site.

201F32.EPS

Figure 32 ◆ Survey marker.

Once contacted, the One-Call system then notifies each of its members. Upon notification, the individual members research their data about the area of interest to determine if their facilities are affected. If affected, they provide temporary markings and information, such as depth and line descriptions, for their facilities at that construction site. *Figure 33* shows a handy pocket Dig Safely card distributed to contractors by the New York One-Call notification center. It gives the procedure for contacting the center by telephone or the internet. All other states have readily available cards similar to the one shown for New York.

Temporary markers used to indicate the various underground utilities on site consist of flags (*Figure 34*), wire, stakes, or paint applied to the route. The choice depends on the circumstances and the length of the line to be marked. The markers used must be durable enough to last throughout the duration of the proposed work. Temporary markers are color coded according to the American Public Works Association (APWA) universal color codes so that the type of the underground utility being marked is easily understood. The APWA universal color codes are as follows:

- *Red* – electric
- *Yellow* – gas and oil
- *Orange* – communication/CATV
- *Blue* – potable water
- *Purple* – reclaimed water
- *Green* – sewer
- *Pink* – temporary markings
- *White* – proposed excavation

Existing pipelines are also identified by permanent markers (*Figure 35*) required by Department of Transportation (DOT) regulations. Signs are required at all road crossings, railroad crossings, water crossings, and intervals along the route of the line. These markers include information about the type of material carried by the pipeline. They also show the name of the company responsible for it and the phone number that can be used to contact the company.

It is very important that you look for these markers before you begin any excavation work. You should be suspicious if you do not see underground utility markers on a site that already has an existing building or is located in a populated area. Even when a building site is located in a rural area, it is very unlikely that there will not be some type of underground utilities in the area.

WARNING!

Always check the plan to see that there are no underground utilities in the area before you do any grading or excavation work. If the plans do not show underground utilities, talk to your supervisor to ensure that there are none located in the area. Hitting underground utilities can be fatal.

201F34.EPS

Figure 34 ◆ Temporary marking flags.

201F35.EPS

Figure 35 ◆ Permanent pipeline markers.

FRONT

BACK

201F33.EPS

Figure 33 ◆ New York Dig Safely card.

1. Mining operations can be set up temporarily for a special project or permanently as a business venture.
   a. True
   b. False

2. During excavation and fill activities, slope control is handled with grade stakes that are placed by _____.
   a. the survey crew
   b. a competent person
   c. the inspector
   d. the contractor

3. For a highway project, the plan sheet shows the center line, the edge of pavement, the right-of-way limits, and the _____.
   a. ditch line
   b. survey lines
   c. reference stakes
   d. profile

4. The purpose of the pre-construction conference is to discuss all of the following *except* _____.
   a. contract negotiations
   b. erosion control
   c. stormwater control
   d. unusual conditions

5. When preparing a job site for excavation or construction, the removal of objects below the original ground is called _____.
   a. clearing
   b. stripping
   c. grubbing
   d. scalping

6. Cycle time is defined as the length of time it takes for a piece of equipment to _____.
   a. load, haul, and dump its load and return for another load
   b. return to the excavation site after it has dumped its load
   c. haul its load to the dump site and return for another load
   d. load fill in all of the vehicles needed for the day's work

7. The EPA requires that operators of construction sites that are one acre or greater establish plans to reduce sedimentation, pollution, and _____.
   a. traffic
   b. erosion
   c. stormwater
   d. riprap

8. Open pits or holes constructed to collect groundwater that are then pumped out mechanically are called _____.
   a. sumps
   b. drainage ditches
   c. holding ponds
   d. borings

9. A temporary opening in which something, such as a pipe or box culvert, is placed and then covered is a _____.
   a. channel
   b. trench
   c. ditch
   d. pit

10. All of the following are steps in the earthmoving cycle *except* _____.
    a. dumping
    b. loading
    c. windrowing
    d. hauling

11. The single largest factor in cycle times for earthmoving operations is _____.
    a. loading
    b. hauling
    c. scraping
    d. excavating

12. The type of dump site used to store material for later use is called a(n) _____.
    a. spoils pile
    b. embankment
    c. stockpile
    d. landfill

13. Waste fill is recycled, treated material approved for use in certain _____.
    a. outdoor products
    b. landscaping projects
    c. playgrounds
    d. landfill areas

14. The binder that provides the highest load-bearing capacity is _____.
    a. cement
    b. lime
    c. fly ash
    d. asphalt

15. When cement is used as a binder, compaction is done after the cement has had time to harden.
    a. True
    b. False

16. When you are working on a job, your first priority is to perform your job _____.
    a. quickly
    b. efficiently
    c. safely
    d. well

17. You have been assigned to a new work site. Walking the site before starting your work, you notice that there are a number of large boulders in the area that are not shown on the plans. You know that you need to notify your supervisor because _____.
    a. the plans are supposed to accurately show the area
    b. these could be a safety hazard to heavy equipment
    c. the property owner must remove them
    d. the engineer may wish to change the building design

18. You are walking the new site before starting your work and you notice that there is some foul-smelling fluid seeping out of the ground. You know that you need to notify your supervisor because it _____.
    a. is probably an underground spring
    b. could be hazardous contamination
    c. needs to be filled before work begins
    d. needs to be shown on the site plans

19. The site plans show that a particularly large tree on the building site needs to be preserved. What do you do?
    a. Find the tree before you start doing any work.
    b. Try to look for the tree while you are working.
    c. Avoid working around any large tree you see.
    d. Ignore it.

20. You are working on an addition to a school and notice red, yellow, orange, and blue markings on the ground. You know that these markings show _____.
    a. grade instructions
    b. existing utilities
    c. ditch locations
    d. property lines

Summary

Earthmoving is the process of digging, loading, hauling, and dumping any material that is needed for, or in the way of, construction. Types of earthmoving include clearing and grubbing, embankment construction, excavation, and backfilling and compaction.

Before construction can begin, other work needs to be done to determine what type and how many pieces of equipment will be needed. This will include analyzing the characteristics of the soil, planning haul routes, putting up temporary signing, and staking.

Clearing and grubbing work is the first activity to be performed. It serves to remove all the vegetation, unwanted structures, and obstructions from the site. Excavation is the digging of the earth to remove material from a site. Fills, backfills, and embankment construction involve the filling in of pits or trenches, or building up of ground in order to raise the grade.

Staking is very important for the equipment operator because it guides the operations. Different kinds of stakes have different information. Types of stakes include right-of-way, slope, grade finish, center line, and ditch line.

The selection of equipment for any type of excavation job will depend on the soil characteristics, the volume of material to be excavated, and distance the material has to be hauled. There are many different types of equipment with different sizes and configurations for almost every type. In order to be efficient, you must select the right type and size of equipment to meet the job at hand.

Soil stabilization and compaction are necessary if the soil is to carry the required weight. Binders such as cement and lime are added to strengthen the soil.

It is very important to always operate your machine in a safe manner. You must be aware of hazardous terrain on the work site and must always check for buried utilities such as gas, electric, water, and sewer lines. If your machine damages buried utilities, it could result in serious injury or death to you or a co-worker, as well as costly project delays.

Notes

Trade Terms Introduced in This Module

Backhaul: The return trip of a piece of equipment after it has completed dumping its load.

Bedrock: The solid layer of rock under the earth's surface. Solid rock, as distinguished from boulders.

Center line: Line that marks the center of a roadway. This is marked on the plans by a line and on the ground by stakes.

Cohesive: The ability to bond together in a permanent or semi-permanent state. To stick together.

Consolidation: To become firm by compacting the particles so they will be closer together.

Core sample: A sample of earth taken from a test boring.

Cycle time: The time it takes for a piece of equipment to complete an operation. This normally would include loading, hauling, dumping, and then returning to the starting point.

Dewatering: Removing water from an area using a drain or pump.

Discing: The mechanical process of using sharp steel discs to cut through and turn over a top layer of soil, usually 4 to 12 inches deep.

Dragline: An excavating machine having a bucket that is dropped from a boom and dragged toward the machine by a cable.

Embankment: Material piled in a uniform manner so as to build up the elevation of an area. Usually, the material is in long narrow strips.

Expansive soil: A clayey soil that swells with an increase in moisture and shrinks with a decrease in moisture.

Gradation: The classification of soils into different particle sizes.

Groundwater: Water beneath the surface of the ground.

Impervious: Not allowing entrance or passage through; for example, soil that will not allow water to pass through it.

Inorganic: Derived from other than living organisms.

Pay item: A defined piece of material or work that the contractor is paid for. Pay items are usually expressed as unit costs.

Pit: An open excavation that usually does not require vertical shoring or bracing.

Organic: Derived from living organisms such as plants and animals.

Riprap: Loose pieces of rock that are placed on the slope of an embankment in order to stabilize the soil.

Select material: Soil or manufactured material that meets a predetermined specification as to some physical property such as size, shape, or hardness.

Shoring: Material used to brace the side of a trench or the vertical face of any excavator.

Soil test: A mechanical or electronic test used to determine the density and moisture of the soil, and therefore the amount of compaction required.

Spoils: Material that has been excavated and stockpiled for future use.

Stormwater: Water from rain or snow.

Test boring: To drill or excavate a hole in the earth in order to take a sample of the material that rests in different layers beneath the surface.

Test pit: See *test boring.*

Trench: A temporary long, narrow excavation that will be covered over when work is completed.

Water table: The depth below the ground's surface at which the soil is saturated with water.

Windrow: A long, straight pile of placed material for the purpose of mixing or scraping up.

Additional Resources

This module is intended to be a thorough resource for task training. The following reference works are suggested for further study. These are optional materials for continued education rather than for task training.

Avoiding Enforcement Actions: How to Effectively Manage Your Construction Site and Survive Regulatory Scrutiny, Grading and Excavation Contractor magazine. July/August 2005. Carol L. Forrest. Santa Barbara, CA: Forester Communication.

Caterpillar Performance Handbook, Edition 27. A CAT® Publication. Peoria, IL: Caterpillar, Inc.

Chemical Soil Stabilization, January/February, 2003. Janis Keating. *Erosion Control* magazine.

Excavating and Grading Handbook, 1987. Nicholas E. Capachi. Carlsbad, CA: Craftsman Book Company.

Heavy Equipment Repair, 1969. Herbert L. Nichols, Jr. Greenwich, CT: North Castle Books.

United States Environmental Protection Agency (EPA) website, National Pollutant Discharge Elimination System (NPDES).

http://cfpub1.epa.gov/npdes/stormwater/const.cfm?program_id=6

Figure Credits

Dale Chadwick, rdchdwck@comcast.net, 201F01

P&H Mining Equipment, 201F02

Deere & Company, 201F03

Sundt Construction, Inc., 201F04, 201F06

Kittelson & Associates, 201F05

John Hoerlein, 201F07

Reprinted courtesy of Caterpillar Inc., 201F10, 201F11, 201F30, 201F31

Komatsu America Corporation, 201F12–201F15

U.S. Environmental Protection Agency, 201F20

Trench Shoring Services, 201F21

Young Corporation, 201F22

Port Industries, Inc., 201F23

Topaz Publications, Inc., 201F32–201F35

NCCER makes every effort to keep these textbooks up-to-date and free of technical errors. We appreciate your help in this process. If you have an idea for improving this textbook, or if you find an error, a typographical mistake, or an inaccuracy in NCCER's Contren® textbooks, please write us, using this form or a photocopy. Be sure to include the exact module number, page number, a detailed description, and the correction, if applicable. Your input will be brought to the attention of the Technical Review Committee. Thank you for your assistance.

Instructors – If you found that additional materials were necessary in order to teach this module effectively, please let us know so that we may include them in the Equipment/Materials list in the Annotated Instructor's Guide.

Write: Product Development and Revision
National Center for Construction Education and Research
P.O. Box 141104, Gainesville, FL 32614-1104

Fax: 352-334-0932

E-mail: curriculum@nccer.org

Craft _____ Module Name _____

Copyright Date _____ Module Number _____ Page Number(s) _____

Description _____

_____

_____

_____

(Optional) Correction _____

_____

_____

(Optional) Your Name and Address _____

_____

_____

22202-06

Dump Trucks

22202-06
Dump Trucks

Topics to be presented in this module include:

Overview

Many heavy equipment operators started at a young age with a toy dump truck and a sand pile. But operating a construction dump truck is not child's play. These powerful trucks move soils and other materials and tow other equipment. It takes a skilled operator to load, move, and unload them safely.

A skilled operator can recognize several different types of dump trucks and choose the best one for the task at hand. Safe operators have a working knowledge of all of the instruments and controls, perform daily inspections and maintenance, and know the rules of the road for both public roads and on-site haul roads.

NOTE

Prior to operating any dump trucks on public roads, trainees must have a commercial driver's license (CDL).

Objectives

When you have completed this module, you will be able to do the following:

1. Describe why dump trucks are widely used in the construction industry.
2. State the types of dump trucks and their uses.
3. Describe the function and operation of the dump hoist, power takeoff unit, auxiliary axle, engine retarder, differential lockout, air brake system, and manual transmission.
4. Demonstrate and state the steps of the pre-operational safety inspection.
5. Perform the proper warmup, operation, and shutdown procedure.
6. State the duties and responsibilities of a dump truck operator.
7. Identify the controls of a dump truck.
8. Safely operate a dump truck.
9. Back up with a trailer attached to the dump truck.

Trade Terms

Articulating	Tandem-axle
Auxiliary axle	Trailer
Articulating	Cab guard
Auxiliary axle	Clutch
Cab guard	Governor
Clutch	Hoist
Governor	Hydraulic
Hoist	Lug
Hydraulic	psi
Lug	rpm
psi	Reservoir
rpm	Split load
Reservoir	Tag axle
Split load	Tandem-axle
Tag axle	Trailer

Required Trainee Materials

1. Pencil and paper
2. Appropriate personal protective equipment

Prerequisites

Before you begin this module, it is recommended that you successfully complete *Core Curriculum Level One*; *Heavy Equipment Operations Level One*; *Heavy Equipment Operations Level Two*, Module 22201-06.

This course map shows all of the modules in the second level of the *Heavy Equipment Operations* curriculum. The suggested training order begins at the bottom and proceeds up. Skill levels increase as you advance on the course map. The local Training Program Sponsor may adjust the training order.

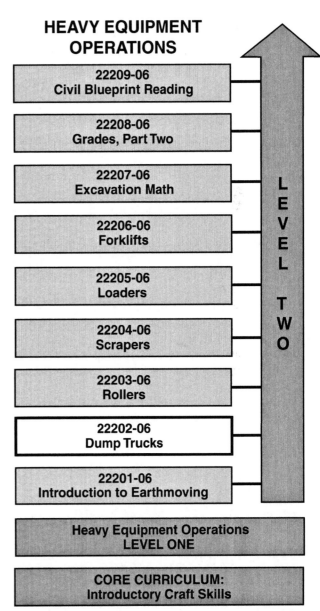

HEAVY EQUIPMENT OPERATIONS

22209-06 Civil Blueprint Reading
22208-06 Grades, Part Two
22207-06 Excavation Math
22206-06 Forklifts
22205-06 Loaders
22204-06 Scrapers
22203-06 Rollers
22202-06 Dump Trucks
22201-06 Introduction to Earthmoving

LEVEL TWO

Heavy Equipment Operations LEVEL ONE

CORE CURRICULUM: Introductory Craft Skills

202CMAP.EPS

1.0.0 ◆ INTRODUCTION

Dump trucks are among the most widely used vehicles in construction work. They are built on a heavy-duty chassis and have a large bed that is used to move large amounts of loose material, such as sand, soil, gravel, or asphalt. Dump trucks can be quickly loaded using a chute or a piece of heavy equipment, such as an excavator (*Figure 1*). The beds can be raised and tilted back with a **hoist** that is mounted on the truck, so the driver can quickly unload the cargo without help.

Dump trucks can also be used to tow dump **trailers**, equipment trailers, and other heavy equipment. Because of their great size and weight, dump trucks can be outfitted with a snowplow and then be used to clear snow from roads, parking lots, and work areas. In addition, they may be fitted with a spreader that scatters sand, salt, or other material onto icy surfaces.

Because of the dump truck's usefulness, large size, and power, the job of a dump truck operator is complex. These trucks can be a danger to operators, workers, and the public when used improperly or unsafely. As a dump truck operator, you need to know how to maintain safe control of your vehicle at all times. This includes not only driving your truck but also safe loading and unloading. When you operate your truck at a construction site, you will be responsible for your safety as well as the safety of your co-workers. When operating your truck on a public road, you will be responsible for the safety of the public. It is up to you to be sure that your truck is well maintained and in good working order before you use it.

You may work on jobs where you never drive your dump truck off the job site. But when you drive your dump truck on a public road—even for a short distance—you must have a commercial driver's license (CDL). The CDL program sets forth minimum national standards that drivers of commercial vehicles must obey. States may choose to add to those standards, so it is your job to know and follow local laws. When you take your truck onto a public road, it is your responsibility to ensure that the vehicle meets all state and local requirements and that you are licensed and trained to operate the vehicle safely.

2.0.0 ◆ TYPES OF DUMP TRUCKS

Dump trucks may be divided into two major categories: on-road trucks and off-road trucks. On-road trucks vary in size and load capacity and may be operated on public roads. Off-road trucks are usually much larger than on-road vehicles and are prohibited by law from routine operation on public roads because of their great size and weight. You as the driver will have the responsibility of knowing which trucks may be operated on a public road.

Dump trucks, just as all motor vehicles, are rated by the amount of weight that their axles may safely carry. You can find the truck's weight capacity in the operator's manual. It is the driver's responsibility to ensure that the maximum load limit for the vehicle is not exceeded. Operating an overweight vehicle can damage the vehicle and place you at risk for an accident.

Roads and bridges are also rated for the weight they can safely support. As trucks became heavier, the federal government introduced the Bridge Formula to protect our nation's bridges and roadways from damage due to excessive weight. The Bridge Formula specifies the maximum weight that each axle on a truck can legally carry. Although the federal government sets the maximum weight, it allows each state to make laws that are even stricter than the federal standards, so you need to know the laws for the states in which you work.

Figure 1 ◆ Excavator loading a dump truck.

 WARNING!
Never operate an overweight dump truck. Not only can it damage the vehicle, but it is also a safety hazard. Carrying too heavy a load negatively affects the maneuverability of the truck and stresses the vehicle's brakes, suspension, axles, and tires, placing you at risk for an accident.

202F01.EPS

2.1.0 On-Road Dump Trucks

The on-road dump truck designation does not mean that all trucks of this type are operated on public roads, since these types of dump trucks may also be driven off-road. Rather, the designation means that the truck falls within the legal size limits for vehicles driven on public roads. When on-road vehicles are driven on public roads, they must be in good working order and adhere to all local laws and regulations, including proper licensing and vehicle registration, state and local safety inspections, and traffic laws. Before you drive a vehicle on a public road, you must ensure that the vehicle is roadworthy and that you have proper documentation to prove that both you and the vehicle are permitted to be on the road. If you are stopped by the police or are involved in an accident, you will be held responsible for illegal operation of a vehicle on a public road.

> **NOTE**
>
> In 1982, the federal government increased the maximum width of vehicles traveling on public roads from 96 inches to 108 inches (8' to 8'-6"). Some states still limit the legal width of on-road vehicles to 96 inches. The legal length of on-road vehicles varies greatly from state to state.

2.1.1 Standard Dump Trucks

Standard dump trucks are the most commonly used dump trucks. These vehicles haul light, medium, and heavy loads. One characteristic that all standard dump trucks have in common is rigid frame construction. This means that the cab and the dump body are made from one frame and operate together.

Figure 2 shows a typical dump truck. Many trucks are outfitted with a load cover that is used to prevent debris from flying off the load while the truck is moving. Using the cover is especially important when the truck is driven at highway speeds. Standard rigid frame dump trucks are equipped with a **cab guard**, which protects the operator from falling objects during loading.

The key difference in these vehicles is the number of axles. Increasing the number of axles increases the amount of weight that a truck can safely carry. See *Figure 3*. Keep in mind that although a truck with multiple axles may safely carry a great amount of weight, you need to ensure that the road and bridges that you drive on can support that weight.

Standard dump trucks include the following axle configurations:

- *Single-axle* – Trucks with one rear axle that are used to haul light loads. These trucks normally have six wheels—two on the front axle and four on the rear axle.
- *Dual-axle* – Trucks with two rear axles that are used to haul heavy loads. These trucks normally have ten wheels—two on the front axle and eight on the rear axles. (Some states refer to any truck with dual-axle or above as **tandem-axle**, while some refer to a tandem-axle as a dual-axle only.)
- *Tri-axle* – Trucks with three rear axles that are used to haul heavier and larger loads than dual-axle vehicles. These trucks normally have 12 wheels—two on the front axle and ten on the rear axles.
- *Quad-axle* – Trucks with four rear axles that are used to haul the heaviest and largest loads. These trucks are rarely used.

CAB GUARD

CAB

ROLL-OUT LOAD COVER

DUMP BODY

202F02.EPS

Figure 2 ◆ Typical dump truck.

(A) DUMP TRUCK

(B) AUXILIARY AXLE DUMP TRUCK

202F03.EPS

Figure 3 ◆ Axle configurations.

A close look at *Figure 3B* shows that some of the tires at the rear of the truck are not touching the ground. These axles are called **auxiliary axles**, or secondary axles. (Primary axles are drive axles with the tires always touching the ground.) Auxiliary axles are lowered when the dump truck is carrying a load. When the tires on the auxiliary axle are touching the ground, they support part of the weight of the load, thus easing the strain on the other axles and permitting the truck to carry more weight than it can with just the primary axles.

2.1.2 Transfer Dump Trucks

Transfer dump trucks are commonly used in many western states because these states have different weight restrictions than other states. A transfer dump truck is a standard dump truck that tows a second dump bed on a trailer. When the bed of the standard dump truck is empty, the dump bed on the trailer is slid into the bed of the truck and then emptied using the truck's hoist mechanism. See *Figure 4*.

2.1.3 Pup Trailers

Pup trailers are similar to the trailers used in transfer dump trucks. See *Figure 5*. The chief difference is that the trailer has its own hoist mechanism so it can be emptied without using a standard dump truck.

2.1.4 Bottom Dump Trucks

Bottom (belly) dump trucks are trailers that are towed behind a truck tractor. See *Figure 6*. As the name implies, the load is dumped through doors in the bottom of the trailer. These trailers are used

202F04.EPS

Figure 4 ◆ Dump truck with transfer trailer.

202F05.EPS

Figure 5 ◆ Dump truck with pup trailer.

when the load needs to be dumped in a row. To dump the load, the driver drives forward while opening the doors and the load is deposited in a row. A good use for this type of dumper would be to dump materials, including asphalt, gravel, and dirt, in a row on a roadbed. The material handler can follow the trailer to quickly spread, lay, and compact the asphalt.

202F06.EPS

Figure 6 ◆ Bottom dump truck.

2.1.5 Side Dump Truck

Side dump trucks are trailers that are towed behind a truck tractor. This type of trailer dumps its load by tilting the trailer body on its side. This type of dumper can only be used when the dumpsite is long enough to permit the trailer access to it. The advantage to this truck is that it can be unloaded quickly.

2.2.0 Off-Road Dump Trucks

Off-road dump trucks are used on construction sites, quarries, and mining operations. These vehicles are usually very large, very wide, and designed to carry extremely heavy loads. As the name implies, these trucks are not permitted on public roads, therefore they are not restricted in size or weight. The heavy-duty construction of these types of vehicles means that they can withstand the tougher conditions of heavy work and short hauls.

NOTE

Don't get confused between on- and off-road dump trucks. On-road dump trucks may be used on- or off-road, but off-road dump trucks are too big to be used on a public road.

Safe operation of off-road trucks is vital because of their immense size, which makes these vehicles difficult and dangerous to operate. Drivers must pay special attention when operating these trucks to prevent unsafe situations. These trucks are often designed to carry their own weight in the dump bed and often have limited visibility to the sides and rear—some off-road trucks are so large that they are equipped with video cameras in various locations on the outside of the vehicle and monitors in the cab so the driver can monitor blind areas. The driver must use skill and knowledge to safely operate these machines.

Off-road dump trucks are categorized by their frames. See *Figure 7*. They can have a rigid frame (like a standard dump truck) or an **articulating** frame. Articulating means there is a joint that allows movement between two parts.

2.2.1 Rigid Dump Truck

Rigid dump trucks are built so the dump body and cab are on one continuous frame, similar to the standard dump truck body. The rigid frame makes maneuvering this truck difficult in tight areas, and it permits one or more tires to leave the ground when the truck is driven over rough terrain.

2.2.2 Articulating Dump Truck

Articulating dump trucks are designed so that the cab is joined to the dump bed with an articulating joint. This joint permits better maneuverability than rigid frame trucks have and allows for vertical movement between the cab and the bed. This vertical maneuverability permits all of the vehicle's wheels to remain on the ground while driving over rough terrain, thus improving safety.

3.0.0 ◆ INSTRUMENTS AND CONTROLS

Dump trucks are designed so the driver can perform most duties from the cab. Two typical cab configurations are shown in *Figure 8*. The layout and operation of the vehicle's instruments and controls will vary between manufacturers, models, and uses of the truck. It is critical to your safety and that of your co-workers and the public that you understand the operation of the controls before you use the truck.

Since each truck is slightly different, you must study the manual for your vehicle. Manuals should be considered part of your equipment and of no less importance than other components such as your seat belt. If you do not understand the written instructions, ask your supervisor or an experienced co-worker for help. Some common features are discussed briefly in the following sections.

3.1.0 Hoist

The hoist mechanism is used to raise and lower the dump bed. See *Figure 9*. Its control is normally located behind or beside the driver's seat. Generally, the dump body can only be raised when the transmission is in neutral, in first gear, or below a certain ground speed. On some truck models, moving the

(A) ARTICULATED TRUCK

(B) RIGID-BODY DUMP TRUCK

202F07.EPS

Figure 7 ◆ Off-road dump trucks.

202F08.EPS

Figure 8 ◆ Typical cab layouts.

dump body control automatically shifts the transmission into neutral. A dump body control normally has four positions: raise, hold, float, and lower. The control positions are as follows:

- *Raise (1)* – Pull the lever up to raise the dump body and empty the load.
- *Hold (2)* – Move the lever down to the hold position and the dump body will not move.
- *Float (3)* – Push the lever down to the float position and the dump body will seek its own level. This is the primary and default position.

- *Lower (4)* – Push the level all the way down to lower the dump body. When the lever is released, it will return to the float position.

3.2.0 Differential Lock

Many dump trucks that are operated in snowy or muddy work environments are equipped with a differential lock feature. To understand differential lock, you must first understand a differential's function.

HOIST —
MECHANISM   202F09.EPS

Figure 9 ◆ Hoist mechanism.

A differential is part of a drive axle assembly. It permits the wheels on the axle to turn at two different speeds and allows vehicles to turn corners smoothly. That is, the outside wheel turns faster than the inside one, because the outside wheel needs to cover more distance than the inside wheel in the same time. When the drive wheels of a vehicle are stuck on ice or in snow or mud and you see one wheel spinning and the other stopped, you are looking at the differential at work.

The differential lock enables the differential so that both wheels turn at the same speed. This allows the truck to move as long as one of the drive wheels has some traction. However, since the drive wheels will be locked, you will not have normal steering control of the vehicle, so before using this feature, ensure that the wheels on the steering axle are pointing forward.

WARNING!

Never attempt normal operation of a vehicle with differential lockout engaged. It can cause an accident.

3.3.0 Power Takeoff

Some dump trucks are equipped with a power takeoff (PTO) unit. A PTO unit is a mechanical link to an engine or transmission to which a cable, belt, or shaft may be connected to power another device. PTO units are usually located on the front of the truck. Sometimes they are used to control the hydraulics for operating the dump bed hoist. The control for the PTO is inside the cab, either on

the dashboard or on the floorboard. Operation is different depending on the types of transmission (standard or automatic) and truck model. Refer to the manufacturer's instructions for details about PTO operation.

3.4.0 Auxiliary Axles

Auxiliary axles are sometimes called secondary axles and are used only when the truck is loaded. They permit the truck to safely carry more weight than the vehicle would without the auxiliary axles by dividing the load weight over more axles. In dump trucks, these axles are usually mounted at the back of the truck, either in front of or behind the drive wheels. When the axle is in front of the drive wheels, it is called an auxiliary axle. When it is behind the drive wheels, it is called a **tag axle**.

Most auxiliary axles are pneumatically raised or lowered using a switch located in the cab. When your truck is equipped with auxiliary axles, it is your responsibility to know how and when to use them—improper use can not only damage the vehicle but can also cause an accident. Remember, in the eyes of the law, the truck operator is responsible for the safe and proper operation of the vehicle.

When a dump truck is empty, the operator can raise these axles to improve the vehicle's gas mileage and maneuverability as well as to reduce toll expenses, since many toll roads base fees on the number of axles on the road. When the vehicle is loaded, auxiliary axles are lowered to spread the load weight over more axles, thus reducing the stress on the other axles and the truck's suspension. When making low speed turns, you may need to raise the auxiliary axle for a short time to improve maneuverability, but you must lower the axle when you come out of the turn. Check the manufacturer's instructions for recommended operation.

CAUTION

Never lower the auxiliary axle when the vehicle is traveling at higher than the manufacturer's recommended speed. This action can severely damage the auxiliary axle tires.

3.5.0 Air Brake System

Most large trucks use air brake systems. The air brake system is a separate endorsement on the CDL test. If you have not taken the air brake part of the test, do not operate a vehicle with air brakes on a public road. An air braking system is very different from what you are used to in your personal vehicle. Your personal vehicle probably uses **hydraulics** to create stopping power, but trucks use air pressure to create stopping power.

In an air braking system (*Figure 10*), an air compressor, which is powered by the truck's engine, supplies air to the **reservoir** tanks for storage until braking is needed. The reservoir has a safety relief valve and the air compressor has a **governor**, so the pressure from either one cannot get too high. The compressed air is stored in the reservoir tanks until you push on the brake pedal. Then the air is released into the brake chambers where it creates a mechanical force to slow the turning of the wheels.

The normal pressure in the reservoir tanks is usually between 60 and 100 pounds per square inch (**psi**). Monitor the air pressure gauge while you're driving to be sure that you have adequate pressure. Many trucks have an alarm or buzzer that will sound when the air pressure is low. Older trucks may have a wig-wag, which is a flagging device that drops from the ceiling of the cab into the driver's view, to indicate low pressure. You should make a habit of glancing at the air pressure gauge to check the reading while you are driving and after applying the brakes.

Air brake systems must be treated differently than other braking systems. First, it takes time for the air to move from the reservoir tanks to the brake chambers. So stopping distances are much longer than with other systems. Second, frequent use of the brakes drains the air from the reservoir, reducing stopping power, so you must avoid pumping the brakes or driving with your foot resting on the brake. Third, the reservoirs in air brake systems gather condensation that can take up room in the tank, displacing air and lowering the availability of compressed air. Further, the water in the tank can freeze and damage the brake system, so the air reservoirs need to be drained periodically, usually after each shift.

Your instructor will help you learn how to use the air brakes in your vehicle. When you are on the job and using a different vehicle, practice using the brakes at low speed and in a safe area until you are sure of your abilities.

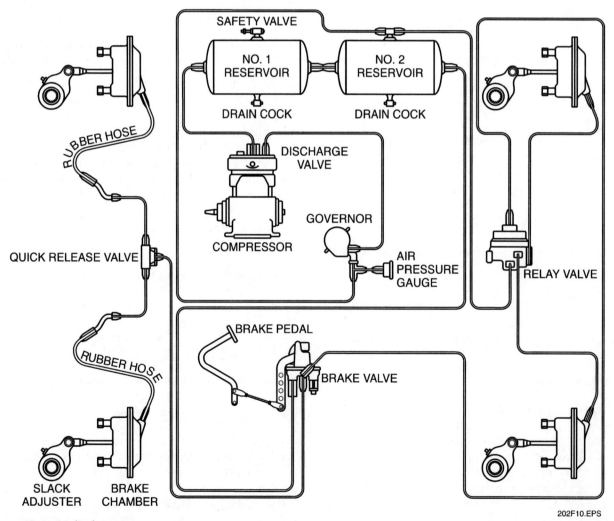

Figure 10 ◆ Air brake system.

202F10.EPS

3.6.0 Manual Transmission

The trucks you drive will most likely have manual transmissions, so you will need to operate the **clutch** and gearshift to place the truck in drive. The Roadranger® ten-speed is one of the most commonly used manual transmission units on dump trucks, although several other types are available. Regardless of the type of transmission, follow the manufacturer's shifting procedure. The steps in shifting will be the same although the pattern may be different. *Figure 11* shows the typical gearshift patterns.

Refer to *Figure 11*. The gear range control switch is located on the gearshift. This switch changes the transmission gears from low (first to fifth gear) to high (sixth to tenth gear). To use low gears, the switch is set to the down position. Low gears provide your truck with more power than the higher gears, so always start your truck from a stop in first gear. Lower gears are used to climb and descend hills and to slow the speed of the truck so that it can stop. Shifting to low gears before stopping helps to spare the brakes from wear and possible overheating. The following procedure describes how to upshift:

Step 1 Set the gearshift lever to the neutral position and start the engine.

Step 2 Wait for the air system to reach normal pressure (usually 60 to 100 psi), but check the manufacturer's recommendations.

Step 3 Check the range control button. If the button is up, push it to the down position.

Step 4 Push the clutch pedal to the floor with your left foot. Shift the gearshift to first

gear, and then slowly lift your foot off the clutch while slowly pressing the accelerator with your right foot until the truck moves.

CAUTION

When operating a clutch, always push the clutch to the floor before shifting and then slowly let out the clutch to drive. Never rest your foot on the clutch pedal while the engine is running—this is called riding the clutch and can cause premature wear of the clutch. Unless you are operating the clutch, your left foot should rest on the floor of the truck.

Step 5 Shift progressively through the gears at manufacturer's recommended **rpm** until you reach fifth gear. Then shift the range selector to high range by pulling the selector knob up and shift into sixth gear. As the lever passes through the neutral position, the transmission will automatically shift from low range to high range.

Step 6 Shift progressively through the upper gears at manufacturer's recommended rpm.

The following describes the procedure for downshifting:

Step 1 Move the shift lever from tenth through each successive lower gear to sixth. When in sixth gear, locate the range control button with your hand.

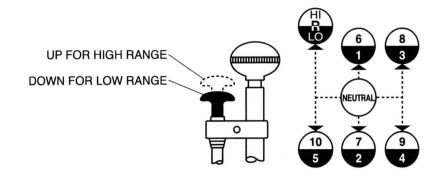

UPSHIFT
- START WITH RANGE SELECTOR IN DOWN POSITION
- SHIFT 1-2-3-4-5
- RAISE RANGE SELECTOR HANDLE
- SHIFT 6-7-8-9-10

DOWNSHIFT
- SHIFT 10-9-8-7-6
- PRESS RANGE SELECTOR HANDLE DOWN
- SHIFT 5-4-3-2-1

202F11.EPS

Figure 11 ◆ Typical gearshift patterns for a Roadranger® transmission.

Step 2 While in sixth gear, push the range control button down, and move the lever to fifth gear. As the lever passes through the neutral position, the transmission will automatically shift from high range to low range.

Step 3 Shift downward through each of the remaining gears.

 CAUTION

Never shift from high range to low range at high speeds. Never make range shifts with the vehicle moving in reverse gear. Always leave the vehicle parked in low gear.

3.7.0 Engine Retarder

Because dump trucks are very large and on-road trucks are designed to operate at highway speeds, they are difficult to stop. Using brakes to stop the vehicle causes a great amount of wear on the braking system. In addition, when a truck is driving downhill, constantly applying the brakes can cause them to overheat and fail. For this reason, on-road trucks are often equipped with an engine retarder, which works by reducing the power coming from the engine, thus slowing the truck. The most famous engine retarder is called the Jake Brake®. It is activated with a switch that is located in the cab of the truck. The Jake Brake® is often blamed for the familiar BRRMP noise that is often associated with trucks, although the manufacturer of the Jake Brake® system maintains that the noise is actually caused by a defective or modified exhaust system. Regardless of the cause, the noise is usually attributed to the engine retarder, and it annoys many people, so Jake Brake® use is prohibited in many areas (see *Figure 12*).

4.0.0 ◆ INSPECTION AND MAINTENANCE

Dump trucks are complicated machines. Unless you are specially trained as a service technician, you will probably not be performing servicing such as oil changes. Rather you will perform simple preventive maintenance tasks that are fast and easy to do. But don't underestimate their importance. A few minutes of your time at some simple tasks can extend the life of your vehicle and help keep you and others safe. Most preventive maintenance will be performed either before you use your truck or at the end of your shift.

4.1.0 Pre-Operational Check

You must inspect your vehicle before, during, and after each shift. Federal regulations may require that you provide written documentation of the inspections. Depending on the use of your vehicle, OSHA may require that the vehicle be inspected before each shift. At a minimum, your employer will have a written checklist that you must complete before you use the truck. Even if OSHA does not require it, your employer may require that you inspect your truck before you use it. A professional truck driver needs to be sure that a vehicle is in good working order whether it is operated off-road or on a public road.

A careful pre-operational inspection is recommended, but as a minimum you must ensure that the following are operational on both on-road and off-road vehicles: service and parking brakes; steering mechanism; headlights, parking lights, brake lights and all reflectors; tires; horn; windshield and windshield wipers; rear-view mirror; and coupling devices (if used). Make sure any problems you find are repaired before you use the truck. The following requirements cover a typical pre-operational check that you should perform before using your truck:

- Check under the truck for fluid leaks, loose wires or parts, and other damage.
- Examine the windshield and mirrors for dirt, cracks, scratches, or debris that will obstruct your view. Clean or repair as needed.
- Open the engine compartment and look for leaking fluid and worn wiring, insulation, or hoses. Check for loose electrical connections. Repair if necessary.

202F12.EPS

Figure 12 ◆ Engine braking prohibited sign.

- Inspect drive belts for serviceability and tightness. Check the level of oil, coolant, hydraulic fluid, and transmission fluid. Add fluids as required. *Figures 13* through *16* show the service points for these checks on a John Deere 250D off-road dump truck.
- If your truck is equipped with windshield washer fluid, check its level and top as needed.
- Check battery connections for tightness and corrosion, and clean or tighten as required. If possible, check electrolyte level in the battery and add water if necessary. Most batteries are maintenance free, so this is probably not necessary. Check that the battery vents are clear of debris. Clean if needed.
- Examine the air filter and clean or replace as needed.
- Check the pressure and condition of each tire. Check the manufacturer's recommendations for tire inflation.
- Clean all grease fittings and lubricate according to the manufacturer's recommendation. In cold areas of the country, this step will be performed at the end of a shift so the equipment will be warm and will more readily receive the lubricant.
- Inspect the braking system according to the manufacturer's instructions and adjust as necessary. If the truck is equipped with air brakes, ensure that the air reservoirs have been drained. In cold areas of the country, reservoirs are bled at the end of a shift to prevent the condensation from freezing and damaging the brake system.

CAUTION

Temperature affects tire pressure. Tires that are correctly inflated in a warm shop will become under-inflated as the truck is operated in freezing weather, which causes excessive tire wear. Check the tire pressure at the temperature in which the truck will operate.

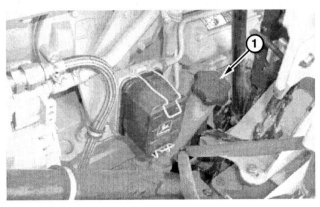

1 - Dipstick and Fill Cap

202F13.EPS

Figure 13 ◆ Check and fill engine oil.

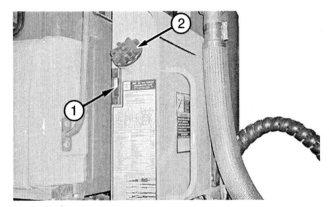

1 - Sight Glass
2 - Fill Cap

202F15.EPS

Figure 15 ◆ Check and fill hydraulic oil.

1 - Expansion Tank
2 - Filler Cap

202F14.EPS

Figure 14 ◆ Check and fill cooling system.

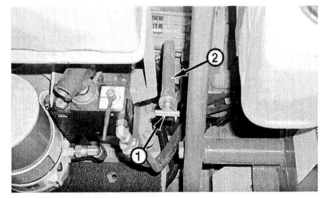

1 - Transmission Dipstick
2 - Fill Tube

202F16.EPS

Figure 16 ◆ Check and fill transmission fluid.

- Look in the dump bed and clean it if necessary. Ensure that the tailgate is operational and all pins, hooks, and chains are in place for safe and effective operation.
- In the cab, remove any trash or other items that can obstruct your view through the windows and windshield or your ability to reach the truck's controls.
- Ensure that all safety devices, such as first aid kit, fire extinguisher, and breakdown triangles are available and operational. See *Figure 17*.
- Set the parking brake and ensure that the gearshift is in neutral, or in park if the transmission is automatic. Adjust all mirrors and the seat. Fasten your seat belt.
- Start the engine and listen for any odd or unusual noises. Investigate as needed.
- Check the reading of all indicators and gauges.
- Check the foot pedals for proper operation.
- Check the operation of the windshield wipers.
- If your vehicle is equipped with communications, check its operation.
- Ask a co-worker to help you to check that the headlights, turn signals, reverse lights, backup alarm, and brake lights are operational.
- Ask a co-worker to be your safety spotter and check the operation of the hoist.

Perform your preventive maintenance tasks intelligently. When you are operating your truck in a dusty environment or on short hauls, perform lubrications more frequently. If your truck is pulling to one side excessively, stop and look at the tires and check the tire pressure with a gauge. When you sense a problem, check it out even if you checked it during your pre-operational check.

NOTE

It is normal for tires that are warm from use to have a higher air pressure than tires that are cold, because air expands as it is heated.

4.2.0 Post-Operation Procedure

The shutdown procedure is shorter than the warmup procedure, but is essential for proper cooling of all parts. Following these basic steps, as well as any special procedures in the manufacturer's recommendations, will increase the life of the engine.

Step 1   Stop the truck, preferably on a level location well out of the flow of traffic.

Step 2   Place the gearshift in neutral position.

202F17.EPS

Figure 17 ◆ Safety devices.

Step 3   Set the parking brake. Chock the wheels when necessary.

Step 4   If your truck has a diesel engine, run it at idle for two to three minutes before shutting the engine down. This permits the engine to cool and allows for gradual and uniform cooling of all engine parts. Turn the key to the OFF position.

Step 5   Perform a walk-around inspection and note leaks and broken or missing parts.

Step 6   Purge the air reservoirs of fluid by following the manufacturer's recommendation and your employer's policy.

Step 7   Clean all grease fittings and lubricate according to the manufacturer's recommendation and your employer's policy.

Step 8   Lock the cab and remove the keys.

Step 9   Report any deficiencies to the proper authority at your company.

 NOTE

Safely parking and securing your vehicle is as important as driving safely. Apply the service brakes to stop the truck. Move the gearshift to neutral and engage the parking brakes. Park on level surfaces if possible. If you cannot park on a level surface, block the wheels.

5.0.0 ◆ SAFETY GUIDELINES

Driving a loaded dump truck can be dangerous whether you drive on public roads or on a work site. As a professional driver, you must be a responsible, qualified, and authorized operator. To be responsible, you must recognize and avoid actions that place you or others at risk. Never perform any unsafe maneuver or operations, even if told to do so by another person. To be qualified, you must understand the manufacturer's written instructions, have training, including actual operation of the vehicle, be properly licensed, and know the safety rules for the job site. Current federal regulations state that at the time of your initial assignment and then at least annually, your employer must instruct you in the safe operation and servicing of equipment that you will use. Most employers have rules about operation and maintenance of equipment—make sure you know them. Finally, your employer must authorize you to operate a vehicle.

5.1.0 Personal Responsibility

You are the most important part of any safety program. When you are behind the wheel of a loaded dump truck, you are responsible for guiding several tons of material around a work site or on a public road. To be a responsible driver, you must avoid actions that can weaken your skills or judgment on the job. Never use drugs or alcohol on the job. They can impair your alertness, judgment, and coordination. When you need to take prescription or over-the-counter medications, seek medical advice about whether you can safely operate machinery. Only you can judge your physical and mental well being—don't take chances with your life or someone else's.

The key to staying safe is to develop good habits while you are learning a new skill. This way safety becomes second nature to you. The following are some good habits to practice:

- Read your operator's manual thoroughly. Know your machine.
- Always follow your job site's safety precautions.
- Wear your hard hat and seat belt. Use other personal protective equipment (PPE) as needed.
- Do not haul people in the dump bed or on the running board.
- Never get under the dump bed when it is raised unless it is securely blocked.
- Keep windshield, windows, and mirrors clean at all times.
- Use a proper three-point mounting and dismounting technique when entering or leaving the vehicle. Never enter or leave the truck by grabbing the steering wheel.
- Exit the cab facing the vehicle, and keep a firm grip on the handholds until you are on the ground.
- Never use cellular phones or other devices that can distract you from your driving.
- Never get under the truck unless the wheels are securely blocked.
- Clean slippery materials off your shoes. Clean shoes will help prevent slipping on steps or having your feet slip from the clutch and brake pedals.
- Set the parking brake and block the wheels when parking trucks.
- Drive defensively.
- Always observe local laws regarding the weight and height limits of vehicles.

A large part of operating a dump truck on a public highway is looking out for other vehicles and pedestrians. Always look ahead of you and think ahead. Practice safety first when driving on public highways and you will have no trouble staying clear of potential accidents.

While you are working on a job, you will need to coordinate your vehicle's movements with other vehicles. Usually one person at a site is appointed signalperson who will direct the movement of vehicles in congested areas. Find out who this is and then watch for and obey his or her signals. When the signalperson is not in view, exercise caution when you are moving your truck. Look and think ahead to avoid situations that could disrupt the smooth flow of traffic. Yield to vehicles approaching from the right, but never take the right of way without first assessing conditions. When operating your vehicle near workers on foot, keep them in sight.

When fueling your truck, always ground the fuel nozzle to the fuel tank filler neck to prevent sparks, and do not reenter the vehicle during fueling to avoid creating static electricity, which can be an ignition source. Never fill the tanks while smoking or when near an open flame. If you spill fuel, immediately clean it up. When re-fueling has been completed, replace and tighten the fuel tank cap.

5.2.0 Responsible Operation

As a professional truck driver, you have a responsibility to your employer to use your vehicle properly. Perform your pre-operation inspection and servicing before you operate the truck. Allow the truck to warm up before driving it, especially if you are in a cold climate. When you are driving your truck, observe all gauges and indicators, and look, listen, and feel for defects, strange noises, and changes in engine or braking power. Allow diesel engines to idle for a few minutes before shutdown to allow the engine to cool slowly.

Learn to shift gears smoothly, and shift as needed to keep the engine operating at the manufacturer's suggested rpm. Use lower gears for low speed, such as on a work site, and high gears for highway speeds. Engines **lug** when driven at too low of a speed for the gear in use. This is hard on an engine and can cause the engine to stall unexpectedly. Lower your speed and gear on rough roads, and avoid driving over curbs, rocks, and other obstacles. Drive your vehicle at moderate speeds, especially when it's fully loaded. Accelerate slowly to avoid spinning the tires, and allow ample stopping time to avoid sliding the tires. Both wear tires excessively.

When you get into the cab, adjust your seat and mirrors, and then fasten your seat belt. Before starting the engine, ensure that the park brake is set and that the gearshift is in neutral (park for automatic transmissions). Follow the vehicle starting procedure recommended by the manufacturer.

At the end of your shift, safely park the vehicle in a firm level spot out of the way of other traffic. If you must park on an incline, position the vehicle across the slope and then follow the vehicle shutdown procedure.

5.2.1 Pinch Points

Pinch points result from the motion between two or more mechanical parts of the equipment. Being caught in a pinch point can cause injury or even death. On dump trucks, pinch points exist in several areas. These are shown in *Figure 18* and marked with a PP. These are the main areas where you are most likely to be caught:

- Between the cab guard and the back of the cab
- Between the dump bed and the frame
- Between the tailgate and the sides or bottom of the dump bed
- Between the rear wheels and the side of the frame
- Between the cowling of the hood and the truck body
- Around exposed rotating machinery (such as belts and gears) that are accessible to the hands

As the operator of the dump truck, it is your responsibility to work safely and make sure co-workers also follow proper safety procedures. There is never a valid reason for getting caught in a pinch point. Follow these tips to work safely around pinch points:

- In general, do not allow workers to be located in a pinch point area, either on the machine or around the machine, while it is in operation.
- Do not stand or allow others to stand or work under a raised dump bed. If work must be performed on the truck with the dump bed raised, the dump bed should be empty and some type of prop or chock placed between the bed and the frame according to the manufacturer's instructions.
- Keep fingers out of the area between the tailgate, tailgate latch, and the sides of the dump bed. Always stand clear when the bed is being lowered because the gate will slam shut as the bed comes down.
- Do not stand on the frame behind the cab as the bed is being lowered. The hydraulic action to

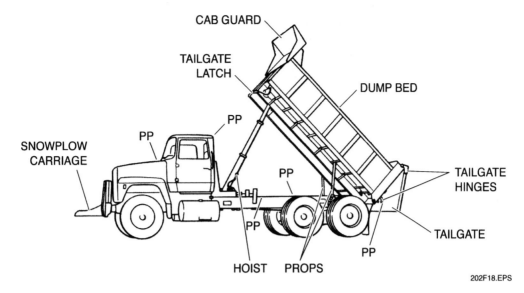

Figure 18 ◆ Pinch points on a dump truck.

lower the bed may be too fast to allow you to get out of the way. Stand on the ground and to the side if you need to operate the dump controls at the back of the cab.

• Do not reach into the area between the rear tires and the frame unless the truck has been shut down and the wheels properly chocked. Any movement by the wheels forward or backward can crush hands and feet, causing serious injury.

5.3.0 Loading and Dumping Safety

The main purpose of your truck is to haul large loads of material in its bed. An improperly loaded dump bed can make your vehicle harder to handle. You must make certain that you and others follow proper loading procedures to ensure a balanced load. To keep yourself and your coworkers safe, follow the safety guidelines below. Never perform any unsafe maneuver or operations, even if told to do so by another person.

Make sure you have the proper heavy equipment for the type of material being hauled. The truck should be in good, safe operating condition. Make sure the frame is sound, the bed is firmly attached, and the tailgate is solid and can be opened easily and locked securely. These items need to be checked before any material is placed in the bed. Avoid the temptation to use a questionable vehicle one last time.

5.3.1 Safe Loading Practices

Overloading a truck causes serious problems for dumping as well as for on-road driving. For example, the weight of the material may be too heavy for the hydraulic hoist to raise the bed, and the material may have to be removed using an excavator or other machine, which could damage the dump bed. Another problem is the force of the weight against the sides and tailgate of the dump bed; this may also cause damage to the equipment and create an unsafe situation. Watch the loading operation and make sure the recommended capacity of the truck is not exceeded.

Each loading site will have its own loading procedure that you must follow. If possible, position the truck for loading so that you can drive away without using the reverse gear—backing a truck is more dangerous than driving forward. If you do need to back the truck, use the route that reduces the time moving in reverse and use a spotter when available.

> **WARNING!**
> Mining occupations can be particularly dangerous, so workers need to be especially alert and safety conscious in the work environment. The Mine Safety and Health Administration (MSHA) cites the following as the four most common accidents that occur during surface hauling operations:
>
> • Trucks going over dump points
> • Vehicles and/or persons being run over by large trucks
> • Workers getting caught in conveyors
> • Haul trucks going out of control

Material must be evenly distributed down the center of the dump bed so the load will be balanced. When your truck is being loaded with a chute, there will most likely be a spotter who will signal to you when to pull forward to evenly distribute the load. Once loading has been completed, pull well away from the loading area, and complete any paperwork before you leave the loading site. Remember to unfurl the load cover if it is needed.

5.3.2 Safe Dumping Practices

Even when the cargo has been loaded properly, many factors can affect the unloading of the truck. You will need to stay alert for problems during the dumping procedure and be ready to stop the procedure at any time conditions warrant.

Most dump areas using rear dump trucks require turning and backing to the edge of the fill. If possible, the fill should be arranged so that any turn is made near the dump spot. The turn spot should be wide enough so that reverse gear is used only once. Turns in reverse should be toward the driver's left to give the driver maximum visibility, and the truck should be level or facing uphill for dumping. Use the low range of the reverse gear when the load is heavy, when the ground is soft or rough, or when complicated steering is required.

When dumping off the edge of a fill, back so that both wheels will be the same distance from the edge, rather than at an angle. Check the stability of the ground at the dump area before driving on it. If one wheel sinks in the ground deeper than the other, it may not be possible to dump the truck safely or to pull out with a load.

The distance from the edge of a fill that can be considered safe is determined by circumstances and the judgment of the operator. If the truck has all-wheel drive, or if the fill is shallow, a close approach can be made. If the fill is soft, slippery, sandy, or otherwise unstable, you should keep the rear wheels six feet or more away from the edge. This can be a difficult judgment for an inexperienced driver, so you may need to ask your supervisor or a more experienced driver for advice.

When the dump bed is raised, the center of gravity for the truck changes to a much higher position, placing the truck at risk of tipping over, especially if the truck is unbalanced. Tipovers can be either to the back or to the side of the dump bed. The following are the leading reasons for backward tipovers:

- *Top-heavy load* – The rear portion of the load is dumped, leaving the rest of the load stuck at the front of the bed, as shown in *Figure 19*.
- *Material stuck together* – Either the material is frozen together or to the bed of the truck.
- *Uneven loading* – This can be either front-to-back or side-to-side.
- *Uneven dumping* – Similar to uneven loading, but is caused by trying to dump part of a load in a particular spot.

When material sticks to the front part of the bed in top-heavy loads, it places a great strain on the hydraulics and causes an unstable condition for the whole truck. The best action to take is to lower the bed to a level position and use a shovel or other hand tool to dislodge the load. Never stand in the dump bed and try to work the material loose while the bed is raised.

When part of a load remains stuck in the bed of the truck, a hazardous condition is created that could cause tipping or damage to the dump bed. This situation is commonly referred to as a **split load**. A split load can be caused by frozen or sticky material that attaches to the bottom and one side of the bed. Compacting the material in the bed when it is loaded may also cause the material to stick on a side and in the corners.

202F19.EPS

Figure 19 ◆ Top-heavy load.

In a split side load, the material splits down the center of the bed and leaves part of the load completely to one side of the bed, as shown in *Figure 20*. This shifts the vertical center of gravity and can tip the unit over to one side. Uneven loading, excessive rear spring deflection, or under-inflated tires can cause this condition. To correct a split load, lower the bed and use a shovel or other hand tool to loosen the load.

Another problem encountered while dumping is allowing the material to pile up against a hinged tailgate. In this case, the upper body of the load usually empties and the load piles up at the back. A load stuck in this position may unbalance the truck so that the front wheels rise off the ground. Until the material is freed, the truck cannot move because there is no steering capability. If the load cannot be dumped, the bed should be lowered and the tailgate opening adjusted or the tailgate removed.

The following actions will help prevent tipover due to materials problems:

- Make sure the bed is empty and clean before loading. Keep the bed washed down with proper fluids when required.
- Test the hoist mechanism with an empty bed to ensure proper operation.
- Adjust the tailgate for loading according to the manufacturer's recommendation.
- Do not dump with the truck on a slope.
- Raise the bed slowly.
- Load the bed evenly. If the bed is loaded unevenly, lower the bed and even out the load with a shovel. Do not do this while the bed is raised.
- Make sure the cab of an articulated dump truck is lined up with the dump bed before dumping.

202F20.EPS

Figure 20 ◆ Split side load.

It is the operator's responsibility to ensure that the equipment is working properly and that the material is being handled safely. Never perform any unsafe maneuver or operations, even if told to do so by another person.

6.0.0 ◆ DUMP TRUCK OPERATION

Because the dump truck is one of the most often used pieces of heavy equipment in the earth moving and construction business, many accidents occur each year involving dump truck operation, both on the road and at the job site. You can avoid most accidents by following required safety procedures and using common sense. Prevention of accidents involving dump trucks is primarily dependent on the person operating the equipment. While the actions of other drivers affect on-road safety, you can take precautions to avoid accidents that may cause injury and damage. Manufacturers design many features into their equipment that make your job safer and easier; however, avoiding unsafe situations and using forethought, good judgment, and skill is solely up to the operator.

This section describes procedures and actions that can be used to ensure safe operation of a dump truck while driving on the road and the job site.

6.1.0 Over-the-Road Operation

Dump truck operators spend a lot of time driving on public highways. Your personal safety and the safety of the occupants of other vehicles depend on your skill, knowledge, and understanding of the trucks you drive. You represent your employer when you drive these vehicles on the road, so project the best possible image through safe, courteous operation of the equipment.

Even with all the safety features that have been built into the design of dump trucks, there are still dangers. You cannot assume that everyone on the road is a good driver. Watch for unsafe situations created by other drivers, and be extremely careful in the operation of your equipment. To help prevent accidents on the road, observe the following safety rules at all times:

- Become acquainted with the operation and maintenance manuals for your truck.
- Know the location and functions of all the controls, gauges, and warning devices.
- Always wear your seat belt. Also, make sure any passengers wear their seat belts. Do not move the equipment until everyone is buckled in.
- Pay close attention at intersections, and wait before accelerating when a red light turns green.

- Never accelerate through a yellow signal.
- If it is necessary to pass a vehicle, pass only in designated passing zones and only after you have checked blind spots for clearance.
- Yield the right-of-way, even when it is yours. Make allowance for other drivers. Obey all state and local traffic regulations at all times. Some department policies and procedures may be more restrictive.
- Always drive at a speed that permits full control of the vehicle and allows for factors such as road, weather, and traffic conditions.
- Do not have loose articles in the cab that might obstruct your vision in all directions or prevent you from maintaining complete control of the vehicle.
- Perform a pre-operational inspection before using the truck.
- If you get drowsy, pull over and take a break. Get out and walk around or change drivers.

To be a safe dump truck operator you must know and practice safe operating principles and procedures on a continual basis. Several areas of operation require particularly close attention, including climbing and descending hills, runaway vehicles, recovering from a skid, and negotiating curves.

6.1.1 Climbing and Descending Hills

A truck with a gross weight of 16,000 pounds may use the same engine as a car weighing only 3,500 pounds. Brakes, although larger, are not increased in size in proportion to the extra weight that must be controlled. Hills, therefore, present problems for trucks both in climbing and in getting down safely.

It is important not to depend solely on your brakes when descending steep hills. They can overheat, and if the vehicle starts to slide with the brakes set, the wheels will lock and you will lose control. Before entering a downhill grade, apply the brakes to reduce your speed, and then shift the transmission into a lower gear. Using a lower gear gives the same effect as using the brakes. A good rule for driving downhill is to use one gear lower than you would use to drive up the same grade.

It is risky to change gears while climbing or descending steep hills. The safest procedure is to select the proper gear before starting to climb or descend. If it is necessary to go to a lower gear during the climb, be sure to shift before the engine lugs to a point where it might stall.

When descending long, steep hills, you may want to come to a complete stop and inspect your vehicle before proceeding. Check your tires for

proper inflation and brakes for proper travel and operation. If everything is working correctly, start down the hill in the lowest gear. Do not use your accelerator. Shift gears as needed to match speed gains as you pick up momentum.

The following are safety tips for descending a steep downhill grade:

- Reduce speed with minimal use of the foot brake.
- Check your speedometer and tachometer frequently. Never exceed the recommended engine speed in any gear.
- Take extra care when weather and road conditions are unfavorable. Never exceed the advised truck speed for any downgrade.
- Keep the truck under complete control at all times.

If you stop on an uphill grade, you will eventually have to move. Starting a loaded truck on a positive grade without any rollback can be difficult. You may need to release the parking brake as you are accelerating in first gear to prevent rollback.

If the steepness of the grade is causing the truck to slow, downshift when the engine rpm reaches the shift point for the next lower gear. Then continue to downshift to match the power demands for the grade until there is no longer a loss of power.

6.1.2 Runaway Vehicles

If brakes fail to hold your vehicle and it starts to run out of control down a hill, the last resort is to ditch the vehicle. Running it off the road against a bank at a gradual angle will slow and stop the vehicle. This must be done promptly before the runaway vehicle has gained too much speed. Proper action in such an emergency may prevent a much more serious accident. Remaining belted in the cab is your safest position.

6.1.3 Controlling a Skid

Statistics show that about one-fourth of all vehicle accidents involve a skid. The reason a skid occurs at all is that the tires have lost their traction on the road surface. When this happens, you begin to lose control of the vehicle's direction. Most skid problems are caused by slippery road conditions like rain, snow, ice, or wet leaves.

If your vehicle should start to skid, stay calm. Avoid hard braking, since slamming on the brakes will lock the wheels, cause further loss of traction, and increase the skid. Steer in the direction of the skid and as the vehicle begins to correct, straighten the front wheels and be ready to correct in the other direction. Many skids need more than one correction to regain control. Avoid over-steering—turning the steering wheel too far whips the rear end of the truck into a skid in the opposite direction. Keep the clutch engaged or the selector lever in drive. Holding the vehicle in gear helps reduce speed and gives you the most control. Avoid lifting your foot from the accelerator suddenly—some expert drivers will even accelerate slightly to get out of a skid.

It is better to try to prevent a skid than to have to recover from one. Adjusting your speed to the conditions in the road will reduce the chance of a skid. Driving within your sight distance and maintaining an adequate distance from the vehicle in front of you will reduce the need for sudden stops and the possibility of a braking skid.

6.1.4 Curves

Taking a curve at high speed with a full load is dangerous. Curves must be taken more slowly when a truck is carrying a load than when the truck is empty. When rounding a curve, centrifugal force will pull the truck toward the outer edge of the curve with the only resistance being the friction of the tires on the road. An increase in load will raise the center of gravity, increase the force that causes the truck to slide sideways, and also increase the risk of tipping over. If traction is good, the truck will tilt slightly, compressing the outside springs and tires, and the truck will stay on the road. If the outward force is greater than the vehicle can handle, the truck will tip over. The only safe action is to reduce your speed before going into the curve and accelerate gently as the vehicle is coming out of the curve.

6.2.0 Backing Safety

Backing is a hazardous maneuver under any circumstance. As the driver, you are always responsible for knowing what is behind you. If possible, do not use a backing maneuver any time you can drive forward. Whenever possible, use a spotter to help guide the truck to reduce the chance of injury or property damage. The rear-view mirrors show only what is to the side of the truck. Anything that is low or immediately behind the vehicle cannot be seen. It is a good practice to get out of the cab and look behind the vehicle before backing.

Use your mirrors when backing, rather than just looking over your shoulder. This gives you the proper visual orientation. The following tips should help you back safely:

- Back slowly. Be sure there is sufficient clearance when backing into narrow spaces.
- Remain properly seated when backing the vehicle, using your mirrors or a spotter.
- Avoid long backing runs. It is much safer to turn around and drive forward.
- Avoid backing downhill.

6.3.0 Using Trucks with Bottom Dump Trailers

In some parts of North America, bottom dumps are used for hauling and spreading, especially for asphalt and aggregate material. These trucks come in many different designs and sizes, but they all dump their load through gates in the bottom of the dump bed. Under certain conditions, there are advantages to using bottom dumps. For example, they can place material in a windrow, which can be worked much easier than dumped piles.

There are precautions that need to be taken with bottom dump trailers:

- Follow the directions of a spotter when starting the dump. If the material is asphalt or aggregate that needs to be placed at exact locations, do not rely on your sense of distance to judge when the dump should start. Being off by half a truck length could cause problems for other equipment.
- Do not allow the material to flow out of the bottom of the bed faster than you can move to maintain an even row. Allowing the material to be dumped in one spot may cause the trailer to become stuck on the pile.
- Never dump while turning. The motion of the bottom of the trailer bed against the side of the dumped material could cause the doors or mechanism to jam.

6.4.0 Operation of Off-Road Trucks

Trucks that operate in mines, quarries, pits, or other types of excavations where the use of public roads is not required are not subject to legal restrictions on size or weight. These trucks, with capacities of up to 50 cubic yards, are very large compared to over-the-road equipment. Specialized semi-trailers can carry up to 100 tons. Off-road truck sizes continue to increase as larger and better-made components are developed. The size of these trucks makes visibility difficult, so drivers must stay alert to changes in the environment. A typical off-the-road truck is shown in *Figure 21*.

Off-road trucks usually have specially designed dump beds and lifting mechanisms; this means the usual precautions taken with a standard dump truck may not apply. For example, many off-road trucks have dump beds that have no tailgate. Once the bed starts to rise, the material is free to fall out the back. Therefore, you want to make sure everyone is clear from the back of the truck before you start to dump.

6.5.0 Transporting Off-Road Dump Trucks

Except for certain types of fixed plant operations in mining, quarrying, and industry, a dump truck operator's work is usually mobile. Equipment operators move with their machines to various job sites. If you work for a construction company as a driver, you will probably work at a dozen job sites in one year. These sites could be located in the same town or city, or they could be in another state, a different area of the country, or even overseas.

When an off-road dump truck needs to be moved to another site, it can be a time-consuming and difficult task. Before the truck can be moved, a great deal of planning must take place so the move will be safe and efficient. Because some of these activities require contact with the public, safety is especially important during the transporting of the equipment.

First, a route must be selected. This requires the ability to read and understand maps, general knowledge of size and weight limits, and knowledge of registrations and permit requirements. Two very important factors in the moving plan involve the identification of points of low clearance and bridge weight restrictions. Local laws or posted limits must be obeyed for weight restrictions. Most interstate highways have underpass clearances of at least 15 feet. However, local road systems may have lower limits.

Once a route is selected, it may be necessary to obtain permits in all jurisdictions in which the truck will travel. Slow-moving vehicles represent a potential danger because they disrupt normal traffic flow. Local law enforcement personnel need to be aware of this in advance. Not only do overweight vehicles require permits before they are allowed on public roads, they may also require a safety escort.

Once the planning and permitting is completed, the vehicle must be loaded correctly for a safe trip. Next, you or someone who works specifically at hauling heavy equipment must transport it. When the destination is reached, the equipment must be unloaded and checked before it is put into service.

6.6.0 Towing

Occasionally, you may need to tow your off-road vehicle, or use it to tow another vehicle or equipment. Always check the operator's manual for your vehicle before attempting to tow it. Never use shortcuts such as a chain or cable to tow a vehicle—even when it is for a short distance within the job site.

 WARNING!
Towing dump trucks is dangerous; improper towing adds to the danger and can cause serious injuries or even death. Follow the correct procedure recommended by the vehicle manufacturer in the operator's manual.

7.0.0 ◆ SNOW REMOVAL

In some areas, dump trucks are used to plow snow and spread sand or salt over icy surfaces. See *Figure 22*. These trucks will be seasonally equipped with snowplows and/or spreaders. Some snowplows are fixed at one level and angle and can be adjusted only by manually moving and resetting the blade. But most snowplow systems allow the operator to adjust the level and angle of the plow from controls in the cab. Other snowplow systems are very sophisticated and are partially automatic, so that the plow will ride over some obstructions without operator intervention. Study the manufacturer's instructions for the equipment used on your truck.

202F21.EPS

Figure 21 ◆ Typical off-road dump truck.

Figure 22 ◆ Dump truck with snow plow.

7.1.0 Cold Weather Safety

When you work in cold snowy weather, you need to take additional precautions to protect yourself. First, always dress suitably. Wear appropriate protective clothing; especially warm gloves, hats, and socks. Fingers, ears, and toes are the first extremities to freeze. Wear moisture-repellent clothing and dress in layers. Layers of lightweight clothing will provide you with more warmth than a single heavyweight garment. Dressing in layers also allows you to remove clothing as you warm up from working.

Know the symptoms of cold exposure and frostbite. Prolonged exposure to cold weather can cause drowsiness and affect your judgment, so take breaks in a warm area when necessary. If you can't get to a warm area, exercise such as jogging in place will help to warm you. Keep warm liquids handy, but avoid drinking alcohol.

Sunlight is intensified when reflected from snow and can cause eye fatigue or temporary blindness. Wear good quality sunglasses to protect your eyes from the strong light. When sunglasses are not needed, remove them to maximize your vision.

7.2.0 Effective Snowplowing

Snow presents an additional hazard to your work since it covers many obstacles. Some obstructions will dislocate or damage the plow and perhaps damage the truck. Always use extreme caution when plowing around the following:

- Bridge and pavement expansion joints
- Headwalls of culverts
- Cattle guards
- Signposts and guardrails
- Hard-packed snow or ice

- Road shoulders
- Raised pavement markers, curbs, and islands
- Fire hydrants

Operation of the snowplow should be done with extreme caution. Working in poor environmental conditions means that snowplow operators need to make every effort to protect themselves and their equipment from accident and injury. Only experience will allow you to develop a feel for plowing different kinds of snow under different conditions. For the mechanical operation of the plow, follow the instructions provided in the manufacturer's instructions.

Before you begin plowing, perform your normal pre-operational inspection and then check to see that the plow is operational. When driving, use extreme caution because the blade has a low clearance even when it is raised. Drive according to road, snow, and visibility conditions. Be alert for hazards. When you arrive at the plow area, adjust your blade to the prescribed height. If you have not been given a prescribed height, set the blade to approximately one-inch clearance. Set the blade angle as shallow as conditions allow. See *Figure 23*. The shallower the angle, the wider the plow lane. A light, dry snow requires a shallow angle, but wet, heavy snow requires an acute angle.

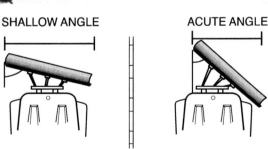

Figure 23 ◆ Snowplow angles.

NOTE

Be sure that the plow angle is not set so that it fails to clear path for the truck wheels.

When plowing a roadway, plow in the direction of traffic flow so that the snow is removed to the shoulder of the road. For two-lane roads, plow from the centerline out to shoulders and make sure your blade overlaps the centerline slightly on the first pass. Plow to the low side of ramps or curves when possible. If plowed to the high side, the snow will melt across the highway and ice over. Be alert for vehicles approaching from behind. Rear-end collisions are common. Plow away from wind whenever possible and practical. Clear snow past intersections and raise the blade before making a turn. Use caution when making your turnaround.

1. Dump trucks are widely used in the construction industry for all of the following reasons *except* _____.
   a. dump trucks can haul large heavy loads of material around a job site or on a public road
   b. dump trucks are designed to be quickly unloaded by the driver without any assistance
   c. dump trucks have uses other than hauling, such as towing trailers and plowing snow
   d. dump trucks are easy to drive, so no special licensing is required to drive on a public road

2. If you don't know the maximum weight capacity of your truck, you should _____.
   a. ask another dump truck driver
   b. ask your immediate supervisor
   c. look it up in the truck's manual
   d. load the bed until the rear tires compress

3. Dump trucks are rated by the _____.
   a. maximum turning radius
   b. size of the box
   c. size of the tires
   d. amount of weight that their axles can safely carry

4. A standard dump truck with a single-axle configuration is used to haul _____.
   a. only light loads
   b. light and heavy loads
   c. medium loads
   d. the heaviest loads

5. A good use for a bottom dump truck is _____.
   a. moving soil to a dumping site
   b. hauling away excavated soil
   c. depositing sand for a foundation
   d. dumping asphalt for a road bed

6. A dump body can only be raised when the transmission is in neutral.
   a. True
   b. False

7. Differential lock is helpful if the truck is stuck in mud and one tire has traction.
   a. True
   b. False

8. A power takeoff unit is often used _____.
   a. to maintain traction on wet roads
   b. for more engine power to go uphill
   c. to provide power to another device
   d. to help slow down a speeding truck

9. When your dump truck is empty, you should raise the auxiliary axles for all of the following reasons *except* to _____.
   a. improve the truck's gas mileage on highways
   b. permit the truck to safely carry a heavy load
   c. decrease the vehicle toll costs on highways
   d. save wear and tear on the auxiliary axle tires

10. When operating a dump truck equipped with air brakes, you should _____.
    a. use them like other braking systems
    b. rest your foot on the brake pedal to maintain pressure
    c. monitor the system air pressure while driving
    d. pump the brakes to avoid locking them

11. To shift gears on manual transmission, you need to operate the gearshift and _____.
    a. clutch
    b. differential
    c. choke
    d. power takeoff

12. It's a good idea to shift the transmission into low gear before stopping to save wear on the _____.
    a. clutch
    b. brakes
    c. retarder
    d. PTO

13. To save wear on the brakes, slow the truck using the _____.
    a. power takeoff
    b. differential lockout
    c. engine retarder
    d. air brake systems

14. On a dump truck, brake lights must be operational _____.
    a. on off-road trucks operated on a public road
    b. on on-road trucks operated on a public road
    c. when the truck is operated on a public road
    d. whenever an on- or off-road truck is used

15. In cold areas, it is common to _____ at the _____ of the work shift.
    a. drain air brake system reservoirs; end
    b. drain air brake system reservoirs; beginning
    c. pressurize the air reservoirs; end
    d. pressurize the air reservoirs; beginning

16. When parking your truck for the night, you need to _____.
    a. place the gear shift in low gear when parked on a hill
    b. chock the wheels when the truck is parked on an incline
    c. select a well-lit area when the truck is parked near traffic
    d. set the parking brake when parked on a steep incline

17. Your employer must provide instruction on the safe operation of your equipment _____.
    a. when you have an accident
    b. at least once a year
    c. at monthly tailgate meetings
    d. only when you are first hired

18. When leaving the truck cab, you should exit _____.
    a. while holding onto the steering wheel
    b. backwards, using the handholds
    c. from the passenger side of the truck
    d. facing front and quickly jumping

19. In a correctly loaded dump truck, the load is _____.
    a. balanced by placing it down the center of the bed
    b. piled as high as the sides of the truck allow
    c. piled as high as possible
    d. compacted to use the most bed space possible

20. When you are dumping at the edge of a fill, it is best to _____.
    a. back your truck so that it is at an angle to the edge
    b. keep the wheels the same distance from the edge
    c. dump the load at the fill edge and then shovel it in
    d. always get the truck as close to the edge as you can

21. When dumping, turns in reverse should be toward the driver's left for _____.
    a. maximum turning radius
    b. maximum visibility
    c. greater stability
    d. better traction

22. If you lose your brakes going down a grade, which of the following is an acceptable procedure to slow down the vehicle?
    a. Push in the clutch and then shift the transmission into reverse.
    b. Turn off the engine and slam on the brakes to lock the wheels.
    c. Steer the truck against the side of the bank at a gradual angle.
    d. Quickly raise the dump bed to decrease power to the engine.

23. Because off-road dump trucks are bigger than on-road trucks, _____.
    a. they are much safer to operate
    b. they have multiple blind spots
    c. the driver is higher up and has good visibility
    d. the driver is unlikely to be injured in an accident

24. When you need to work in cold snowy weather, you should dress in layers of clothing.
    a. True
    b. False

25. The plow shown in *Figure 1* is set to _____.
    a. plow heavy wet snow
    b. plow dry light snow
    c. clear ice from the road
    d. clear snow or ice

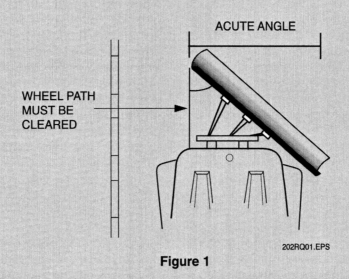

ACUTE ANGLE

WHEEL PATH MUST BE CLEARED

202RQ01.EPS

Figure 1

Summary

Dump trucks are among the basic pieces of heavy equipment in the construction contractor's fleet. Driving a dump truck requires skill and knowledge. The operator must be able to handle the operation of the truck on public highways as well as on job sites and perform difficult maneuvers such as backing up and dumping the load. The driver must be able to do these tasks safely and efficiently.

The dump truck consists of a dump bed and cab mounted on a rigid frame. The engine can either be gasoline or diesel powered. The transmission can be automatic or manual. Manual transmissions usually have 10 or more gears. The configuration of the axles can either be single, dual, or multiple in the rear, with single or double tires mounted on each side. The dump bed of the truck is operated by a hydraulic hoist mounted on the frame. Most trucks come with an air brake system, which you must know how to use and maintain in proper working order.

As a dump truck operator, you are responsible for its preventive maintenance and safety. You must perform daily walk-around inspections to make sure the truck is operating properly and safely. Proper preventive maintenance will prolong the life of your truck. Repair any problems you find or report the problem to your supervisor. Do not drive a truck that is not operating properly or is unsafe. Follow the procedures in your operator's manual for proper inspection and operation activities.

Dump trucks may tip over if loaded improperly or when dumping material that is frozen or stuck together. The tipping usually occurs when the dump bed is being hoisted and the material is being unloaded. Tipping can also take place if a large load causes the center of gravity to be too high and the truck is moving too fast around a curve.

Dump trucks are often used for snow removal. A snowplow can be attached to the front of the truck and a spreader can be loaded in, or mounted on, the bed. These attachments are only mounted on the truck when needed and are stored the rest of the time.

Notes

Trade Terms Introduced in This Module

Articulating: Movement between two parts by means of a joint.

Auxiliary axle: An additional axle that is mounted behind or in front of the truck's drive axles and is used to increase the safe weight capacity of the truck.

Cab guard: Protects the truck cab from falling rocks and load shift.

Clutch: Device used to connect or disconnect two parts of the transmission.

Governor: Device for automatic control of speed, pressure, or temperature.

Hoist: Hydraulic or mechanical lifting device.

Hydraulic: Powered or moved by liquid under pressure.

Lug: Effect produced when engine is operating in too high a transmission gear. Engine rotation is jerky, and the engine sounds heavy and labored.

psi: Pressure in pounds per square inch.

rpm: Revolutions per minute.

Reservoir: Storage tank.

Split load: Load that results when material being dumped does not flow evenly from the dump bed. A split load is more likely to occur with wet sand, clay, or asphaltic materials.

Tag axle: Auxiliary axle that is mounted behind the truck's drive wheels.

Tandem-axle: Usually a double-axle drive unit, but some states call all multiple axle units tandem-axle.

Trailer: Towed vehicle that rests on its own wheels.

Additional Resources

This module is intended to be a thorough resource for task training. The following reference works are suggested for further study. These are optional materials for continued education rather than for task training.

Moving the Earth, Fourth Edition. New York, NY: McGraw-Hill.

Truck Driver's Guide to CDL, First Edition. New York, NY: Prentice Hall Press.

Trucking's Web Resources for Journalists and Communicators, 2005. *twna.org*.

Figure Credits

Deere & Company, 202F01, 202F07B, 202F09, 202F13–202F16

Topaz Publications, Inc., 202F02, 202F12, 202F23 (photo)

International Truck and Engine Corporation, 202F03, 202F06

Clement Industries, Inc., 202F04, 202F05

Reprinted courtesy of Caterpillar Inc., 202F07A, 202F08, 202F21

Courtesy of Amerex Corporation, 202F17 (fire extinguisher)

Peterson Manufacturing Company, 202F17 (warning triangles)

First Aid Only, Inc., 202F17 (first aid kit)

Associated General Contractors, 202F18

South Middleton, PA Department of Roads, 202F22

NCCER makes every effort to keep these textbooks up-to-date and free of technical errors. We appreciate your help in this process. If you have an idea for improving this textbook, or if you find an error, a typographical mistake, or an inaccuracy in NCCER's Contren® textbooks, please write us, using this form or a photocopy. Be sure to include the exact module number, page number, a detailed description, and the correction, if applicable. Your input will be brought to the attention of the Technical Review Committee. Thank you for your assistance.

Instructors – If you found that additional materials were necessary in order to teach this module effectively, please let us know so that we may include them in the Equipment/Materials list in the Annotated Instructor's Guide.

Write: Product Development and Revision
National Center for Construction Education and Research
P.O. Box 141104, Gainesville, FL 32614-1104

Fax: 352-334-0932

E-mail: curriculum@nccer.org

Craft _____ Module Name _____

Copyright Date _____ Module Number _____ Page Number(s) _____

Description _____

_____

_____

_____

(Optional) Correction _____

_____

_____

(Optional) Your Name and Address _____

_____

_____

_____

22203-06

Rollers

22203-06
Rollers

Topics to be presented in this module include:

Overview

All structures must have a solid foundation. Rollers provide surface compaction so that the structures have a firm level foundation to carry their weight. Without good compaction of the subsurface, the structure will settle and crack. Rollers also compact soils or gravel to build roads. Specialty rollers are used to smooth asphalt for the final finish for roads.

There are several types of rollers designed for different types of compaction. An operator must know the functions and features of each type. In addition to safe operation of the equipment, an operator must understand various properties of soil and how to test their work.

Objectives

When you have completed this module, you will be able to do the following:

1. Describe the uses of a roller.
2. Identify the components and controls on a typical roller.
3. Explain safety rules for operating a roller.
4. Perform prestart inspection and maintenance procedures.
5. Start, warm up, and shut down a roller.
6. Perform basic maneuvers with a roller.
7. Describe the attachments used on rollers.

Trade Terms

Compaction
Density
Foot
Ground contact pressure (GCP)
Lift
Mat
Pad
Puddling
Sheepsfoot
Settling
Tamping roller
Vibratory roller

Required Trainee Materials

1. Pencil and paper
2. Appropriate personal protective equipment

Prerequisites

Before you begin this module, it is recommended that you successfully complete *Core Curriculum; Heavy Equipment Operations Level One; Heavy Equipment Operations Level Two*, Modules 22201-06 and 22202-06.

This course map shows all of the modules in the second level of the *Heavy Equipment Operations* curriculum. The suggested training order begins at the bottom and proceeds up. Skill levels increase as you advance on the course map. The local Training Program Sponsor may adjust the training order.

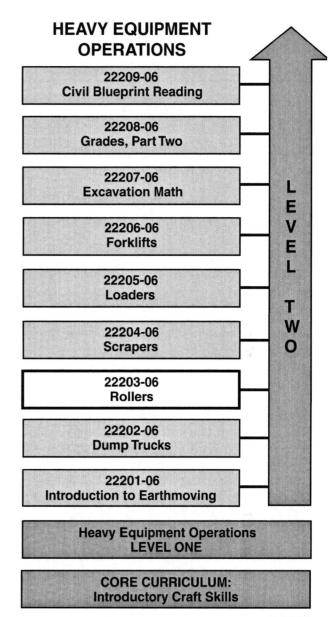

HEAVY EQUIPMENT OPERATIONS

22209-06
Civil Blueprint Reading

22208-06
Grades, Part Two

22207-06
Excavation Math

22206-06
Forklifts

22205-06
Loaders

22204-06
Scrapers

22203-06
Rollers

22202-06
Dump Trucks

22201-06
Introduction to Earthmoving

LEVEL TWO

Heavy Equipment Operations
LEVEL ONE

CORE CURRICULUM:
Introductory Craft Skills

203CMAP.EPS

1.0.0 ◆ INTRODUCTION

The practice of rolling roads during construction and for continuing maintenance goes back to Roman times. Horses or slaves dragged large stone rolls to create a smooth and fairly solid surface for vehicles. Steam-powered engines were used to roll roads in America after the civil war. The first steamrollers were imported from England. American inventor Abbott Q. Ross patented his steamroller in 1871 and Anders Lindelof patented his design in 1873. The largest producer of steamrollers, Buffalo-Springfield, manufactured the last steamroller in 1935, as steamrollers were replaced by gasoline-powered rollers.

The surfaces that buildings and roads are built on must be able to carry the weight of the structure and its contents or the road and the traffic that will travel on it. To do this, each building and road is designed with certain foundation specifications. Specialized equipment such as rollers and compactors are used to make the underlying soil, asphalt, or gravel as smooth, level, and solid as possible.

Soil consists of irregular particles; air and water fill the spaces between the particles. When weight is placed on soil, the particles are packed closer together and the air and water are pressed out. This process can occur naturally and is called **settling**. However, it is slow and inconsistent. If an area settles after construction, the structures above can fail. For example, if the ground settles after a building is completed, the foundation can crack. Repairing this type of damage is extremely expensive.

The **density** of the soil is a measure of how closely the particles are packed. Denser soils can bear more weight. **Compaction** is the engineered process of increasing the density of the soil. It mimics the natural processes, but is faster and more consistent. The underlying soil is compacted before construction so that it will not change shape or volume under the weight of the structure, pavement, or traffic.

Soil can be compacted by pressure, vibration, kneading, impact, or a combination of these methods. Different machines with specialized rollers compact soil using one or more of these methods. For example, a **vibratory roller** supplies pressure and vibration, while pneumatic rollers supply kneading and pressure. Various compaction techniques are used, depending on site conditions and soil types.

There are several reasons a site must be compacted. In addition to providing a base for buildings or roads, compaction provides a stable base for all site operations. Many sites require loose fill to level the land. If the fill is not compacted, rain water can soak into the spaces in the soil and turn it into mud. Equipment will get stuck in the mud, causing expensive delays or equipment damage. If the mud is deep enough, it will make the site impassable and stop work. Compacting loose fill minimizes the spaces in the soil so that rain cannot penetrate the soil and turn it into mud.

Some fills may be compacted by a process called **puddling**. This is an enhancement of the natural settling process. Water is added to the area until the soil is semiliquid. It is then allowed to dry and settle. A small vibrator speeds up the drying process of any puddled fill. Vibration brings excess moisture to the surface. This method is particularly effective on gravel fills.

2.0.0 ◆ TYPES OF ROLLERS

Rollers are used for compacting and smoothing materials such as earth, gravel, and asphalt. They compress the materials to the desired density to provide strength so that the specified loads can be carried without causing settling. There are several types of compaction rollers, each used for different types of work.

Early compaction equipment included a horse-drawn steel roller filled with rocks. Various designs have been developed based on these early models. Several companies manufacture rolling equipment in different configurations and sizes. Although there are many different models and makes of equipment, currently there are four basic designs. The four basic types of rollers are shown in *Figure 1* and include the following:

- Pneumatic tire
- Steel-wheel
- Vibratory
- **Sheepsfoot**

Each of these types can be further subdivided into two or more size classifications. For example, pneumatic tire rollers are manufactured in three distinct size categories, based on tire diameter.

Static forces are those produced by the heavy weight of the roller. Dynamic forces are those that use a combination of weight and energy to produce a vibratory or tamping effect on the soil. Rollers and compactors use both static and dynamic forces to compact the soil and achieve the required soil density.

STEEL WHEEL (SHEEPS-FOOT) ROLLER

STEEL WHEEL ROLLER

SMOOTH-DRUM, RUBBER TIRE
VIBRATING ROLLER

PNEUMATIC TIRE ROLLER

203F01.EPS

Figure 1 ◆ Various types of rollers.

2.1.0 Pneumatic Tire Roller

Pneumatic tire rollers (*Figure 2*) achieve excellent results in almost every type of compaction. However, some types of wet soils may stick to the tires, limiting their effectiveness. Scrapers are often attached to scrape mud from the tires. The pneumatic tire roller first appeared in the early 1930's as a towed wagon with a rock-filled box. The primary design remains a weight-filled box mounted on two rows of tires. Some models are self-propelled, while others are towed.

The tires are the compaction devices on these rollers. Typically there are two axles with a set of tires on each axle. The tires are smooth with no treads. There are four to five tires on one axle and five or six on the other. Compaction is varied by changing the air pressure in the tires. The normal range is between 60 and 120 psi.

203F02.EPS

Figure 2 ◆ Pneumatic tire roller.

Rubber-tired rollers provide uniform coverage and equal wheel loading. The rear wheels are staggered to roll the spaces left by the front wheels in order to achieve uniform coverage. This offset adds a kneading action as the material is pressed toward the spaces between tires. The kneading action aligns the aggregate particles to their most stable positions. Equal wheel loading on uneven terrain is provided by articulated axles.

The tires are designed to provide a specific **ground contact pressure (GCP)** on the soil. The amount of GCP is controlled by both wheel loading and the tire inflation pressure. GCP values were standardized in the United States in 1967. Compactors carry decals that identify their maximum allowable wheel load and the ground contact pressures that can be attained with the various combinations of wheel loadings and tire inflation pressure.

Rubber-tired rollers are used to compact all types of surfaces. In most cases, the rubber-tired roller is followed by finish steel-wheel rolling. When used as an intermediate roller, the rubber-tired roller contributes to compaction and particle alignment.

NOTE

Pneumatic tire rollers are used in asphalt road construction. The tendency of the asphalt binder to stick to the rubber tires increases as the temperature of the tires decreases. Heavy mats made of rubber or plywood can be mounted around the tire area to prevent heat loss and keep the tires at or near mat temperature. This will reduce the tendency of the binders to stick to the tires and increase rolling efficiency.

2.2.0 Steel-Wheel Roller

Steel-wheel rollers (*Figure 3*) are rollers as well as compactors. Their rolls are machined to provide a smooth, concentric surface. They are designed primarily for pavement course rolling, but are also used for rolling gravel roads, road bases, and some subgrades.

There are three basic types of steel-wheel rollers: three-wheel, two-wheel, and single wheel. The three-wheel roller has one wide guide roll and two narrow drive rolls. The two-wheel, or tandem, has two rolls of equal width. The single-wheel roller has one steel rolling wheel and rubber drive tires. The drive tires are typically mounted on the rear of the machine. There is also a unique version of the tandem roller, called the three-axle tandem, which is similar to the tandem in appearance, but has two

guide rolls on a walking beam. The walking beam allows the forces exerted by the rolls to be varied. The three-axle tandem is generally the heaviest of all steel-wheel rollers, but there are different size categories in each style.

Steel-wheel rollers have hydrostatic drive, hydrostatic steering, steering wheels, and a pressure type fog-spray system for wetting the rolls. A self-contained portability wheel on the 4- to 6-ton tandems facilitates towing without the use of a trailer. Hydrostatic drive has proven to be the single biggest improvement in recent years. It provides variable speed control, together with dynamic braking action.

2.3.0 Vibratory Compactors

Vibratory compactors combine static weight with a generated, cyclic force. This dynamic type of compaction was developed to provide high densities in granular soils. The vibrations move the particles, and the static pressure forces them into a compact structure. *Figure 4* shows a vibratory steel-wheel roller. Vibratory compactors work best in granular soils and are ineffective in silt and clay.

CAUTION

The vibratory compactor must be in motion when the vibrator is on.

Rollers are manufactured in many styles and sizes. Many vibratory rollers are also available to compact deep-lift asphalt bases, complete with spray systems and smooth tires.

Static rollers depend on the weight of the machine to compress materials. A vibratory roller combines static and dynamic forces to generate higher compaction. The vibration or dynamic force is created by rotating a series of off-balance weights within a steel compacting drum (*Figure 5*). This rotation develops a centrifugal force that is sufficient to lift and drop the steel drum as it turns. The vibrating amplitude is one-half the total lift and drop distance.

The vibrating amplitude can be changed by adjusting the off-balance weight. The weight can be increased or moved further from the center axle to increase the amplitude. Decreasing the weight, moving it closer to the center axle, or adding a balancing weight will decrease the vibrating amplitude. Some machines have fluid-filled chambers that can be adjusted to increase the weight or balance the off-balance weight to increase or decrease the vibrating amplitude.

Figure 3 ◆ Steel-wheel roller.

Figure 4 ◆ Vibratory steel-wheeled roller.

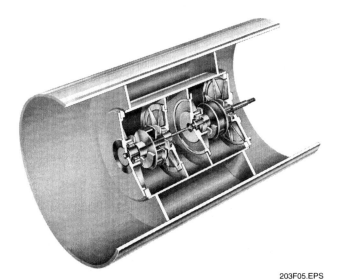

Figure 5 † Cutaway of vibratory roller.

The other measure of vibratory motion is frequency. Frequency is simply the number of times the lift and drop cycle is repeated each second. Increasing the speed that the off-balance weight is rotating will increase the frequency. Decreasing the rotational speed will decrease the frequency. As the machine moves across the surface of the material, the lift and drop cycle repeats 28 to 67 times per second. The vibratory motion increases the compacting force up to six times the static weight of the drum assembly.

Most soils vibrate more easily at a specific frequency known as the resonant frequency. In most cases, vibration at this resonant frequency will produce faster compaction. In some vibratory compactors, the frequency can be varied to achieve the best results. Many have a monitoring system that indicates the vibrating amplitude.

2.4.0 Steel-Wheeled Sheepsfoot

The sheepsfoot compactor, or **tamping roller** (*Figure 6*), offers fast compaction and high production. Self propelled units have good maneuverability and pulling power. Other units can be towed behind a bulldozer so the area can be graded and compacted in one pass. Alternately, a blade can be mounted on the front of a self-propelled compactor. Cleaner bars can be added to the back to remove the dirt caught between the feet.

The steel drum rollers have projecting pads or feet. The feet are normally 7 to 9 inches long. There are two basic styles, sheepsfoot and tapered.

Sheepsfoot rollers compact a little at the surface, but provide greater compaction under the feet. In deep or soft fill, the wheel actually rides higher with each pass. This is known as walking out. The tow-type sheepsfoot is increasingly being replaced by large, self-propelled units. However, the towed unit still fills a compaction need. With its slower speed and lower cost, it can provide economical compaction in smaller fill areas.

Figure 6 ◆ Sheepsfoot roller.

3.0.0 ◆ IDENTIFICATION OF EQUIPMENT

Rollers and compactors are designed to perform certain tasks. Because the tasks are similar, the machines designed to complete them are also similar in appearance and operation, regardless of the manufacturer. Optional equipment may be available to improve efficiency or increase the usefulness of the machine, but the basics are the same.

There are approximately 16 companies that manufacture rollers, and describing the operation of each model is impossible. The information presented in this module is largely based on the operation and maintenance requirements of a Caterpillar 815F soil compactor. Where significantly different design features occur between manufacturers, those differences will be noted and briefly described. The Bomag Model BW5AS steel-wheel roller and the Vibromax single-drum roller model VM106 D/PD are used as comparison machines. The term roller will be used to include both rollers and compactors.

It is important to read and understand the instruction manual that applies to the make and model of the equipment you are operating. If the manual is not available, ask your supervisor for it. Do not operate any piece of equipment without understanding how it works and what to do to ensure safety.

The Caterpillar 815F soil compactor is a 46,000 pound diesel-powered articulated compactor with bulldozing capability. The engine is capable of developing 240 horsepower. The transmission has four forward and four reverse gears. *Figure 7* shows the primary components of the Caterpillar 815F roller/compactor.

A diesel engine provides power to move the roller and drive the hydraulic pumps. On two-wheel smooth rollers and pneumatic rollers, the engine is usually mounted in the center of the machine between the rollers. On articulated rollers, the engine is mounted in the rear section, which also contains the operator's cab. Usually, power is supplied to the rear roller or wheels, which move the machine, although some units have four-wheel drive. The front or guide roller is used to steer the machine.

Several of the machine systems are hydraulic. One or more hydraulic pumps may be interconnected. The hydrostatic drive provides a stepless change of gear ratios and a smooth transition when changing directions. The service brakes may be part of the hydrostatic drive. In this configuration, the forward/neutral/reverse (FNR) lever on the steering column is used for braking. Steering is typically accomplished using two hydraulic rams. On vibratory rollers, the vibration system may also be hydraulically activated. The hydraulic system is also used to control any attachments.

The following provides a brief description of the controls and indicators that are common to most rollers. Consult the instruction manual of the particular make and model you are operating for information specific to that machine.

203F07.EPS

Figure 7 ◆ Roller with dozer blade.

3.1.0 Operator's Cab

The operator is enclosed in a rollover protective structure (ROPS) and/or a falling objects protective structure (FOPS), also known as the cab. Most of the machine controls are located in the cab. It is the central hub for roller operations. An operator must understand the controls and instruments before operating a roller. Study the operator's manual to become familiar with the controls and instruments and their functions.

Figure 8 is an overhead view of the interior of the operator's cab on a Caterpillar 815F. Some switches and controls are located outside of the main cab. However, the controls for normal operations can be reached from the operator's seat. A steering wheel, accelerator, and brakes are used to maneuver the machine. The transmission is controlled with a lever on the left side of the steering wheel. The instrument panel is in front of the steering wheel. A joystick control lever to the right of the steering wheel controls the dozer blade. The engine is started with a key switch. An additional switch panel to control ancillary machine functions is located to the right of the operator's seat.

Figure 9 shows the cab for an Ingersoll-Rand DD-138 steel-wheel vibratory roller. This is a typical open-cab design which provides greater visibility of the leading roller edge. The driver is protected from falling objects by the FOPS/ROPS.

> **NOTE**
> Some rollers feature an open operator's platform covered with a ROPS. The open design allows greater visibility in all directions. However, the open design also subjects the operator to greater levels of noise, fumes, heat, and cold. Take appropriate safety precautions to protect yourself from harmful exposures.

203F09.EPS

Figure 9 ◆ Operator's cab on an Ingersoll-Rand DD-138.

TRANSMISSION
SPEED SELECT
LEVER

LEFT BRAKE PEDAL

INSTRUMENT PANEL

STEERING WHEEL

JOYSTICK BLADE CONTROL

ACCELERATOR PEDAL

RIGHT BRAKE PEDAL

203F08.EPS

Figure 8 ◆ Operator's cab on a Caterpillar 815F.

3.2.0 Instruments and Indicators

Operators must pay close attention to the instrument panel when operating a roller. The gauges and lights show the status of the machine's systems. These are electronically monitored. *Figure 10* shows the instrument panel on a Caterpillar 825G. The left-hand panel contains four gauges. In the center is the speedometer and digital readout. The panel on the right contains a series of warning lights.

Figure 11 shows a typical four-gauge panel. These gauges include the water temperature gauge, the transmission oil temperature gauge, the hydraulic oil temperature gauge, and the fuel level gauge. The following sections will describe each of these gauges and their functions.

3.2.1 Engine Coolant Temperature Gauge

The engine coolant temperature gauge (1, *Figure 11*) indicates the temperature of the water or coolant flowing through the cooling system. Refer to the operator's manual to determine the correct operating range for normal roller operations. Temperature gauges normally read left to right, with cold on the left and hot on the right. If the gauge is in the white zone, the coolant temperature is in the normal range. Most gauges have a red segment. If the needle is in the red zone, the coolant temperature is excessive. Some machines may also activate warning lights if the engine overheats.

CAUTION

Operating equipment when temperature gauges are in the red zone may severely damage it. Stop operations, determine the cause of problem, and resolve it before continuing operations.

If the engine temperature gets too high, stop operation immediately. Get out of the machine and investigate the problem. There are several checks that the operator can perform. First check the engine coolant level. Add more fluid if it is too low. Check that the fan belt is not loose or broken. Replace it if necessary. Check that the radiator fins are not plugged with dirt or debris. Clean them if necessary. These are the three primary causes of the engine overheating. If initial troubleshooting fails to resolve the problem, stop operations and take the machine out of service.

WARNING!

Engine coolant is extremely hot and under pressure. Never remove a radiator cap while the engine is hot. Check the operator's manual and follow the procedure to safely check and fill engine coolant.

3.2.2 Transmission Oil Temperature Gauge

The transmission oil temperature gauge (2, *Figure 11*) indicates the temperature of the oil flowing through the transmission. This gauge also reads left to right in increasing temperature. It has a red zone that indicates excessive temperatures. When the weather is colder, allow the transmission oil to warm up sufficiently before operating the machine.

3.2.3 Hydraulic Oil Temperature Gauge

The hydraulic oil temperature gauge (3, *Figure 11*) indicates the temperature of the oil flowing through the hydraulic system. This gauge also

203F10.EPS

Figure 10 ◆ Instrument panel for the Caterpillar 825G series roller.

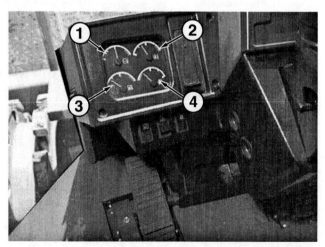

203F11.EPS

Figure 11 ◆ Typical four-gauge panel for a roller.

reads left to right in increasing temperature and has a red zone that indicates excessive temperatures. Check the operator's manual for the normal operating range. You must allow the hydraulic oil to warm up sufficiently before operating the machine under a load. Cycle the hydraulic system under low rpm to warm it up.

NOTE
If the hydraulic system is sluggish, the hydraulic oil has not warmed sufficiently. Allow the machine to idle before operating it under a load.

3.2.4 Fuel Level Gauge

This gauge (4, *Figure 11*) indicates the amount of fuel in the roller's fuel tank. Some models may have a low fuel warning zone and low fuel warning light. Stop and refuel if the machine is running low. Avoid running out of fuel on diesel engine rollers because the fuel lines and injectors must be bled of air before the engine can be restarted.

3.2.5 Digital Display

Some machines have a combined digital display. In *Figure 12*, the digital display is on the lower half of the panel (1). The display can be toggled to show the hour meter, tachometer, speedometer, odometer, or diagnostic codes.

The hour meter indicates the total hours of operation. It shows the period of time the machine has been running. Periodic maintenance is scheduled based on hours of service, which will be covered in more detail later in this module.

203F12.EPS

Figure 12 ◆ Digital display and warning lights.

The tachometer indicates the engine speed in revolutions per minute (rpm). The speedometer shows the machine's ground speed. Typically they can be set for either miles per hour (mph) or kilometers per hour (kph). Diagnostic codes are shown when the machine is not functioning correctly. They are used by service mechanics to identify problems with the machine's systems.

3.2.6 Indicators

There is a series of warning lights (2) above the display window as shown in *Figure 12*. When lit, these lights show that the machine systems are not operating under normal conditions. Generally, they indicate that there is something wrong with the machine. If any of the warning lights start flashing, stop the machine and resolve the problem before continuing operations. The exception is the parking brake light. If it is lit, the parking brake is activated. However, if it starts flashing the parking brake is not functioning properly.

Typical warning lights show that the following systems are not functioning properly:

- Engine oil pressure
- Parking brake (when flashing)
- Brake oil pressure
- Electrical system
- Low fuel light
- Primary steering
- Supplemental steering

CAUTION
A flashing indicator light requires immediate action by the operator. Stop the machine and investigate the cause. If you continue operations, you may cause serious damage to the equipment. Resolve the problem before continuing operations.

There are often increasing levels of warning alarms on a machine. At the first level, alert indicators will light up. The operator must take action in the near future to correct the problem. At the second level, the alert indicators will flash. The operator must take action immediately to correct the problem and avoid machine damage. At the third level the action indicators will flash and a warning alarm will sound. The operator must immediately shut down the machine to avoid machine damage or operator injury.

3.3.0 Controls

Controls and their locations can vary between makes and models of rollers. Vehicle movement controls on rollers can be similar to those of cars and trucks. A steering wheel is used in combination with throttle and brake foot pedals to control vehicle movement. The dozer blade is controlled with a joystick or levers. Switches and levers activate ancillary and specialty machine functions. The controls for a Caterpillar 815F will be described in this module. Review the operator's manual to fully understand the controls for the machine you are operating.

3.3.1 Disconnect Switch

Some models are equipped with disconnect switches located outside the cab. These switches disconnect critical functions so that the machine cannot be operated. These switches must be activated before the machine can be started. Turn them on before mounting the machine. They offer an additional level of safety and security. Typically, unauthorized users will not activate these switches and will not be able to operate the equipment.

Some models have a fuel shutoff switch that must be turned to the ON position before the machine can be operated. The fuel shutoff switch physically prevents fuel from flowing from the fuel tank into the supply lines. This prevents unwanted fuel flow during idle periods or when transporting the machine, significantly reducing the potential for fuel leaks.

Alternatively, some machines have a battery disconnect switch, as shown in *Figure 13*. When the battery disconnect switch is turned off, the entire electrical system is disabled. This switch should be turned off when the machine is not in use overnight or longer to prevent a short circuit or active components from draining the battery. Before mounting the machine, check that the switch is in the ON position.

NOTE

Some machines have security panels or vandal guards. These panels can be locked into a no-access position when the machine is not in use. They prevent access to the machine's controls so it cannot be started or operated. Engage any vandal guards when leaving the machine unattended. Move the guard down to the stowed position to operate the machine. Lock the cab door, secure and lock the engine enclosure. Always engage the security systems when leaving the roller unattended.

3.3.2 Seat and Steering Wheel Adjustment

When first entering the cab, the operator should adjust the seat and steering wheel. While seat and steering wheel adjustments are not directly involved in roller operations, correct positioning of these items aids in safe operation. The operator should adjust the seat and steering wheel position and then fasten the seat belt before operating the roller.

Most seats can be moved up or down and forward or backward using a series of levers (*Figure 14*). Some provide an adjustable shock absorber function using springs or hydraulics. The angle of the back of the seat can be set by lifting the seat recline lever. Pull up on the fore/aft lever and slide the seat to adjust it. Release the lever to lock the seat in place. The seat should be adjusted so that the operator's legs are almost straight when the accelerator or brake pedals are fully depressed and the operator's back is flat against the back of the seat. Lift the seat cushion angle lever and adjust the seat to the desired position. Releasing the lever locks the seat in position. The stiffness of the seat suspension can be set to different positions by rotating a knob. The correct adjustment is based on the weight of the operator, and approximate setting is often shown on a scale. The seat height can be raised or lowered by lifting the seat height lever. The armrest can be adjusted by rotating a knob on the side of the armrest.

The steering wheel on some rollers can be adjusted up or down and in or out as desired after the seat is correctly positioned. In *Figure 15*, moving the handle (1) upward unlocks the steering column (2). The steering column can then be tilted and positioned correctly. Release the handle to lock the steering column into position. Because seat and steering wheel adjustment devices vary widely for various makes and models, refer to the operator manual for specific instructions.

Many rollers are designed to be driven in both directions. To increase operator comfort and improve visibility, some seats are mounted so that they face the side. Side-mounted seats allow the operator to see equally well in either direction. Other units have dual swivel seats like the ones shown in *Figure 16*. The driver rotates the seat and steering column depending on the direction of travel. This allows the operators to see the drum edge and rolling pattern more clearly. Some smaller models feature a seat that can be moved around the operators platform to access the controls from several different positions.

Figure 13 ◆ Battery disconnect switch.

Figure 14 ◆ Seat adjustment controls.

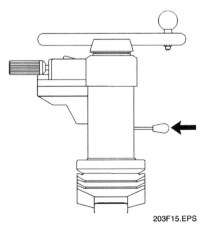

Figure 15 ◆ Steering wheel adjustment controls.

Figure 16 ◆ Dual-seat configuration.

3.3.3 Engine Start Switch

The ignition switch functions can vary widely between makes and models of rollers. Some only activate the starter and ignition system. Others may activate fuel pumps, fuel valves, and starting aids. In some cases, the starter and starting aids are engaged by other manual controls. The multi-purpose ignition switch for the Caterpillar 815F series roller is shown in *Figure 17*. This switch is located to the right of the steering wheel on the console.

The engine start switch has the following three positions:

- *Off (1)* – Turning the key to this position stops the engine. It will also disconnect power to electrical circuits in the cab. However, several lights remain active when the key is in the OFF position, including the hazard warning light, the interior light, and the parking lights.

- *On (2)* – Turning the key to this position activates all the electrical circuits except the starter motor circuit. When the key is first turned to the ON position, it may initiate a momentary instrument panel and indicator bulb check.

- *Start (3)* – Turn the key to this position to activate the starter and start the engine. Release the key when the engine starts. This position is spring-loaded to return to the ON position when the key is released. If the engine fails to start, the key must be returned to the off position before the starter can be activated again. To reduce battery load during starting, the ignition switch of some rollers may be configured to shut off power to accessories and lights when the key is in the start position.

four transmission speeds indicated on the collar. Rotate the transmission collar on the control lever to select the transmission speed. The four speeds can be used for either forward or reverse. Do not skip gears when downshifting.

Some rollers have joystick controls. One configuration is to have two joysticks, one for steering and one for travel control (*Figure 19*). All major machine functions are operated with button switches located at the tips of both joysticks. This allows the operators to easily engage and disengage drum vibration and manually control the water-spray system.

On some vibratory rollers, the vibratory action can be adjusted. *Figure 20* shows a typical control for a steel-wheel vibratory roller. The vibrations per minute (vpm) can be varied from 1,225 to 2,050. The high and low amplitude is selected via a switch below the vibration control.

3.4.0 Attachments

Typically, rollers are used only for their primary purpose and do not have interchangeable attachments like forklifts or loaders. Some models do feature interchangeable rollers. The roller can be changed from a smooth steel-wheel roller to a segmented-pad roller. But generally, roller attachments are limited to a water spray units, a dozer blade, or a scarifier.

Some roller units are attachments. They are portable and towed by tractors or other equipment. These rollers or compactors are typically towed by a dozer. Thus grading and compacting can occur in a single pass.

Water spray units are typically used when rolling asphalt. They can be mounted on steel-wheel or pneumatic tire rollers but come as standard equipment on many models. They are used when rolling asphalt to keep the drum moist. Asphalt is less likely to stick to the wet surface of the roller.

3.4.1 Dozer Blade

The blade on a roller can be controlled with levers or with a joystick. The blade can be moved in several directions. The joystick may be used in conjunction with switchers, triggers, or buttons. The four ways to move the blade are as follows:

- *Lift* – Lift lowers or raises the blade. Lowering the blade allows you to change the amount of bite or depth to which the blade will dig into the material. Raising the blade permits you to travel, shape slopes, or create stockpiles. Most controls have a float position, which permits the blade to adjust freely to the contour of the ground. The float position is commonly used in reverse to smooth the surface.

203F17.EPS

Figure 17 ◆ Engine start switch.

3.3.4 Vehicle Movement Controls

Roller movement is controlled in a manner similar to that of cars and trucks. The throttle and brakes are operated by foot pedals. A steering wheel is used to turn the vehicle. The transmission is controlled with a lever on the right of the steering column. Moving the steering wheel to the left or right steers the vehicle to the left or right. Some steering wheels are equipped with a knob. This gives the operator greater control of the wheel when steering with one hand. This is important when the operator is using one hand to steer and the other to operate the dozer blade with the joystick.

The transmission is controlled via a lever on the left side of the steering column (*Figure 18*). The lever has three positions; forward, neutral, and reverse. Move the lever to select the direction of travel. Move the lever upward to set the transmission in forward. The center position is neutral. Move the lever downward for reverse. There are

Figure 18 ◆ Transmission control.

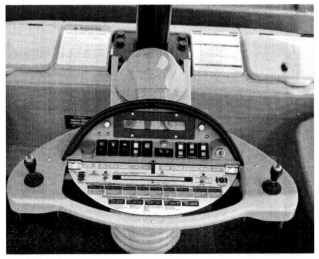

Figure 19 ◆ Dual joystick controls.

Figure 20 ◆ Vibratory controls.

- *Angle* – Angle adjusts the blade in relation to the direction of travel. When moving a load, the blade should be perpendicular to the line of travel. For filling a ditch, the blade should be angled to permit the load to be pushed off to the side.

- *Tilt* – Tilt changes the angle of the blade relative to the ground. This permits the blade to cut deeper on one side than on the other. This process is very useful for performing side hill work where the blade tends to hang lower on the downhill side. It is also useful for crowning roads and grading slopes and curves.

- *Pitch* – Pitch is the slope of the blade from top to bottom. The greater the slope, the more the blade tends to dig in. On most machines the blade pitch must be changed manually. Because of the difficulty in controlling the pitch, it is only changed for unusual situations.

A typical control arrangement is shown in *Figure 21*. The right and left movement of the joystick angles the blade right and left. Backward and forward movement of the joystick raises and lowers the blade. The center position is the float or hold position. The blade will remain in the same position when the controls are in the hold position.

The controls for blade movement on a roller vary. There can be a second lever or joystick to control this movement. Alternatively, the blade could be adjusted with pins that must be moved manually. In a two-lever configuration, moving the lever forward tilts the blade forward. Pulling the lever backward tilts the blade backward. When the lever is released, the lever will return to the center hold position and the blade will stop moving. Always read the operator's manual to understand the controls for the machine you will be operating.

Figure 21 ◆ Blade control.

4.0.0 ◆ SAFETY GUIDELINES

Safe operation is the responsibility of the operator. Operators must develop safe working habits and recognize hazardous conditions to protect themselves and others from injury or death. Always be aware of unsafe conditions to protect yourself from injury and the roller from damage. Become familiar with the operation and function of all controls and instruments before operating the equipment. Read and fully understand the operator's manual.

The majority of accidents on the job site are the result of a combination of factors such as inattention, inexperience, and carelessness. It is vitally important for everyone on the site to be aware of and follow the highest safety practices.

The construction industry is always at the top of all most dangerous profession lists. The heavy equipment sector is one of the most dangerous parts of the industry. When on site and operating heavy equipment, it is of utmost importance for you to be aware of your surroundings. On a busy construction site there are many people crossing the path of heavy equipment during the course of a business day. Everyone is concentrating on their jobs, but it must be remembered that safety should be the number one consideration.

The first aspect of safe operation is to prepare yourself to work safely. The following are general rules that should be followed to prepare for safe operation:

- Read and understand the operating manual of the machine you will be operating.
- Read and understand your company's safety manual and safety regulations.
- Inspect the machine for potential safety problems at the beginning of your shift and as possible throughout the day.
- Perform the daily maintenance procedures required by the operator's manual. A well maintained machine is safer than a poorly maintained machine.
- Make yourself aware of any physical properties at the site, such as overhead wires, that may be a safety hazard.
- Wear appropriate clothing. Leave jewelry at home.
- Wear eye and ear protection, a hardhat, and safety shoes.
- Remove or tie down loose tools, tie down safety equipment, and remove or tie down building materials.
- Use your seat belt.
- Read and understand all safety labels or stickers placed on the machine. Keep these signs clean so they can be read. Replace any labels or stickers that are missing or not readable.

The second aspect of safe operations is to avoid dangerous behavior. The following general rules will help you stay clear of risky actions.

- Do not allow others to ride on the machine.
- Do not drill holes into, or weld anything to, the structure of the machine.
- Never use accessories not specifically designed to be used with your machine.
- Do not disable alarms, lights, or related safety devices and make sure they are working properly.
- Do not smoke while checking machine fluid levels. Some fluids and related vapors are highly explosive. Do not smoke when refueling or when servicing the ether cold weather start system.
- Make sure all shields, guards, and access panels are in place before operating the machine.
- Never drill into or weld to the structure of an ROPS. Weakening of the structure by doing so can result in injury or death of the operator and bystanders.
- Keep objects and hands away from moving fan blades. Contact with the blades will severely cut body parts and cut or throw objects.
- Do not allow oil and grease to accumulate on the machine, as it creates both a fire hazard and a slip-and-fall hazard. Clean spills from the machine immediately and clean with high pressure water or steam every 1,000 hours.
- Dispose of dangerous fluids and material properly, following all applicable environmental regulations.

Finally use common sense and advance planning to protect yourself and others from hazards. Use the following general rules to operate the equipment:

- Stay seated and use the seat belt while operating the machine.
- Protect yourself from dust, dirt, and fumes.
- Check the engine coolant level only after the engine has been stopped and you can safely remove the filler cap with your bare hand. Vent the pressure and remove the cap slowly.
- When moving the machine to a colder climate, remember to check or replace fluids that may be affected by the temperature change.
- Know the maximum height of your machine.
- Frequent replacement of the same fuse may indicate an electrical problem that should be investigated.
- Always replace fuses with the same type and size.

NOTE

Many construction workers are at risk for hearing loss due to noise exposure over the course of their career. With an average exposure of 80 decibels, a worker has a 3 percent chance of hearing loss. But an average exposure of 85 decibels increases potential hearing loss to 15 percent. At 90 decibels, it further increases to 29 percent. Rollers produce between 80 and 105 decibels. A general rule of thumb is that you need hearing protection when operating a roller from an open cab or when you have to raise your voice to be heard.

4.1.0 Specific Roller/Compactor Safety Rules

In addition to the general hazards from operating heavy equipment on a construction site, there are specific hazards related to operating rollers. The following safety rules apply specifically to rollers:

- Know the limits for operating the roller on slopes. Do not turn on a slope or move across a slope. Only operate up and down slopes.
- Be careful when operating equipment close to the edge of a hill or embankment. Roller compaction can cause the slope to collapse. The roller could become unstable and overturn.
- Machines that produce vibration can cause the walls of trenches or embankments to collapse. Make sure these walls are properly braced.
- Do not operate the vibrator when the machine is stopped. Excessive vibration can cause the ground to subside and cause the machine to overturn.
- Articulated machines that have a center pivot do not have clearance for personnel in the pivot area. Keep all personnel clear of the pinch area.
- The transmission of some rollers will not hold a parked machine on any grade. Always apply the secondary parking brake and move the transmission control lever to the neutral position after the machine has stopped. Do not use the secondary brake system to slow and then stop the machine. It is to be used to hold the machine in a parked position only. Do not use the machine if it can be moved when the secondary brake system is on.
- Use extreme caution when raising or lowering scrapers. The scrapers are under spring tension. The lower edge of the scraper can become very sharp due to wear.

 **Roller Rollover**

A highway construction worker died after the roller she was operating slipped off the edge of the road surface and rolled over. The worker was compacting a road bed. The earth under the rear tires gave way when she backed up near the edge. She was thrown from the open cab door, pinned under the ROPS, and died from the injury. The machine was equipped with a ROPS but not a seat belt.

The Bottom Line: Always wear your seat belt. Do not operate a roller that is not equipped with a ROPS and a seat belt.

Source: *Electronic Library of Construction Occupational Safety and Health.*

- Do not operate a water spray system with an empty reservoir. This can damage the pump.

4.2.0 Tire Safety Rules

Tire pressure is an important aspect of roller operation for rubber-tired rollers. Air, nitrogen, or a liquid mixture may be used to inflate rubber tires. Some models require inflation with air and a mixture of calcium chloride and water. Refer to the operator's manual for your specific model for complete instruction. Remember that most heavy equipment tires are mounted on a split rim and, unlike automobile tires, can explode with devastating force under certain conditions.

Use the following guidelines to be safe when working with rubber tires on a roller:

- Check the tire pressure at the beginning of each shift and at intervals of ten hours thereafter. Inspect each tire for excessive wear and damage to both the tire and rim.
- When inflating the tire, park the machine to position the filler valve at the top of the tire.
- Never inflate a completely deflated tire. Leave that to a trained tire technician.
- When inflating a tire, never stand directly in front of the tire. Position yourself to the side of the valve.
- Never expose tire rims to a direct flame or high heat like that from a torch or welder.

5.0.0 ◆ BASIC PREVENTIVE MAINTENANCE

Routine maintenance is a regular effort to lubricate and service the equipment. This helps prevent breakdowns and keeps the machine is good working order. Consistent attention to these details keeps you and your machine operating efficiently, which improves productivity and safety.

These tasks must be performed on a regular basis. Some tasks, like inspection and lubrication, should be done daily. Other tasks must be done weekly, monthly, or annually. Maintenance is scheduled based on the hours of service. The operator must keep track of the hours of service and schedule maintenance. Usually a service log is kept with the machine. Some preventive maintenance procedures can be easily performed with the right tools and equipment.

CAUTION

Roller service is normally based on hours of service. A service schedule is contained in the operator's manual. Failure to perform scheduled maintenance could damage the machine.

Maintenance requirements are basically separated into two main categories: as required and scheduled. The scheduled maintenance is further divided into nine cycles:

- Daily (10 hours), or once per shift
- Weekly (50 hours)
- Bi-weekly (100 hours)
- Monthly (250 hours)
- Three months (500 hours)
- Six months (1,000 hours)
- One year (2,000 hours)
- Two years (3,000 hours)
- Four years (6,000 hours)

There is a list of inspections and required service activities in the operator's manual.

5.1.0 Daily Inspection and Maintenance

Perform all maintenance procedures according to the instruction given in the machine's operating and/or service manual. Before performing any maintenance on the machine, you must read and understand all safety rules in the operating and/or service manual. Although not all of the maintenance tasks are performed by the operator, it is the operator's responsibility to ensure they have been done.

Before beginning work each day, you must inspect your machine. Some companies provide a daily inspection checklist. Include all items on the daily or 10 hour maintenance schedule as well as any as-required maintenance. Complete the inspection and daily maintenance before starting the engine. This will help prevent a breakdown during operation.

The daily inspection is often called a walk-around. The operator should walk completely around the machine checking various items. Look around and under the machine for leaks, damaged components, or missing bolts or pins. If you see evidence of a leak, locate the source and get it repaired. If leaks are suspected, check fluid levels more frequently. Items to be checked and serviced on a daily inspection include the following:

- Engine precleaner screen for debris
- Engine compartment for debris
- Engine for damaged belts
- Hydraulic system for leaks

- Cooling system for leaks, debris, damaged hoses
- Transmission for leaks
- Fuel tank for water and sediment
- Covers and guards secured in place
- Front and rear differentials for leaks
- Tires for proper inflation and damage
- Blade control linkage for damage and/or wear (if applicable)
- Steps, walkways and handholds for debris and damage
- ROPS for damage
- Windows for cleanliness
- Cab for cleanliness
- Seat belt for damage or wear
- Roller cleaner bars
- Pad foot tips (*Figure 22*)
- Lights for damage
- Indicators, gauges, and brakes
- Back-up alarm for proper operation
- Roller and cleaner bar for damage

203F22.EPS

Figure 22 ◆ Check pad foot tips.

 WARNING!

Do not check for hydraulic leaks with your bare hands. Use cardboard or another device. Pressurized fluids can cause severe injuries to unprotected skin. Long-term exposure can cause cancer or other chronic diseases.

During your daily inspection, check the fluid levels and add fluids as needed. Make sure that the machine is level to ensure that fluid levels are accurate. The following fluids should be checked:

- Engine oil (*Figure 23*)
- Coolant
- Transmission oil (*Figures 24* and *25*)
- Hydraulic oil (*Figures 26* and *27*)

Some maintenance is performed as required. This means that it should be done whenever it is needed. Include as-required maintenance in your daily walk-around. These tasks should also be performed during or at the end of the work shift, as needed. Typical as-required maintenance includes the following:

- Clean air intake
- Service primary air filter element
- Service secondary air filter element (*Figure 28*)
- Check ether starting aid
- Check fuses (*Figure 29*)
- Check circuit breakers

203F23.EPS

Figure 23 ◆ Check the engine oil and coolant levels.

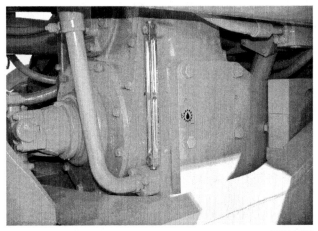

203F24.EPS

Figure 24 ◆ Transmission fluid sight gauge.

Figure 25 ◆ Transmission fluid fill port.

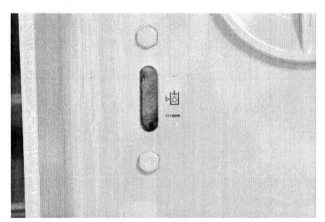

Figure 26 ◆ Hydraulic fluid sight gauge.

Figure 27 ◆ Hydraulic fluid fill port.

- Inspect windshield wipers
- Fill windshield washer fluid
- Inspect cutting edge and end bits
- Inspect tamping tips
- Inspect cleaner bar
- Inspect oil filter

> **NOTE**
>
> Roller tires can be inflated with air or a liquid mixture. Tires inflated with liquid ballast provide added machine stability. The liquid is a mixture of water and calcium chloride. The latter provides antifreeze protection. No matter what is inside the tires, check the tire pressure daily to make sure they meet the recommended pressure in the operator's manual.

Figure 28 ◆ Clean the secondary air filter.

Figure 29 ◆ Check the fuses.

The operator's manual usually has detailed instructions for performing periodic mainte- nance. If you find any problems with your machine that you are not authorized to fix, inform the foreman or field mechanic before operating the machine.

5.2.0 Servicing a Roller

A roller should be serviced by a trained mechanic. When servicing a roller, follow the manufac- turer's recommendations and service chart. Any special servicing for a particular piece of equip- ment will be highlighted in the manual. Normally, the service chart recommends specific intervals, based on hours of run time, for such things as changing oil, filters, and coolant. In addition, some machines provide computer diagnostic information. The machine's systems are moni- tored by a computer. The computer will display diagnostic codes which provide service informa- tion. The codes and required service are explained in the machine's service manual.

Hydraulic fluids should be changed whenever they become dirty or break down due to overheat- ing. Continuous and hard operation of the hydraulic system can heat the hydraulic fluid to the boiling point and cause it to break down. Filters should also be replaced during regular servicing.

Before performing maintenance procedures, always complete the following steps:

Step 1 Park the machine on a level surface to ensure that fluid levels are indicated cor- rectly.

Step 2 Lower all equipment to the ground. Operate the controls to relieve hydraulic pressure.

Step 3 Engage the parking brake.

Step 4 Lock the transmission in neutral.

5.2.1 Preventive Maintenance Records

Accurate, up-to-date maintenance records are essential. Each machine should have a record that describes any inspection or service that is to be performed and the corresponding time intervals. Typically, an operator's manual and some sort of inspection sheet are kept with the equipment at all times.

6.0.0 ◆ BASIC OPERATION

Operation of a roller requires constant attention to the controls and instruments, the rollers, and the surrounding environment. Rollers often work in conjunction with other machines. Operators must be aware of other operations going on around the equipment.

Operating a roller is fairly straightforward. It is simply a matter of starting the machine, disen- gaging the parking brake, shifting into gear, and depressing the accelerator to move the machine. Rollers move slowly; typical speeds are between three and five miles per hour.

The initial operations you must learn include the proper ways to start and shut down the machine and how to move forward and back- ward. There are standard rolling patterns for spe- cific applications like rolling asphalt. This section also include information on basic dozing and land clearing as well as compaction.

6.1.0 Suggestions for Effective Roller Operation

Before starting roller operation, make sure you are familiar with the area of operations and the soil conditions. Check the area for both vertical and horizontal clearances. Make sure that the path is clear of electrical power lines and other obstacles. Make note of soft areas, trenches, embankments, or other unstable areas. Mark any hidden dangers.

Check your machine to make sure that it is adjusted for the work that you will be performing. Where appropriate, check that the following are set correctly:

- Water spray unit
- Tire inflation
- Ballast or other weights

6.2.0 Preparing to Work

Preparing to work involves getting yourself orga- nized in the cab and starting your machine. Mount your equipment using the grab rails and foot rests. Always maintain three points of contact when mounting equipment. Keep grab rails and foot rests clear of dirt, mud, grease, ice, and snow. Adjust the seat to a comfortable operating posi- tion. The seat should be adjusted to allow full pedal travel with your back against the seat back. This will permit the application of maximum force on the brake pedals. Make sure you can see clearly and reach all the controls. If your machine has dual seats, adjust both seats before operating the machine.

Operator stations vary depending on the manufacturer, size, and age of the equipment. However, all stations have gauges, indicators, switches, levers, and pedals. Gauges tell you the status of critical items such as water temperature, oil pressure, and fuel level. Indicators alert you to low oil pressure, engine overheating, and electrical system malfunctions. Switches activate starting aids and turn accessories such as lights on and off. Typical instruments and controls were described previously. Review the operator's manual so that you know the specifics of the machine you will be operating.

The startup and shutdown of an engine is very important. Proper startup lengthens the life of the engine and other components. A slow warm up is essential for proper operation of the machine under load. Similarly, the machine must be shut down properly to cool the hot fluids circulating through the system. These fluids must cool so they can cool the metal parts of the engine before it is switched off.

6.2.1 Start-Up

There may be specific start-up procedures for the piece of equipment you are operating, but in general, the start-up procedure should follow this sequence:

Step 1 Be sure the transmission control is in neutral.

Step 2 Use a lever or knob to engage the parking brake (2, Figure 30), depending on the roller make and model. Some rollers have a parking brake test switch (1, Figure 30).

NOTE

When the parking brake is engaged an indicator light on the dash will light up or flash. If it does not, stop and correct the problem before operating the equipment.

Step 3 Depress the throttle control slightly.

Step 4 Turn the ignition switch to the start position. The engine should turn over. Do not operate the starter for more than 30 seconds at a time. If the engine fails to start, turn the key to the OFF position and wait 2 to 5 minutes before cranking again.

Step 5 Warm up the engine for at least 5 minutes. In colder temperatures warm up the machine for a longer period.

Step 6 Make sure all the gauges and instruments are working properly.

Step 7 Shift the transmission into forward and rotate the gear control to low range.

Step 8 Release the parking brake and depress the service brakes.

Step 9 Check all the controls for proper operation.

Step 10 Check service brakes for proper operation.

Step 11 Check the steering for proper operation.

Step 12 Manipulate the controls to be sure all components are operating properly.

Step 13 Shift the transmission to neutral.

Step 14 Reset the parking brake.

Step 15 Make a final visual check for leaks, unusual noises, or vibrations.

If the machine you are using has a diesel engine, there are special procedures for starting the engine in cold temperatures. Many diesel engines have glow plugs that heat up the engine for ignition. Follow the manufacturer's instructions for starting the engine in cold temperatures.

Some units are also equipped with an ether starting aid. Depress the switch (Figure 31) while cranking the engine. Depress the switch every two seconds until the engine is running smoothly. Review the operator's manual so that you fully understand the procedures for using these aids.

203F30.EPS

Figure 30 ◆ Parking brake.

Figure 31 ◆ Starting aid switch.

As soon as the engine starts, release the key; it should return to the ON position. Adjust the engine speed to approximately half throttle. Let the engine warm up to operating temperature before moving the roller.

> **WARNING!**
> Ether is used in some older equipment to start cold engines. Ether is a highly flammable gas and should only be used under strict supervision and in accordance with the manufacturer's instructions.

6.2.2 Checking Gauges and Indicators

Keep the engine speed low until the oil pressure registers. The oil pressure light should initially light and then go out. If the oil pressure light does not turn off within 10 seconds, stop the engine, investigate, and correct the problem.

Check the other gauges and indicators to see that the engine is operating normally. On most rollers, the instrument panel will run a self test when the machine is first started. Check that the coolant temperature, hydraulic oil temperature, and oil pressure indicators are in the normal range. If there are any problems, shut down the machine and investigate or get a mechanic to look at the problem.

6.2.3 Shutdown

Shutdown should also follow a specific procedure. Proper shutdown will reduce engine wear and possible damage to the machine.

Step 1 Find a dry, level spot to park the roller. Stop the roller by decreasing the engine speed and placing the direction lever in neutral. Depress the service brakes and bring the machine to a full stop.

Step 2 Place the transmission in neutral and engage the parking brake.

Step 3 Release the service brake and make sure that the parking brake is holding the machine.

Step 4 Lower the dozer blade so that it is resting on the ground.

Step 5 Place the speed control in low idle and let the engine run for approximately five minutes.

Step 6 Turn the engine start switch to the OFF position and remove the key.

Step 7 Release hydraulic pressure by moving the control levers until all movement stops.

Step 8 Engage any vandalism covers and other security measures.

> **CAUTION**
> Rollers tend to be very heavy. Parking on a slope or incline is not advised. If you must park on a slope, park so that the machine is facing uphill. Provide a support behind the machine such as a concrete barrier.

6.3.0 Basic Maneuvering

To maneuver the roller you must be able to move forward, backward, and turn. Basic maneuvering was covered in detail in *Heavy Equipment Operations Level One*. The first part of this section serves as a review.

6.3.1 Moving Forward

The first basic maneuver is learning to drive forward. Adjust the seat and fasten the seat belt before operating the machine. To move forward, follow these steps:

Step 1 Before starting to move, use the joystick to raise the blade to clear any obstructions.

Step 2 Depress the service brakes and release the parking brake. Move the shift lever to low forward.

Step 3 Release the service brakes and depress the accelerator pedal to start the roller moving.

Step 4 Drive the machine forward for the best visibility. Steer the machine using a steering wheel.

NOTE
Slow down or brake before changing direction. This will increase the service life of the engine.

6.3.2 Moving Backward

To back up or reverse direction, slowly come to a complete stop. Engage the parking brake. If necessary adjust the seat or switch seats for optimal visibility.

Check to make sure that the travel area is free from pedestrians or obstacles. Depress the service brakes and release the parking brake. Move the shift lever to reverse. Release the service brake and slowly depress the accelerator to begin to move backwards.

NOTE
Many rollers use dynamic braking, which uses the engine to slow the machine. To slow the machine, move the transmission lever slowly toward and then to the neutral position. The machine will coast to a stop. Once the machine is stopped move the shift lever to reverse and increase engine speed to move in the desired direction.

6.3.3 Steering and Turning

How you steer a roller depends on the make and model; however, most have a steering wheel. Some roller steering wheels have a knob that allows the wheel to be turned with one hand. This allows the operator to use the other hand to control the joystick to move the blade or operate other controls.

The steering wheel on a roller operates in the same manner as the steering wheels on a car or truck. Moving the wheel to the right turns the roller to the right. Turning the wheel to the left moves the wheels to the left.

6.4.0 Transporting the Roller

Typically a roller is only driven around a site. If it must be transported from one job site to another,

it is usually loaded and hauled on a transporter. Transport the roller on a properly equipped trailer or other transport vehicle. Before beginning to load the equipment for transport, make sure the following tasks have been completed:

• Check the operator's manual to determine if the loaded equipment complies with height, width, and weight limitations for over-the-road hauling.

• Check the operator's manual to identify the correct tie down points on the equipment (*Figure 32*).

• Be sure to get the proper permits, if required.

Once the above tasks are completed and the loading plan determined, carry out the following procedures:

Step 1 Position the trailer or transporting vehicle. Always block the wheels of the transporter after it is in position but before loading is started.

Step 2 If equipped, raise the blade slightly. Drive the roller onto the transporter. Whether the roller is facing forward or backward will depend on the recommendation of the manufacturer. Most manufacturers recommend backing the roller onto the transporter.

Step 3 Move the transmission lever to neutral, engage the parking brake, turn off the engine, and remove the key.

Step 4 Manipulate the hydraulic controls to relieve any remaining hydraulic pressure.

Step 5 Lock the door to the cab and any access covers. Attach any vandalism protection.

203F32.EPS

Figure 32 ◆ Roller tie down points.

Step 6 Engage the battery disconnect or fuel shutoff switches.

Step 7 Secure the machine with the proper tie-down equipment as specified by the manufacturer. Place chocks at the front and back of all four tires.

Step 8 Cover the exhaust and air intake openings with tape or a plastic cover.

Step 9 Place appropriate flags or markers on the equipment if needed for height and width restrictions.

WARNING!

The machine may shift while in transit if it is not properly tied down. If the machine shifts in transport it could cause personal injury or death. Follow the manufacturer's safety procedures.

Unloading the equipment from the transporter would be the reverse of the loading operation.

7.0.0 ◆ WORK ACTIVITIES

Operation of the roller is fairly straightforward, but it requires an understanding of the compaction process. With this knowledge, you will have no trouble operating the roller. A basic description of compaction and soils is included in this section.

7.1.0 Equipment Selection

There are four types of compactors, as previously described: rubber-tire, steel-wheel, vibratory-steel-wheel, and segmented-pad. They supply static and dynamic forces to compact soil,

depending on equipment configuration. Therefore, choosing the right equipment is an important part of the compaction process.

Equipment selection is based on the type of material to be compacted. Some types of compactors are more efficient for compacting a certain soil type while other methods are completely ineffective. *Table 1* compares the effectiveness of several types of compactors on different types of soil.

Soil may be compacted with various types of equipment by pressure, kneading, vibration, impact, or a combination of these methods. The steel-wheeled and segmented-pad rollers supply pressure with some kneading. Vibratory rollers supply both pressure and vibration.

The weight of a smooth steel-wheeled roller is applied along a straight line across the direction of travel. A high spot may carry the full weight of the roll, while a low spot may be bridged over and receive no compaction. With adequate weight, the soil is squeezed from high to low spots, equalizing irregularities in spreading and producing a smooth surface. However, high spots that are too hard to yield will support the roller and prevent compaction.

Segmented-pad rollers compact mostly with their feet from the bottom up. As the soil is compacted, the roller rises and walks out of the ground. These rollers are good on fine-grained plastic soils and are least efficient in sandy and gravely types. Avoid excessive weight, as the feet may shear the soil and damage its structure.

Rubber-tired rollers are suitable for use in any type of soil, but weight and tire pressure must be proper for the soil type. Results are affected by the shape of the tires, their air pressure, and the total wheel or axle load. They are not affected by tire pressure alone, as is often assumed. Very heavy units of 50 tons or more may be effective at compressing rock fills.

Table 1 Compaction Selection Chart

		Materials			
		Vibrating Sheepsfoot Rammer	Static Sheepsfoot Grid Roller Scraper	Vibrating Plate Compactor Vibrating Roller Vibrating Sheepsfoot	Scraper Rubber-tired Roller Loader Grid Roller
	Lift Thickness	Impact	Pressure (with kneading)	Vibration	Kneading (with Pressure)
Gravel	12+	Poor	No	Good	Very Good
Sand	10+/-	Poor	No	Excellent	Good
Silt	6+/-	Good	Good	Poor	Excellent
Clay	6+/-	Excellent	Very Good	No	Good

Vibration is most effective in sand and gravel soils, but may increase the effectiveness of a roller in any soil. It is particularly effective in bringing excess moisture to the surface.

Trenches and other small areas may be compacted by impact or by pneumatic or gasoline-powered hammers, jump rammers, or vibrators. Gravel fills may be compacted by puddling.

7.2.0 Compaction Method Considerations

Since a building or roadway is only as stable as the ground on which it is built, soil compaction is one of the most important jobs on any project. Fill that is not properly compacted will settle naturally over time, giving the effect of shrinking. Clay soils that are improperly compacted may absorb moisture and swell. In both cases, the resulting movement of the ground can damage roadways and structures, requiring costly repairs.

Uneven compaction across a fill area can cause the greatest damage. An entire project that settles evenly may cause very little harm, but when part of a foundation settles more than another part, it can severely damage the building foundation (see *Figure 33*). Uneven settling of a roadbed can ruin its entire surface.

The following should be considered in compaction of any area:

- Moisture content
- Layers
- Contractual requirements

7.2.1 Moisture Content

Moisture content is the single most important variable in soil compaction. Moisture acts as a lubricant, allowing soil particles to easily slide across one another to fill in air gaps. In soil that contains too little moisture, the particles resist sliding. In soils that contain too much moisture, the particles float and permit the gaps to be filled with water. Water cannot be compressed, and when it drains it will leave air gaps. Since both conditions are unacceptable, a great deal of effort goes into determining the correct amount of moisture for each mixture. This amount is called the optimal moisture content.

Soil is commonly sprayed with water as it is being placed in a fill so it may be compacted to the specified density, but often soil from a cut contains too much moisture to be compacted properly. In this case, you may need to use a grader or other equipment to turn the soil to allow it to dry out. Then when the moisture content is acceptable, the

area can be graded and compacted to specification. However, in naturally wet locations or wet weather, the soil may never dry out sufficiently to meet compaction specifications.

> **NOTE**
>
> The hand test is one method to determine if the soil has sufficient moisture to be properly compacted. Pick up a handful of soil, squeeze it in your hand, and then drop it. The soil should easily form a ball. It should break into a couple of pieces when dropped. If the soil is powdery and does not retain its shape, it is too dry. If the soil shatters when dropped, it is too dry. If the soil leaves traces of moisture on your fingers when squeezed, it is too wet for compaction. If it stays in one piece when dropped, it is also too wet.

7.2.2 Layers

Fill is typically placed in layers, or lifts, with the contract specifying the maximum thickness of each lift. As each lift is placed, it is compacted to the desired density, and another lift is added until the specified depth is reached. Common lift thicknesses are 4, 6, 8, and 10 inches. Compacting equipment is designed for particular soil types and lift thicknesses, so it is important to check the manufacturer's specifications when selecting equipment. For example, a very heavy rubber-tired roller works well for coarse rock fills, but not on cohesive clay, which needs a sheepsfoot drum.

7.2.3 Contractual Requirements

Compaction is usually specifically addressed in the construction contract. It is important for you to know and follow the exact contractual requirements

203F33.EPS

Figure 33 ◆ Foundation crack.

during compaction. Violating contractual specifications places your employer at risk for legal liability should something go wrong on the project due to compaction errors. Compaction can be specified in the following ways:

- *Method only* – Specifies the type and weight of the equipment used, the thickness of the lifts, and the number of passes needed. The owner would likely specify testing to ensure that the density is acceptable, but the contract would need to be amended if more compaction is required.
- *Method and result* – Specifies the method and the required density. If the specified density cannot be attained with the specified method, then the contract must be amended to use other equipment or change the density requirement.
- *Suggested method and result* – Suggests a method and specifies a resultant density. This method allows the contractor to use any method as long as the specified density is attained. This can be problematic because circumstances beyond the contractor control, such as too much moisture, could prevent reaching the specified density. In this case, the contract may be written so that the contractor is not responsible for failing to meet the specified density when the suggested method is used.
- *Result only (performance specification)* – Specifies the density only. This can be problematic to the contractor when circumstances, such as too much moisture, prevent reaching the specified density.

7.3.0 Checking Quality

Each compacted layer must meet design specifications to ensure that the final product will stand up under projected loads. It must also meet smoothness requirements. Each stabilized layer must be checked for the following:

- Compaction or density
- Thickness
- Mix uniformity
- Gradation
- Loadbearing capacity
- Moisture content
- Binder content

The amount of compaction needed to bring the soil to the engineers' requirements is really just a measure of the density of the soil. There are tests that will determine if the compaction performed meets the specifications of the designers.

Because it is doubtful the machine operator will be required to do this test, we'll discuss these tests

very briefly in the following paragraphs just to familiarize the operator with the technology. There are two types of tests primarily in use today: the sand test and nuclear testing.

> **NOTE**
>
> On-board density meters are available for many machines. These meters measure the density of the soil and provide the information to the operator through a display panel. The sensors measure the reaction of the drum while it is compacting. As the soil stiffens, less energy is transmitted to the ground and more is reflected back into the equipment. The operator receives feedback from instruments that show if compaction has reached the desired level. These readings can be as accurate as other testing methods.

7.3.1 Sand Cone Test

This is the oldest testing method. Many engineers consider this the most accurate testing method. To perform this test, dig a round hole with a volume of one-tenth of a cubic foot (*Figure 34*). Weigh the soil taken out of the hole. Send the soil removed from the hole to a laboratory. At the laboratory, the soil is dried and weighed again to determine how much of the total weight was water. An additional sample is taken and analyzed the same way so the lab can plot a moisture density curve of the site.

The specific volume of the hole is determined using a calibrated jar of dry sand. The sand is calibrated so that the volume can be easily determined from the weight. Weigh the jar full of dry sand. Fill the hole with sand using the jar and cone device shown in *Figure 35*. Weigh the container

203F34.EPS

Figure 34 ◆ Digging a test hole.

and remaining sand again. Subtract that from the original weight to determine the weight of sand in the hole. Use the conversion factor to convert the weight of the sand into the volume of the hole.

The dry weight of the soil removed from the hole is divided by the volume of sand needed to fill the hole. This gives the density of the compacted soil in pounds per cubic foot. This result is compared to the theoretical maximum density, which gives the relative density of the soil that was just compacted.

7.3.2 Nuclear Testing

Nuclear testing involves either placing the testing machine on the ground to get a reading or inserting a probe attached to the machine into a small hole drilled into the soil. Either method sends impulses into the soil that are reflected back to the device and recorded. Denser soils absorb more impulses. The more a soil is compacted the fewer impulses are returned. The lab technician then creates a moisture density curve of the site in the same way as the sand cone test.

 WARNING!

Nuclear equipment should only be used by trained personnel. Exposure to nuclear materials can cause serious illness.

Nuclear devices are being replaced with other types of equipment to avoid the risks posed by nuclear materials. The electrical density gauge (EDG) measures the physical properties of compacted soils used in road beds and foundations (*Figure 36*). This device is battery operated and can determine the wet and dry density, gravimetric moisture content, and percent compaction. The kit includes a console, four electrodes, a hammer, soil sensor and cables, template, temperature probe, and battery charger.

7.4.0 Leveling and Compacting

A roller equipped with a dozer blade can be used to level and compact an area using a single machine. The dozer blade works to skim soils from high spots and fill in lower areas. The soil is compacted with each pass. First check the ballast to bring the machine to the desired compacting weight. Lower the blade to ground level. Place the transmission in second gear. If the engine slows, lower it to first gear. Doze soils to the desired fill height. When the end of the row is reached, reverse the direction of the machine and place the blade control in the float

position. Straddle the previous tracks made with the compactor. Repeat the forward and backward pattern for three or four passes or until the desired compaction is achieved.

203F35.EPS

Figure 35 ◆ Soil density testing equipment.

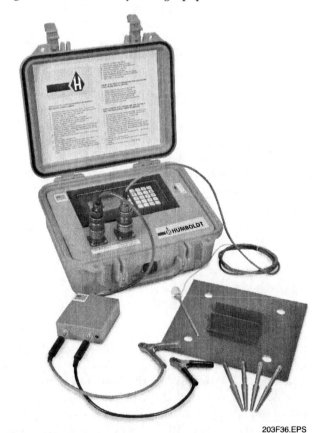

203F36.EPS

Figure 36 ◆ Electrical density gauge.

7.5.0 Backfilling

Backfilling around culverts or structures must be done carefully so that they are not damaged. A roller with a dozer blade can backfill a trenched area and then compact the soil above. The soil should be placed against the structure or pipe in layers, taking care not to concentrate the weight of the fill in one place. Also, be careful not to push against the structure with the material or the blade. To reduce pressure on the material and provide better visibility, raise the blade as you approach the structure.

When backfilling a trench that has pipe or conduit in it, take care to protect the pipe from bending or breaking due to uneven or excessive pressure caused by the material being pushed into the trench. Material is usually placed in layers, depending on the compaction required around the pipe. The typical range for each layer is six inches to several feet. Make sure there is sufficient cover on top of the pipe or any buried structure before driving the roller over it.

7.6.0 Compacting Cement

The stabilized area must be compacted after the stabilizer has finished. Only adequate compaction produces the required density. If cement is to be used, the compaction must be done immediately after mixing because the cement in the mixture will begin to set. A cement mixture also requires that a vibratory roller compaction machine be used to obtain the needed results.

When compacting a lime-stabilized area, the compaction must be done when the moisture content has dropped to a point within the specified range. This may be done immediately or after a short time.

The number of passes made by the compactor is important to the amount of compaction achieved. The number of passes will depend on the moisture content of the material, the layer thickness, the compactor type, and the degree of compaction called for in the specifications. With proper compaction, a final curing stage may be eliminated before the next course is applied.

7.7.0 Compacting Asphalt

It is important that newly placed asphalt be compacted to a hard, smooth surface. Compaction is the most important factor in the performance of a hot-mix asphalt (HMA) pavement. Adequate compaction of the mix increases surface life, decreases permanent deformation or rutting, reduces oxidation or aging, decreases moisture damage, increases strength and stability, and decreases low temperature cracking.

Compaction is the process by which the asphalt mix is compressed and reduced in volume. Compaction reduces air voids and increases the unit weight or density of the mix. As a result, the asphalt-coated aggregates in the mix are forced closer together, increasing their interlock and interparticle friction and providing more strength.

Asphalt pavement is usually compacted with static steel-wheel rollers, vibratory steel-wheel rollers, or pneumatic tire rollers. A combination roller with a vibratory drum and pneumatic tires is also used (*Figure 37*).

Factors related to the design of the mix that affect compaction are aggregate properties and cement type. Paving factors that affect the roller's efficiency include temperature of the mix and uniformity of the pavement layer. Generally, the higher the HMA temperature, the easier it is to compact. Surrounding temperature and layer thickness will also affect the mix temperature. Compaction variables that can be controlled during the process and that have an effect on the level of density include the following:

- Roller speed
- Number of passes
- Rolling zone
- Rolling pattern
- Direction and mode

7.7.1 Roller Speed

The more quickly a roller passes over a particular point in the new surface, the less time the weight of the roller dwells on that point. This means that less compactive effort is applied. As roller speed

203F37.EPS

Figure 37 ◆ Combination pneumatic and steel-wheel vibratory roller.

increases, the amount of density gain achieved with each pass decreases. The roller speed selected depends on the paver speed, the layer thickness, and the position of the equipment in the roller train.

7.7.2 Number of Passes

To obtain the target air-void content and uniform density on an asphalt mixture, each point in the pavement must be rolled a certain number of times. The number of passes depends on many variables, including the type of equipment, the mix properties, and the position of the equipment in the roller train. Test strips are performed at the beginning to determine the minimum number of passes. Different roller combinations and patterns may be tested.

7.7.3 Rolling Zone

Compaction must be achieved while the viscosity of the asphalt binder in the mix and stiffness of the mix are low enough to allow for reorientation of the aggregate particles under the roller. A rule of thumb for achieving proper levels of air voids is that the mix should be above 175°F.

To obtain the required level most quickly, the initial compaction should occur directly behind the laydown machine. If the mixture is stable enough, breakdown rolling can be carried out very close to the paver while the material is quite hot. More density can be obtained with one pass when the temperature is 250°F than when it is 230°F, so the front of the rolling zone should be as close to the paver as possible.

7.7.4 Rolling Pattern

Roller passes must be distributed evenly over the width and length of the mat. Proper rolling patterns help ensure that compaction is adequate over all the surface and not concentrated in certain areas. All too often, the center of the paver

lane, the area between the wheel paths, receives most of the compaction effort, while wheel paths and edges receive less.

The pattern to be used on a specific project should be determined in the beginning by considering the following factors:

- Type of mix
- Type of roller
- Desired density
- Layer thickness
- Temperature

During the rolling process, density can be checked with a nuclear density gauge. If necessary, adjustments can be made then. A typical rolling pattern is shown in *Figure 38*.

7.7.5 Direction and Mode

For vibratory rollers, direction of travel and mode of operation are also under the operator's control. The direction of travel is the orientation of the roller drums to the paver. The mode of operation is either full vibratory mode with both drums vibrating or combination mode with only one drum vibrating. The direction and mode generally used depends on the type of roller and the type of mixture.

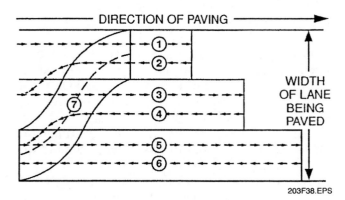

Figure 38 ◆ Typical rolling pattern.

Review Questions

1. The measure of how closely soil particles are packed together is the _____.
   a. psi
   b. plasticity index
   c. compaction
   d. density

2. Soil consists of irregular particles; air and _____ fill the spaces between the particles.
   a. sand
   b. bacteria
   c. water
   d. dirt

3. There are _____ basic roller designs.
   a. two
   b. four
   c. six
   d. ten

4. A pneumatic tire roller uses _____ for compaction.
   a. weight
   b. only tire pressure
   c. vibratory forces
   d. a combination of wheel loading and tire pressure

5. Vibratory rollers are ineffective in _____.
   a. granular soils
   b. clay
   c. sand
   d. wet areas

6. Typically, the feet on segmented-pad rollers are _____ long.
   a. 3 to 4 inches
   b. 7 to 9 inches
   c. 12 inches
   d. 24 inches

7. Typically, roller movement is controlled with _____.
   a. joysticks
   b. levers and knobs
   c. steering wheel and switches
   d. steering wheel and pedals

8. The fuel shutoff switch physically prevents fuel from flowing from the gas pump into the fuel tank.
   a. True
   b. False

9. Water spray units are typically used for rolling _____.
   a. asphalt
   b. gravel
   c. sand
   d. clay

10. Hydraulic fluids should be changed whenever they become dirty or break down due to overheating.
    a. True
    b. False

11. Ensure that fluid levels are indicated properly by _____.
    a. parking on a level surface
    b. changing the air filter
    c. engaging the parking brake
    d. checking the service codes

12. Many diesel engines have _____ that heat up the engine before ignition.
    a. hydrostatic transmission
    b. glow plugs
    c. heater coils
    d. starter switches

13. The sheepsfoot roller supplies pressure, along with _____.
    a. compaction
    b. vibration
    c. impact
    d. kneading

14. The sand-cone test and _____ are used to check soil density.
    a. nuclear testing
    b. water pressure
    c. litmus test
    d. the plasticity index

15. When leveling an area use the _____ to skim soil from the high spot to fill lower areas.
    a. segmented-pads
    b. hydraulic rams
    c. dozer blade
    d. wheel scraper

16. When backfilling a culvert trench be careful not to _____.
    a. damage the culvert
    b. cover the pipe
    c. over-compact the soil
    d. get the soil too wet

17. Generally, higher HMA temperatures _____.
    a. have no effect on compaction
    b. are easier to compact
    c. must be cooled before compaction
    d. cannot be compacted

18. As roller speed increases, the amount of density gain achieved with each pass _____.
    a. increases
    b. decreases
    c. is not affected

19. In HMA compaction, the front of the rolling zone should be _____.
    a. as close to the paver as possible
    b. at least 10 feet behind the paver
    c. in front of the paving zone
    d. as far away from the paver as possible

20. Proper rolling patterns help ensure that _____.
    a. each section is only compacted once
    b. the roller does not have to backtrack
    c. compaction is concentrated in certain areas
    d. compaction is adequate over the entire surface

Summary

Rollers are used for compacting and smoothing materials such as earth, gravel, cement, and asphalt. They compress the material to the desired density to provide strength so that the specified loads can be carried without settlement.

There are four types of rollers; each is used for different types of work. The four types of rollers are rubber-tire, steel-wheel, vibratory, and segmented pad rollers. They provide static and dynamic force to compact soils and paving material.

Vehicle movement is controlled with the steering wheel, accelerator, and brake pedals. If the roller is equipped with a blade, it is controlled with either levers or a joystick, as well as switches. Study the operator's manual to become familiar with the machine you will be operating.

Safety considerations when operating a roller include keeping the roller in good working condition, obeying all safety rules, being aware of other people and equipment in the area where you are operating, and not taking chances. Perform inspections and maintenance daily to keep the roller in good working order.

The type of roller selected depends on the material being compacted. Some rollers are more effective with certain types of soils, while others are primarily used for asphalt. Operators must understand the properties of soil that contribute to compaction. Moisture content is one of several factors that contribute to good compaction. Once the soil is compacted, it must be tested to ensure that sufficient density has been achieved.

Notes

Compaction: Using an engineered process, such as rolling, tamping, or soaking, to reduce the bulk and increase the density of soil.

Density: The ratio of the weight of a substance to its volume.

Foot: In tamping rollers, one of a number of projections from a cylindrical drum that contact the ground.

Ground contact pressure (GCP): The weight of the machine divided by the area in square inches of the ground directly supporting it.

Lift: A layer of material that is a specific thickness; the depth of material that is being rolled by the roller.

Mat: Asphalt as it comes out of a spreader box or paving machine in a smooth flat form.

Pad: On a segmented or sheepsfoot roller, the part of the roller that contacts the ground; also called the foot.

Puddling: A process in which water is added to the soil until is it semi-liquid; the soil is then allowed to dry before being vibrated.

Sheepsfoot: A tamping roller with feet expanded at the outer tips.

Settling: The natural wetting and drying process whereby soil particles become more compact and denser.

Tamping roller: One or more steel drums fitted with projecting feet and towed with a box frame.

Vibratory roller: A compacting device that mechanically vibrates the soils while it rolls. It can be self-propelled or towed.

Additional Resources

This module is intended to be a thorough resource for task training. The following reference works are suggested for further study. These are optional materials for continued education rather than for task training.

Caterpillar Performance Handbook, Edition 27. A CAT® Publication. Peoria, IL: Caterpillar, Inc.

Excavation and Grading Revised, 1987. Nick Capachi. Carlsbad, CA: Craftsman Book Company.

Moving The Earth, Fourth Edition 1998. Herbert L. Nichols Jr. and David A. Day. McGraw-Hill, New York, NY.

Figure Credits

Reprinted courtesy of Caterpillar Inc., 203F01 (sheepsfoot roller, pneumatic tire roller), 203F03, 203F05-203F08, 203F10-203F13, 203F15-203F18, 203F21-203F29, 203F31

BOMAG Americas, Inc., 203F01 (steel wheel roller), 203F02, 203F04, 203F19

JCB Vibromax, 203F01 (smooth drum, rubber tire vibratory roller), 203F37

Ingersoll Rand Construction Technologies, 203F09, 203F14, 203F20, 203F30, 203F32

Contractors Depot and Multiquip, Inc., Table 1

Topaz Publications, Inc., 203F33

National Center for Asphalt Technology, 203F34

Humboldt Manufacturing Company, 203F35, 203F36

CONTREN® LEARNING SERIES — USER UPDATE

The NCCER makes every effort to keep these textbooks up-to-date and free of technical errors. We appreciate your help in this process. If you have an idea for improving this textbook, or if you find an error, a typographical mistake, or an inaccuracy in NCCER's Contren® textbooks, please write us, using this form or a photocopy. Be sure to include the exact module number, page number, a detailed description, and the correction, if applicable. Your input will be brought to the attention of the Technical Review Committee. Thank you for your assistance.

Instructors – If you found that additional materials were necessary in order to teach this module effectively, please let us know so that we may include them in the Equipment/Materials list in the Annotated Instructor's Guide.

Write:	Product Development and Revision
	National Center for Construction Education and Research
	P.O. Box 141104, Gainesville, FL 32614-1104
Fax:	352-334-0932
E-mail:	curriculum@nccer.org

Craft _____ Module Name _____

Copyright Date _____ Module Number _____ Page Number(s) _____

Description _____

_____

_____

_____

(Optional) Correction _____

_____

_____

(Optional) Your Name and Address _____

_____

_____

22204-06

Scrapers

22204-06
Scrapers

Topics to be presented in this module include:

Overview

The scraper was designed with a single purpose – bulk earthmoving. Thin layers of earth are scraped into the pan and can be transported and spread in another location. A skilled scraper operator can even prepare finish grades.

The first step to becoming a skilled scraper operator is understanding all of the controls and instruments. The next step is to understand how to inspect and maintain the machine. This provides a solid foundation to safe scraper operation.

Scrapers can do it all, but often they also work in tandem with other scrapers or bulldozers. Understanding how to work efficiently as a team is an important skill for a scraper operator. This includes understanding hauling costs so you can maximize your efficiency.

Objectives

When you have completed this module, you will be able to do the following:

1. Describe the uses of a scraper.
2. Identify the components and controls on a typical scraper.
3. Explain safety rules for operating a scraper.
4. Perform prestart inspection and preventive maintenance procedures for scrapers.
5. Start, warm up, and shut down a scraper.
6. Perform basic maneuvers with a scraper.
7. Perform the basic earthmoving operations with a scraper.

Trade Terms

Apron	Lug down
Bowl	Pay material
Ejector	Ripping
End shoes	Stockpile
Fixed time	Stripping
Grade checker	

Required Trainee Materials

1. Pencil and paper
2. Appropriate personal protective equipment

Prerequisites

Before you begin this module, it is recommended that you successfully complete *Core Curriculum*; *Heavy Equipment Operations Level One*; *Heavy Equipment Operations Level Two*, Modules 22201-06 through 22203-06.

This course map shows all of the modules in the second level of the *Heavy Equipment Operations* curriculum. The suggested training order begins at the bottom and proceeds up. Skill levels increase as you advance on the course map. The local Training Program Sponsor may adjust the training order.

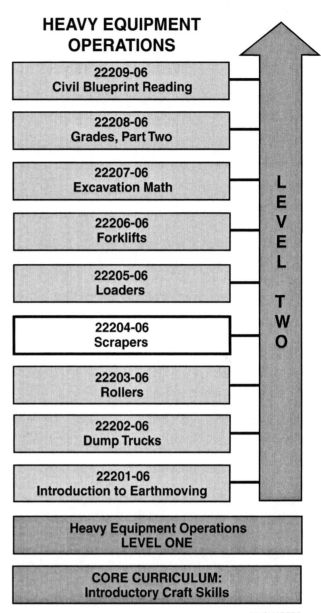

HEAVY EQUIPMENT OPERATIONS

22209-06
Civil Blueprint Reading

22208-06
Grades, Part Two

22207-06
Excavation Math

22206-06
Forklifts

22205-06
Loaders

22204-06
Scrapers

22203-06
Rollers

22202-06
Dump Trucks

22201-06
Introduction to Earthmoving

LEVEL TWO

Heavy Equipment Operations
LEVEL ONE

CORE CURRICULUM:
Introductory Craft Skills

204CMAP.EPS

1.0.0 ◆ INTRODUCTION

Self-propelled scrapers were invented in the late 1930s. They are also called a carry-alls or pans. The scraper is undoubtedly the most efficient piece of equipment that has been developed for bulk, wide-area excavation. They are used to remove dirt and other materials from a site by scraping it into a **bowl**. They are also used to spread dirt and other material that has been loaded into the bowl with a loader.

There are many variations of the scraper. Some of the earlier scrapers were mounted on four wheels and drawn by a tractor or bulldozer. A power take-off from the dozer powered the cables operating the scraper. Today, most scrapers are self-propelled, although towed units are still used. Four-wheeled, self-propelled units are preferred if any length of haul is required. Their deficiency in loading is overcome by using bulldozers or pushers during the loading operation.

1.1.0 Types of Scrapers

Scrapers are manufactured in different sizes for performing different kinds of work. There are small scrapers that are used for small jobs such as industrial land development or for work in tight places. Smaller units can be towed by a dozer or tractor. Larger scrapers are typically used for rough grading on large excavation or construction projects for roads, airports, and dams.

The main characteristics of the smallest and largest self-propelled scrapers in commercial production are shown in *Table 1*.

Two of the more interesting characteristics from this comparison are the capacity of the bowl and the depth of the cut. With the larger scrapers, tremendous amounts of material can be cut and moved in one cycle. Because of their size, scraper operation requires good coordination and quick reflexes.

A number of manufacturers make several different types of scrapers. Generally they fall into the following four categories based on the arrangement of the bowl and the power unit:

• Standard self-propelled
• Tandem
• Elevating
• Towed

1.1.1 Standard Self-Propelled Scraper

A standard self-propelled scraper consists of a bowl unit and a tractor. A cutting edge is mounted at the bottom of the bowl to scrape the earth into the bowl. On larger models, the bowl can hold more than 40 cubic yards of material, and the cutting edge width spans up to 13 feet. An **apron** holds the material in

Table 1 Largest and Smallest Scrapers

Scrapers	Smallest	Largest
Power	175 hp	550 hp
Bowl Capacity	11 cu yds	44 cu yds
Top Speed	21 mph	31 mph
Width of Cut	92"	151"
Maximum Depth of Cut	6.3"	17.3"
Maximum Depth of Spread	14.6"	20"

204T01.EPS

the bowl and an **ejector** pushes it out. *Figure 1* shows a standard self-propelled scraper.

The tractor is connected to the bowl unit by an arched gooseneck. The tractor section includes the engine, drive train, hydraulics, and operator's cab. A standard self-propelled scraper has two axles; the front axle is the drive axle. The engine projects forward of the drive axle so the weight of the tractor is on the drive wheels for better traction. This is known as an overhung configuration.

1.1.2 Tandem and Push-Pull Scrapers

In some conditions, a standard scraper will not have sufficient power or traction to fully load the bowl. There are several methods used to supply more power to the scraper to accomplish this. Tandem scrapers have two engines: one in the tractor and a second at the rear of the scraper unit. These machines can work on steeper grades and rougher terrain.

Even with tandem engines, additional power may be needed. In certain conditions, another machine is used to push the scraper to load it completely. A tractor or bulldozer can be used, or two push-pull scrapers can work together. The two scrapers are connected by a bail and hook. One machine pushes the other and the power of the two

204F01.EPS

Figure 1 ◆ Standard self-propelled scraper.

machines is used to load one machine. When the first machine is full, it is used to pull the second, so both machines are used to load the bowl of the second scraper. The scrapers unhook and travel separately to unload and repeat the process. *Figure 2* shows two tandem scrapers working together.

1.1.3 Elevating Scraper

The elevating scraper has an independently powered auger or elevator mounted in the bowl. The elevator lifts the material to the top of the pile and evenly distributes it into the bowl. As the material is lifted, it is conditioned, or broken up. This blends the material, reduces air voids, and creates a more consistent payload. These scrapers do not need additional power to obtain a full load.

There are some drawbacks to this type, however. In most models, the elevator replaces the apron in the front of the bowl. However, large boulders or tree stumps can damage the elevator or get stuck in the bowl. The weight of the elevator or auger can slow the scraper during the hauling phase. Finally, the elevator must be maintained, which increases machine costs. However, these models are economical to use in many situations.

Figure 3 shows an elevating scraper hauling a load. The elevator on this machine is a paddle wheel, a circular design with blades around a hub. Other elevators have a single spiral blade that rotates like an auger (*Figure 4*). Typically the elevator is hydraulic or electrically powered.

1.1.4 Towed Scraper

The pull-type or towed scraper (*Figure 5*) does not have its own engine. It must be pulled by a crawler tractor, farm tractor, bulldozer, or other equipment. Towed scrapers pre-date self-propelled models, but were replaced by self-propelled units on most construction projects. Recently, however, towed units have gained popularity. Some contractors use a large crawler tractor to pull two towed scraper units to move dirt economically. Several companies produce construction-grade towed scrapers in sizes from 7 to 24 cubic yards.

Figure 2 ◆ Two tandem scrapers.

1.2.0 Uses of the Scraper

Scrapers are highly mobile excavators that can dig, carry, and spread loads. They are used on many different types of excavation and construction projects, such as road building, mining, dam construction, and industrial plant development. Scrapers are also used to build canals, dikes, levees, and sometimes roads. They can be used efficiently where any large amount of material must be moved.

Scrapers are used in various construction projects for the following tasks:

- Cut, load, and spread granular material
- Strip and **stockpile** topsoil
- Perform finish grading

204F03.EPS

Figure 3 ◆ Elevating scraper.

204F04.EPS

Figure 4 ◆ Auger–type elevating scraper.

204F05.EPS

Figure 5 ◆ Towed scraper.

2.0.0 ◆ IDENTIFICATION OF EQUIPMENT

Currently, three companies manufacture self-propelled scrapers and six other companies manufacture pull-type scrapers. This module describes the operations of a tandem scraper, specifically the Caterpillar 627G. It is a popular and versatile scraper. The features and operations are typical of large scrapers. The controls, instruments, and operating procedures for other models may differ slightly. Always review the operator's manual before operating a particular piece of equipment.

2.1.0 Basic Parts of a Scraper

The basic parts of a tandem scraper are shown in *Figure 6*. There are two engines, one on the rear scraper unit and one in front on the tractor. The operator's station is in the front of the tractor. The back section contains the bowl and an arched section called a gooseneck. The front and back sections are connected by a cushioned hitch.

Figure 7 shows a bail and push plate in the front of the machine. These are used to connect two machines when they are working together. The major operating components of the scraper are the bowl, the apron, and the ejector.

The bowl looks like a large scoop and hangs from a frame supported on rubber-tired wheels. The forward edge can be tilted vertically about 20 degrees. A replaceable cutting edge is attached to the lip of the bowl. Typically, the cutting edge is made up of three or four steel blades that project below the adjoining sections. The cutting edge is lowered by hydraulic cylinders until it penetrates the soil to the

desired level. As the scraper moves forward, a thin layer of earth is forced back into the bowl.

The cutting edge is made of steel (*Figure 8*). In harder surfaces, teeth are fitted onto the edge to improve digging. They are replaced when they become dull or worn. When they are worn, some cutting edges can be lowered to extend their working life. After additional wear they must be replaced.

When the bowl is full, the operator tilts it upward. This brings the cutting edge above the ground surface so that cutting stops. The apron is dropped down over the open end of the bowl and rests on the cutting edge. The apron closes the bowl and prevents spillage during hauling. The apron can also be lowered to clamp onto bulky objects like tree stumps or boulders.

The ejector is used to unload material from the bowl. When the scraper reaches the unloading area,

BAIL   PUSH PLATE   204F07.EPS

Figure 7 ◆ Bail and push plate.

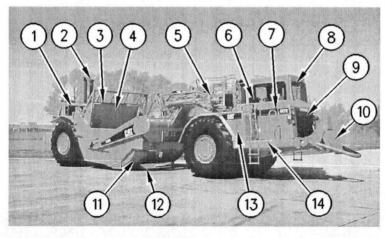

(1) Scraper engine
(2) Precleaner (scraper)
(3) Fuel tank
(4) Ejector
(5) Cushion-hitch
(6) Precleaner (tractor)
(7) Tractor engine compartment
(8) Operator station
(9) Radiator (tractor)
(10) Bail and push plate
(11) Apron
(12) Cutting edge
(13) Hydraulic tank
(14) Battery compartment (tractor)

204F06.EPS

Figure 6 ◆ Basic parts of a scraper.

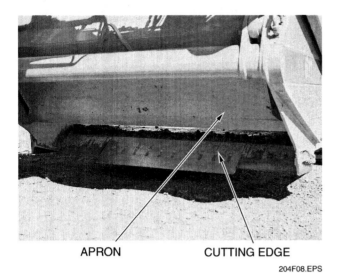

APRON   CUTTING EDGE

204F08.EPS

Figure 8 ◆ Cutting edge.

the bowl is lowered until the cutting edge is 1 to 6 inches above the ground surface. The tilt is generally not sufficient to force the material to spill out of the bowl. The ejector is used to push the material out from the back as the apron is raised. The ejector plate, which is the width of the bowl, is forced slowly forward against the material in the bowl. The apron is lifted gradually as the ejector moves forward to provide a uniform discharge of the material.

The tandem scraper is powered by two diesel engines. The tractor engine provides high power and torque for lugging and loading. The two engines are coordinated via an electronic control module. Both engines have numerous sensors which monitor engine conditions and alert the operator to potential problems. The transmission is electronically controlled. Gears one and two are used for loading operations, and the higher gears are used for hauling.

2.2.0 Controls on a Standard Scraper

The scraper should be operated only from the operator's cab. Although the location of the controls varies with manufacturer and model, all of the primary controls for the scraper are located within easy reach of the operator's seat. Typical controls found in the cab of a tandem scraper are shown in *Figure 9*. Always check the operator's manual for the model you are operating to become familiar with the controls.

Vehicle movement controls on scrapers can be similar to those of cars and trucks (*Figure 10*). A steering wheel is used in combination with foot pedals to control vehicle movement. The throttle foot pedal is used to control the engine speed. There are two pedals, one for each engine. One controls the tractor engine speed, while the other

controls the scraper engine speed. Depress the tractor throttle to increase travel speed, release it to decrease it. A third pedal is for the service brakes. Depress the pedal to engage the service brakes.

NOTE
Both throttle foot pedals can be operated at the same time with one foot.

The scraper, apron, and ejector are controlled with either a joystick or levers. This module will describe joystick controls. Review the operator's manual before operating a machine so that you understand how each of the controls function.

2.2.1 Disconnect Switches

Some loader models have a fuel shutoff switch that must be turned to the ON position before the machine can be operated. The fuel shutoff switch physically prevents fuel from flowing from the fuel tank into the supply lines. This prevents unwanted fuel flow during idle periods or when transporting the machine, significantly reducing the potential for fuel leaks.

Some machines have a battery or electrical system disconnect switch. When the battery disconnect switch is turned off, the entire electrical system is disabled. This switch should be turned off when the machine is left overnight or longer to prevent a short circuit or active components from draining the battery. Before mounting the machine, check that the switch is in the ON position. Tandem scrapers with two engines will have two switches, one on the scraper engine and one on the tractor engine (*Figure 11*).

CAUTION
Never switch the battery disconnect switch to the OFF position while the engine is running. This could seriously damage the electrical system.

2.2.2 Seat and Steering Wheel Adjustment

Upon first entering the cab, the operator should adjust the seat and steering wheel. While seat and steering wheel adjustments are not directly involved in scraper operations, correct positioning of these items can aid in safe operation. The operator should adjust the seat and steering wheel position and then fasten the seat belt before operating the scraper.

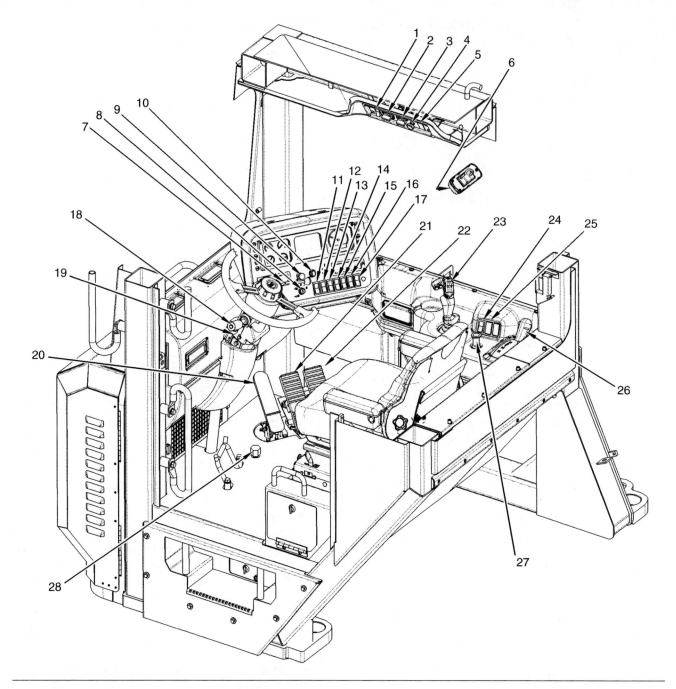

Operator controls in the cab

(1) Heating and air conditioning switch
(2) Temperature control switch
(3) Fan control switch
(4) Window wiper/washer switch
(5) Side light switch (if equipped)
(6) Interior dome light switch
(7) Retarder control
(8) Engine start switch (tractor)
(9) Cigar lighter
(10) 12V Power receptacle

(11) Headlight switch
(12) Warning hazard light switch
(13) Bowl floodlight switch
(14) Scraper transmission neutral/run control
(15) Mode select switch
(16) Tractor-Scraper select switch
(17) Engine start switch (scraper)
(18) Multifunction switch
(19) Steering column tilt and telescope
control

(20) Service brake control
(21) Throttle control (tractor)
(22) Throttle control (scraper)
(23) Joystick control
(24) Throttle lock control
(25) Scraper power limiter control (if
equipped)
(26) Transmission control
(27) Parking brake control
(28) Differential lock control

204F09.EPS

Figure 9 ◆ Operator's cab for a tandem scraper.

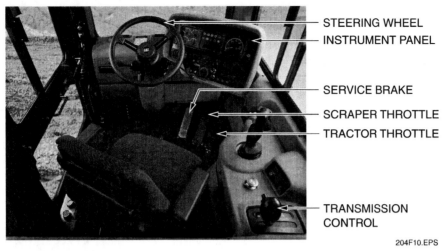

Figure 10 ◆ Vehicle movement controls.

Most seats can be moved up or down and forward or backward. Some provide an adjustable shock absorber function using springs or hydraulics. The seat should be adjusted so that the operator's legs are almost straight when the clutch or brake pedals are fully depressed and the operator's back is flat against the back of the seat.

After the seat is correctly positioned, adjust the steering wheel. Most scrapers have a lever on the steering column to adjust the wheel. On most models the lever will return to the lock position when released. Move the lever up and tilt the steering column to position it correctly. The steering wheel can be moved higher or lower through a telescoping function. Push the lever down to move the wheel up or down. Because seat and steering wheel adjustment devices vary widely for various makes and models, refer to the operator's manual for specific instructions.

2.2.3 Engine Start Switch

The ignition switch functions can vary widely between makes and models of scrapers. Some only activate the starter and ignition system. Others may activate fuel pumps, fuel valves, and starting aids. In some cases, the starter and starting aids are engaged by other manual controls. The engine start switch for the tractor engine is located on the console under the instrument panel.

NOTE

Tandem scrapers have two engines. The engines are usually started separately. Typically, the tractor engine is started first and then the scraper engine is started. Both engines may also have separate battery disconnect switches. The switch for each engine must be turned to the ON position before starting.

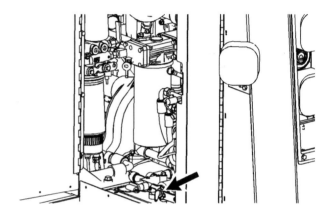

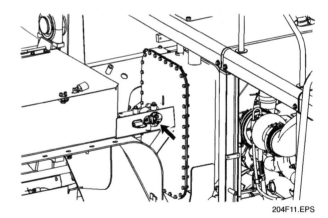

Figure 11 ◆ Battery disconnect switches.

The tractor engine start switch has the following three positions:

- *Off* – Turning the key to this position stops the engine. On tandem scrapers it will stop both engines. It will also disconnect power to electrical circuits in the cab. However, several lights remain active when the key is on the OFF position, including the hazard warning light, the interior light, and the parking lights.

- *On* – Turning the key to this position activates all of the electrical circuits except the starter motor circuit. When the key is first turned to the ON position, it may initiate a momentary instrument panel and indicator bulb check.
- *Start* – The key is turned to this position to activate the starter, which starts the engine. This position is spring-loaded to return to the ON position when the key is released. If the engine fails to start, the key must be returned to the OFF position before the starter can be activated again. To reduce battery load during starting, the ignition switch of some scrapers may be configured to shut off power to accessories and lights when the key is in the start position.

CAUTION

Activate the starter for a maximum of 30 seconds. If the machine does not start, turn the key to the OFF position and wait 2 minutes before activating the starter again.

Start the scraper engine after starting the tractor engine. There is a separate switch on the instrument panel to start the scraper engine. The key switch for the tractor engine must be in the ON position before the scraper engine can be started. Push the scraper engine switch upward to to start the scraper engine. If the engine fails to start, release the switch before attempting to start the engine again.

2.2.4 Vehicle Movement Controls

Scraper movement is controlled in a manner similar to that of cars and trucks. The engine throttle and brakes are operated by foot pedals. A steering wheel is used to turn the vehicle. Moving the steering wheel to the left or right steers the vehicle to the left or right.

The transmission is controlled by a gear selection lever on the right side of the operator's seat (*Figure 12*). The gear selector should be set to N or neutral when you park or start the engine. Move the lever to R to move the machine in reverse. There are several forward speeds. On some models the number of forward gears is set by the service personnel. To change gears, squeeze the trigger and move the gear selection lever to the desired gear. Release the trigger to mechanically lock the control into the current gear.

NOTE

The engine will not start unless the gear selection lever is set to neutral.

On some machines, the transmission will automatically shift between second gear and any higher gear that is selected. Manual shifting is only required between first and second gear. Typically, a top gear selection button is located on the gear selection lever. Once the engine is running and the transmission is in neutral, the operator may select any gear. This will limit automatic upshifts to the highest gear set. Be sure you understand the transmission controls before operating the machine.

2.2.5 Scraper Controls

Scraper functions are controlled with a joystick or a series of levers. This section will describe joystick controls. The joystick is located on the right side of the operator's seat (*Figure 13*). For machines with lever controls, review the operator's manual to familiarize yourself with the operations of those controls.

The joystick combines the functions of several levers and switches into one control unit. This makes it much easier to operate several functions at the same time with one hand. The joystick controls the

TRANSMISSION CONTROL   204F12.EPS

Figure 12 ◆ Gear selection lever.

204F13.EPS

Figure 13 ◆ Joystick.

bowl, ejector, apron, and hitch. The control features on the joystick are shown in *Figure 14*. Moving the joystick forward lowers the bowl. Moving it back raises the bowl. Moving the joystick to the left moves the ejector forward, which pushes the material out of the bowl. Moving the joystick to the right moves the ejector to the back of the bowl to allow additional material to be placed in the bowl. The speed of movement is increased the farther the joystick is moved away from the central position. Some models have automatic features. For example, when the joystick is moved all the way to the right (detent position) it will remain there until the ejector has fully returned, then spring back to the neutral position.

On the top of the joystick are several buttons labeled 3, 4, and 5 in *Figure 14*. The thumb wheel switch (3) controls the apron position. Pushing the thumb wheel all the way to the left and releasing it puts the apron into float mode. This allows the apron to maintain a constant position relative to the bowl. Use the float position when hauling a load. This allows you to adjust the bowl position without having to adjust the apron and prevents materials from spilling out of the bowl as adjustments are made during hauling. Pushing the thumb wheel downward lowers or closes the apron. Pushing it upward raises or opens it.

The transmission hold switch (4) prevents the transmission from shifting. Push the button in to engage this feature. Push the button again to disengage this feature. The transmission hold allows the operator to maintain the converter drive for increased rimpull or hold the current gear for enhanced control.

> **NOTE**
> The tractor and scraper transmissions will both respond at the same time.

The cushion hitch lockout lever (5) locks the cushion hitch for improved control of the cutting edge during loading and dumping. The cushion hitch should be in the ON position at all times except when loading and hauling. Many joysticks have a trigger control. On some models the trigger raises and lowers the bail. On machines with an elevator or auger, the trigger on the joystick activates those functions.

2.2.6 Differential Lock

The differential lock is engaged with a button on the floor of the cab next to the brake and throttles and is also operated with your foot. To engage the differential lock, depress the button. Release the button to disengage it. On some models there are two buttons, one on either side of the pedals. Both buttons have the same function, but are used for different seat configurations.

The differential lock overrides the normal operation of the front axle differential. Torque is transmitted to both wheels, even though one wheel may not have traction. This helps maintain traction when ground conditions are soft or slippery.

Engage the differential lock when the wheels are not spinning. If the wheels start to spin, release pressure on the accelerator until the wheels stop

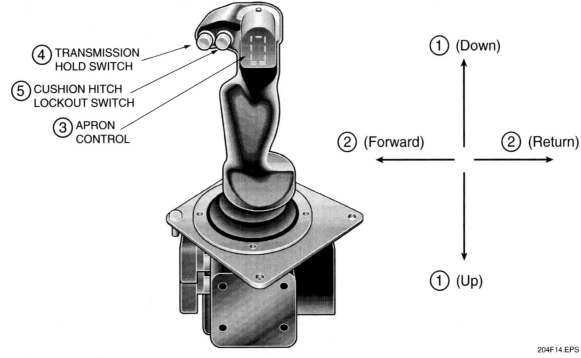

Figure 14 ◆ Joystick configuration.

spinning. Engage the differential lock and increase pressure on the accelerator. Once clear of the area, release pressure on the accelerator and release the switch to disengage the differential lock.

CAUTION

Do not engage the differential lock at high speeds or while the wheels are spinning. This could damage the differential.

2.2.7 Additional Controls

There are several other important controls in addition to those described above. Some of these features offer added control of the machine in specialized situations. These controls are not available on all models. The controls and their functions are as follows:

- **Bail hitch** *control* – Controls the raising and lowering of the bail hitch.
- *Fuel ratio override* – Changes the mixture of fuel to the injectors.
- *Governor override button* – Allows the operator to exceed the governor setting of the engine speed.
- *Heat-start switch* – Control panel switch used to activate glow plug and start engine. Newer machines have an automatic feature that operates glow plugs or ether starting aids.
- *Individual wheel brake lever* – Controls braking action to individual wheels.
- *Parking brake control* – Secures the scraper when stopped or parked (*Figure 15*).
- *Retarder lever* – The hydraulic retarder acts like an internal brake on the driveline. It reduces the need to apply the service brakes. The retarder takes a few seconds to engage and must be activated 3 to 4 seconds before it is needed. Using the retarder will decrease wear on the service brakes and enhance machine control.

204F15.EPS

Figure 15 ◆ Parking brake.

- *Stop button* – A kill switch for the engine.
- *Transmission hold* – A pedal or switch that allows the operator to lock in the transmission to maintain converter drive for increased rimpull or hold the current gear for increased control.
- *Transmission neutralizer* – Neutralizes the transmission when the service brake is applied and locks the transmission in gear.
- *Transmission control lever safety lock* – Locks the transmission in gear.

2.3.0 Instruments

An operator must pay attention to the instrument panel. The instrument panel includes the gauges that indicate engine and transmission temperature. There are several warning lights and indicators that must also be monitored. An operator can seriously damage the equipment if the instrument panel is not closely monitored.

The instrument panel varies on different makes and models of scrapers. Generally they include the instruments and indicators covered in the following sections. A typical instrument panel is shown in *Figure 16*. Most of these instruments and indicators are similar to those in other machines.

The instrument panel on a Caterpillar 627G includes a quad gauge panel (*Figure 17*). The gauges include: engine coolant temperature, transmission/torque converter oil temperature, fuel level, and air supply pressure. On dual-engine scrapers, the display can be switched between the tractor or scraper engine monitoring systems. The function of these gauges is described in the sections that follow. Other types of scrapers may have different gauges. Refer to the operator's manual for the specific gauges on the machine you are operating.

NOTE

On dual-engine machines, if one engine is experiencing trouble, the display will show the data for that engine. When the engine coolant or torque converter oil temperature exceeds the maximum operating temperature, warning lights will come on and the action lamp will flash.

The instrument panel also includes a speedometer, service hour meter, indicator lights, and warning lights, which are also described in the sections that follow. Indicator lights show that various machine functions are engaged. Warning lights indicate that the machine's systems are not functioning properly. Frequently used switches and indicator lights are located on the instrument panel. Other lights and switches are located on the overhead console.

2.3.1 Engine Coolant Temperature Gauge

The engine coolant temperature gauge (1, *Figure 17*) indicates the temperature of the coolant flowing through the cooling system. Refer to the operator's manual to determine the correct operating range for normal scraper operations.

Temperature gauges normally read left to right, with cold on the left and hot on the right. If the gauge is in the white zone, the coolant temperature is in the normal range. Most gauges have a section that is red. If the needle is in the red zone, the coolant temperature is excessive. Stop the machine immediately and investigate the problem. Some machines may also activate warning lights if the engine overheats.

 CAUTION

Operating equipment when temperature gauges are in the red zone may severely damage it. Stop operation, determine the cause of problem, and resolve it before continuing operation.

If the engine temperature gets too high, stop operation immediately. Get out of the machine and follow the proper procedures for investigating and repairing the problem. There are several checks that the operator can perform. First check the engine coolant level (*Figure 18*). Add more fluid if it is too low. Check that the fan belt is not loose or broken. Replace it if necessary. Check that the radiator fins are not fouled. Clean them if necessary. These are the three primary causes of the engine overheating. If initial troubleshooting fails to resolve the problem, stop operation and take the machine out of service.

 WARNING!

Engine coolant is extremely hot and is under pressure. Check the operator's manual and follow the procedure to safely check and fill engine coolant.

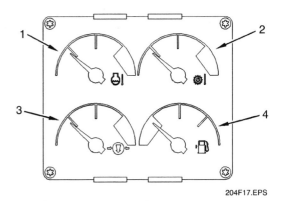

204F17.EPS

Figure 17 ◆ Quad gauge panel.

204F18.EPS

Figure 18 ◆ Check the engine coolant level.

QUAD GAUGE

INDICATOR LIGHTS

MAIN DISPLAY PANEL

SPEEDOMETER/TACHOMETER AND GEAR INDICATOR

INDICATOR LIGHTS

204F16.EPS

Figure 16 ◆ Instrument panel.

2.3.2 Transmission/Torque Converter Oil Temperature Gauge

The transmission or torque converter oil temperature gauge (2, *Figure 17*) indicates the temperature of the oil flowing through the transmission or torque converter. This gauge also reads left to right in increasing temperature. It has a red zone that indicates excessive temperatures. If the gauge is in the red zone, immediately reduce the load on the machine. This should reduce the temperature. If it remains in the red zone, stop the machine and investigate the problem. When the weather is colder, allow the transmission oil to warm up sufficiently before operating the machine.

2.3.3 System Air Pressure Gauge

The air supply pressure gauge (3, *Figure 17*) indicates the air pressure in the air tanks. When the monitoring system is in tractor mode, the gauge will show the air pressure in the tractor's air tanks. When the monitoring system is in scraper mode, it will show the air pressure in the scraper's air tanks.

The brakes are actuated with air pressure. Thus, it is critical that the air pressure remain within the normal operating range. A pushbutton releases a spring-applied mechanism that engages the parking brake. On many models, the brakes will lock on if the pressure decreases below a certain level. However, the operator must be aware of the air pressure in the system and take action if the pressure drops below safe operating levels.

2.3.4 Fuel Level Gauge

This gauge indicates the amount of fuel in the scraper's fuel tank. Most diesel engines have a low-fuel warning zone. Some also have a low-fuel warning light. Avoid running out of fuel on diesel engine scrapers because the fuel lines and injectors must be bled of air before the engine can be restarted.

2.3.5 Speedometer/Tachometer/Gear Indicator

The Caterpillar 627G has a combined tachometer, speedometer, and gear indicator (*Figure 19*). The analog tachometer (1) indicates the engine speed in rpm. It can be set to display the engine speed of either the front-tractor engine or the rear-scraper engine. If the monitoring system is in tractor mode, it will show the engine speed on the tractor. If it is in scraper mode, it will show the engine speed of the scraper. The digital speedometer (2) shows the machine's ground speed. Typically the speedometer can be set for either miles per hour (mph) or kilometers per hour (kph). The transmission gear

and direction indicator (3) will show the actual operating gear of the tractor engine unless the monitoring system is set in scraper mode.

2.3.7 Indicator Lights and Main Display

There are two sets of indicator lights on either side of the instrument panel (*Figure 20*). These lights show that various features are activated under normal operating conditions. They do not indicate that there is something wrong with the machine, merely that a function is active.

Typical indicator lights include the following:

- Turn signals (1, 6)
- High beams (2)
- Apron float (3)
- Tractor mode (4)
- Scraper mode (5)
- Transmission hold (7)
- Ejector return (8)
- Throttle lock (9)

The main display module (*Figure 21*) includes ten alert indicator or warning lights and a digital display window. Warning lights indicate that there is something wrong with the machine or its systems. When they are lit, the system is not functioning properly. For example, if the oil pressure light is lit,

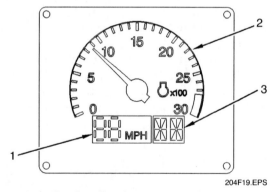

204F19.EPS

Figure 19 ◆ Combined tachometer, speedometer, and gear indicator.

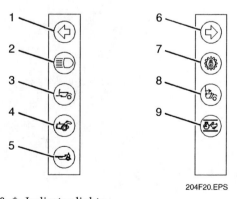

204F20.EPS

Figure 20 ◆ Indicator lights.

the oil pressure on the selected engine is too low. The operator can set the digital display to show data on several machine systems.

Typical alert indicators or warning lights include the following:

- Fuel temperature (1)
- Air pressure (2)
- Engine oil pressure (3)
- Secondary steering (4)
- Engine coolant temperature (5)
- Hydraulic system failure (6)
- Power train system (7)
- Charging system (8)
- Parking brake (9)
- Torque converter oil temperature (10)

> **NOTE**
>
> The parking brake light will light up when the parking brake is engaged. It should go out when the parking brake is disengaged. If the light comes on when the parking brake is not engaged, shut the machine down immediately and investigate the problem.

There are often increasing levels of warning alarms on a machine. At the first level, the alert indicators will light up. The operator must take action in the near future to correct the problem. Failure of these systems will not endanger the operator or the equipment. At the second level, the alert indicator and the action light will come on. The operator must take immediate action and change how the machine is being operated to correct the problem and avoid machine damage. At the third level, the alert indicator and the action light will come on and a warning alarm will sound. The operator must immediately shut down the machine to avoid machine damage or operator injury.

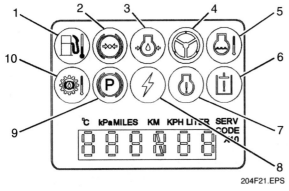

Figure 21 ◆ Main display panel.

2.4.0 Attachments Used on or with Scrapers

There are very few attachments for scrapers because they are designed to perform one function. The one variation for scrapers is the difference in tire type. The four tire patterns typically used are based on tread pattern and tire profile. Tires can cost more that $1,000 each. Considerable thought must be given to the selection of tires for a particular job, as they are expensive and wear differently depending on the particular operation.

The four different types of tires are as follows:

- *Traction tire* – A self-cleaning, directional-bar-type tire that provides maximum traction.
- *Hard-lug or rock-rib tire* – Used if tire is subject to rough operation over rocky terrain.
- *Button-type tire* – Generally used on the non-driving or trailing wheels and is particularly desirable in sand.
- *Balloon tires* – Generally used where work is done in soft areas and where maximum flotation is needed.

3.0.0 ◆ SAFETY GUIDELINES

Safe operation is the responsibility of the operator. Operators must develop safe working habits and recognize hazardous conditions to protect themselves and others from injury or death. Always be aware of unsafe conditions to protect the load and the scraper from damage. Before beginning work in a new area, walk around to locate any cliffs, steep banks, or other obstacles. Become familiar with the operation and function of all controls and instruments before operating the equipment. Read and fully understand the operator's manual.

3.1.0 Operator Safety

Nobody wants to have an accident or be hurt. There are a number of things you can do to protect yourself and those around you from getting hurt on the job. Be alert and avoid accidents.

Know and follow your employer's safety rules. Your employer or supervisor will provide you with the requirements for proper dress and safety equipment. The following are recommended safety procedures for all occasions:

- Only operate the machine from the operator's cab.
- Mount and dismount the equipment carefully using three points of contact, as shown in *Figure 22*.
- Wear a hard hat, safety glasses, safety boots, and gloves when operating the equipment.

- Do not wear loose clothing or jewelry that could catch on controls or moving parts.
- Keep the windshield, windows, and mirrors clean at all times.
- Never operate equipment under the influence of alcohol or drugs.
- Never smoke while refueling.
- Do not use a cell phone and avoid other sources of static electricity while refueling because it could cause an explosion.
- Never remove protective guards or panels.
- Always lower the bowl to the ground before performing any service or when leaving the scraper unattended.

 WARNING!
Never attempt to search for leaks with your bare hands. Hydraulic and cooling systems operate at high pressure. Fluids under high pressure can cause serious injury.

3.2.0 Safety of Co-Workers and the Public

You are not only responsible for your personal safety, but also for the safety of other people who may be working around you. Sometimes, you may be working in areas that are very close to other equipment. In these areas, take time to be aware of what is going on around you. Remember, it is often difficult to hear when operating a scraper. Use a spotter and a radio in crowded conditions.

The main safety points when working around other people include the following:

- Walk around the equipment to make sure that everyone is clear of the equipment before starting and moving it.
- Never let anyone in or near the pivot area in an articulated machine.
- Always look in the direction of travel.
- Know and understand the traffic rules for the area you will be operating in.
- Exercise particular care at blind spots, crossings, and other locations where there is other traffic or where pedestrians may step into the travel path.
- Give the right of way to loaded machines on jobsite haul roads and in pits.
- Maintain a safe distance between other machines and vehicles (*Figure 23*).
- Pass cautiously and only when necessary.
- Stay in gear when driving downhill. Do not coast in neutral.

- Avoid traveling across a hill. Drive up and down the slope.
- In steep slope operation, do not allow the engine to overspeed. Select the proper gear before starting down the slope.

3.3.0 Equipment Safety

Your scraper has been designed with certain safety features to protect you as well as the equipment. For example, it has guards, canopies, shields, roll-over

204F22.EPS

Figure 22 ◆ Mount using three points of contact.

204F23.EPS

Figure 23 ◆ Maintain safe traveling distance.

protection, and seat belts. Know your equipment's safety devices and be sure they are in working order.

Use to the following guidelines to keep your equipment in good working order:

- Perform pre-start inspection and lubrication daily (*Figure 24*).
- Look and listen to make sure the equipment is functioning normally. Stop if it is malfunctioning. Correct or report trouble immediately.
- Use caution when backing up to a hitch. Use a spotter if necessary.
- Keep the machine under control. Do not try to work the machine beyond its capacity.
- Keep the work areas smooth and level, which allows for easier maneuvering and greater stability.

The basic rule is know your equipment. Learn the purpose and use of all gauges and controls as well as your equipment's limitations. Never operate your machine if it is not in good working order.

4.0.0 ◆ BASIC PREVENTIVE MAINTENANCE

Preventive maintenance is an organized effort to regularly perform periodic lubrication and other service work in order to reduce instances of poor performance and breakdowns at critical times. By performing preventive maintenance on your scraper, you keep it operating efficiently and safely, and avoid the possibility of costly failures in the future.

Preventive maintenance of equipment is essential and relatively easy if you have the right tools and equipment. The leading cause of premature equipment failure is putting things off. Preventive maintenance should become a habit, performed on a regular basis.

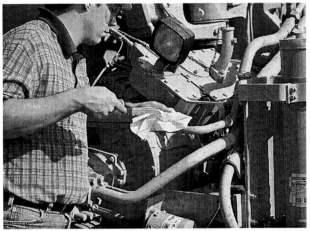

204F24.EPS

Figure 24 ◆ Perform daily maintenance.

CAUTION
Scraper service is normally based on hours of service. A service schedule is contained in the operator's manual. Failure to perform scheduled maintenance could result in damage to the machine.

4.1.0 Daily Inspection Checks

Maintenance time intervals for most machines are established by the Society of Automotive Engineers (SAE) and adopted by most equipment manufacturers. Instructions for preventive maintenance are usually in the operator's manual of each piece of equipment. Common time intervals are: 10 hours (daily); 50 hours (weekly); 100 hours, 250 hours, 500 hours, and 1,000 hours. The *Appendix* is an example of typical periodic maintenance requirements.

The first thing you must do each day before beginning work is conduct a daily inspection. This should be done before starting the engine. The inspection is intended to identify any potential problems that could cause a breakdown and to determine whether or not the machine should be operated. Inspect the equipment before, during, and after operation. The general inspection of the equipment includes looking for the following:

- Leaks (oil, fuel, hydraulic, and coolant)
- Worn or cut hoses
- Loose or missing bolts
- Trash or dirt buildup around the wheels and hydraulic rams
- Broken or missing parts
- Damage to gauges or indicators
- Damaged lines
- Dull, worn, or damaged blades
- General wear and tear

WARNING!
Do not check for hydraulic leaks with your bare hands. Use cardboard or another device. Pressurized fluids can cause severe injuries to unprotected skin. Long term exposure can cause cancer or other chronic diseases.

NOTE
If a leak is observed, locate the source of the leak and fix it. If leaks are suspected, check fluid levels more frequently.

Some manufacturers require that daily maintenance be performed on specific parts. These generally are parts that are the most exposed to dirt or dust and may malfunction if not cleaned or serviced. For example, the service manual may recommend lubricating specific bearings every 10 hours of operation (*Figure 25*), or always cleaning the air filter before starting the engine.

> **NOTE**
> Remember that tandem scrapers have two engines. Make sure to check both engines.

To reduce the possibility of a breakdown or malfunction of a component, pay particular attention to the following areas:

- *Air cleaner* – If the machine is equipped with an air cleaner service indicator, check the indicator. If the indicator shows red, the air filter and intake chamber need to be cleaned. If the machine does not have a service indicator attachment, then you must remove the air cleaner cover, inspect the filter, and clean out any dirt at the bottom of the bowl. Check the operator's manual for the specific procedures to follow.
- *Battery* – Check the battery cable connections. Make sure that the terminals are free of corrosion and the clamps are tight.
- *Cooling system* – Check the coolant level (*Figure 26*). Make sure it is at the level specified in the operator's manual.
- *Drive belts* – Check the condition and adjustments of drive belts on the engine.
- *Engine oil* – Check the engine oil level to make sure it is within the safe starting range (*Figure 27*).
- *Environmental controls* – Your machine may be equipped with lights and windshield wipers. If you are going to operate under conditions where these accessories are needed, make sure they work.
- *Fuel level* – Check the fuel level in the fuel tank(s). Do this manually with the aid of the fuel dip stick or marking vial. Do not rely on the fuel gauge at this point. Also check the fuel pump sediment bowl if one is fitted on your machine.
- *Hydraulic fluid level* – Check the hydraulic fluid levels in the reservoirs (*Figure 28*).
- *Hydraulic lines and couplings* – Check the lines that run between the tractor and the scraper unit. They are subjected to continuous flexing and can easily be damaged.
- *Transmission fluid* – Check the level of the transmission fluid to make sure it is in the operating range (*Figure 29*).
- *Tires* – Check tires for cuts or low pressure. Uneven pressure in tires can cause poor scraper performance.
- *Wires, insulation, and connections* – Check wires and end connections, both in the engine compartment and in the cab.

If you find any problems with the machine that you are not authorized to fix, inform your supervisor or the equipment mechanic. Have the problem corrected before beginning operation.

4.2.0 Servicing a Scraper

When servicing a scraper, follow the manufacturer's recommendations and service chart. Any special servicing for that particular piece of equipment will be highlighted in the manual. Normally, the service chart recommends specific intervals based on hours of run time for such things as changing oil, filters, and coolant. Hydraulic oils should be changed whenever they become dirty or break down due to overheating. Continuous and hard operation of the hydraulic system can heat the hydraulic fluid to the boiling point and cause it to break down. Filters should also be replaced as necessary during regular servicing.

Some machines have computer diagnostics. When the machine needs service, a code will flash on the display panel. These codes are cross-referenced in a book which describes the system that needs service. On some machines, a computer can be plugged into the system for diagnostics as shown in *Figure 30*.

Before performing maintenance procedures, perform the following steps:

Step 1 Be sure the machine is parked on a level surface to ensure that fluid levels are indicated correctly.

204F25.EPS

Figure 25 ◆ Lubricate bearings as required.

Step 2 Be sure all equipment is lowered to the ground and that there is no hydraulic pressure.

Step 3 Be sure the parking brake is engaged.

Step 4 Be sure the transmission is locked in neutral.

4.3.0 Preventive Maintenance Records

Accurate, up-to-date maintenance records are essential for knowing the history of your equipment. Each machine should have a record describing inspection and service that is to be done, along with the corresponding time intervals. Typically, an operator's manual and some type of inspection sheet are kept with the equipment at all times.

204F26.EPS

Figure 26 ◆ Check coolant.

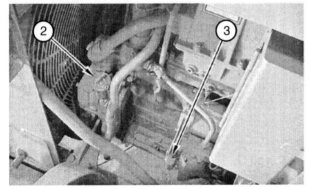

1 - Tractor Engine Fill Port
2 - Scraper Engine Oil Fill Port
3 - Scraper Engine Oil Dip Stick   204F27.EPS

Figure 27 ◆ Check engine oil on both engines.

204F28.EPS

Figure 28 ◆ Check hydraulic oil

1 - Tractor Transmission Oil Fill Port
2, 3 - Sight Glasses
4 - Scraper Transmission Oil Fill Port
5 - Scraper Transmission Sight Gauge   204F29.EPS

Figure 29 ◆ Check transmission fluid.

204F30.EPS

Figure 30 ◆ Computer diagnosis.

5.0.0 ◆ SCRAPER OPERATION

Operation of a scraper requires constant attention to the use of the controls and the surrounding environment. Scrapers have a tremendous amount of power and force. Once that force is put in motion it is not an easy task to stop or redirect it quickly. Do not take risks. If there is doubt about the capability of the machine to do some work, stop the equipment and investigate the situation by discussing it with the foreman or engineer in charge.

Whether it is a slope that may be too steep or an area that looks like the material is too unstable to work, you should know the limitations of the equipment and decide how to do the job ahead of time. Once the operation has begun, it may be too late to stop or change direction. This could cause extra work or result in an unsafe condition.

5.1.0 Suggestions for Effective Scraper Operation

The following are suggestions to improve operating efficiency of your scraper:

- Keep equipment clean. Make sure the cab is clean so that nothing affects the operation of the controls.
- Plan operations before starting.
- Set up the work cycle so that it will be as short as possible. This will reduce the **fixed time**.
- Dump and spread material at the highest practical travel speed.
- Up shift and travel at the highest practical speed for return to cut.
- Do not work the machine over capacity.
- Observe all safety rules and regulations.
- Service equipment regularly to ensure efficient operation.

Use common sense and caution when operating on hills. Reduce engine speed when you are going over a hill. Select the appropriate travel speed before you drive the machine downhill. Do not shift the transmission while traveling downhill or coast in neutral. Use the same gear going downhill as you would driving up the hill. Do not allow the engine to overspeed when you go downhill. Use the retarder or service brakes to reduce engine speed. Use extreme care when turning on a grade; avoid it if possible.

5.2.0 Preparing to Work

Preparing to work involves getting yourself organized in the cab and starting your machine. Mount your equipment using the grab rails and foot rests.

Adjust your seat to a comfortable operating position. Your seat should be adjusted to allow full brake pedal travel with your back against the seat back. This will permit application of maximum force on the brake pedals. Adjust the steering column and the mirrors. Fasten your seat belt.

Operator stations vary, depending on the manufacturer, size, and age of the equipment. However, all stations have gauges, indicators, switches, levers, and pedals. Gauges tell you the status of critical items, such as water temperature, oil pressure, battery voltage, and fuel. Typically, you will have at least a tachometer, voltmeter, temperature gauge, oil pressure gauge, and an hour meter. Read the operator's manual and learn the purpose and normal operating range for each gauge on the machine.

Indicators alert the operator to low oil pressure, engine overheating, clogged air and oil filters, and electrical system malfunctions. Switches are used for disconnecting the electrical system, activating the glow plug, and starting the engine, as well as turning accessories such as lights on and off.

5.2.1 Startup

There may be specific start-up procedures for the piece of equipment you are operating, but in general the start-up procedure should follow this sequence. If the machine has dual engines, start the tractor engine first and then start the scraper engine.

 NOTE
Remember to switch on any disconnect switches before entering the cab.

Step 1 Turn the battery disconnect switches for the tractor engine and the scraper engine to the ON position.

Step 2 Move the transmission control to the neutral position.

Step 3 Engage the parking brake.

Step 4 Move the joystick to the hold position.

Step 5 Turn the key switch to the start position. Release the key when the engine starts.

 NOTE
Never operate the starter more than 30 seconds at a time. If the engine fails to start, wait 2 to 5 minutes before cranking again.

Step 6 Check the oil pressure gauge immediately after the engine starts. If there is no oil pressure, shut off the engine and look for the cause.

Step 7 Allow the engine to warm up. When the temperatures is above 60°F, allow 5 minutes; for colder temperatures, allow 10 to 15 minutes.

Some machines have special features for starting the engine in cold temperatures. These features include glow plugs and ether starting aids. Glow plugs heat the engine to build up heat for ignition. Newer models have automatic starting aids. However, all engines must be allowed to warm up before they are operated.

For dual-engine scrapers, start the scraper engine after starting the tractor engine. The controls used in starting the scraper engine are shown in *Figure 31.*

Step 1 Move the transmission control to the neutral position (1).

Step 2 Engage the parking brake (2).

Step 3 Move the joystick to the hold position (3).

Step 4 Turn the key switch to the ON position (4).

Step 5 Wait until the glow plug light goes out (5).

NOTE

On some machines, if the coolant temperature is below freezing, an air inlet heater will be activated when the key switch is turned to the ON position.

Step 6 Depress the top of the scraper engine start switch (6). Release the switch when the engine starts.

Step 7 Check the oil pressure gauge immediately after the engine starts; if there is no oil pressure, shut off the engine and look for the cause.

5.2.2 Warmup

While the machine is warming up check the gauges and indicators. Keep the engine speed low until the oil pressure registers. If oil pressure does not register within 10 seconds, stop the engine and investigate. Cycle all the controls to circulate warm oil through all system components. Once the machine is warmed up, check that all systems are working properly before operating the machine under a load.

Follow these steps to check that all of the systems are working properly:

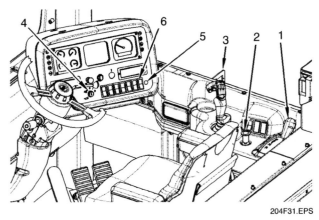

Figure 31 ♦ Controls used for starting the scraper engine.

Step 1 Check brake pressure.

Step 2 Determine that all gauges are functioning properly.

Step 3 Manipulate controls to be sure all components are working properly.

Step 4 Unlock gears and shift gears to low range.

Step 5 Lightly depress the accelerator.

Step 6 Manipulate controls to move the scraper forward and backward, and to make turns.

Step 7 Check the service brake.

Step 8 Shift gears to neutral and lock.

Step 9 Make a final visual check for leaks and listen for any abnormal knocks or noises.

Check the other gauges and indicators to see that the engine is operating normally. Check that all of the gauges are in the normal range. If there are any problems, shut down the engine and investigate, or get a mechanic to look at the problem.

5.2.3 Shutdown

Shutdown should follow a specific procedure also. Proper shutdown will reduce engine wear and possible damage to the machine. Find a dry level spot to park the scraper. Stop the machine by decreasing the engine speed, then depress the clutch and apply the brakes gently. Depress the brake and bring the scraper to a full stop. Place the transmission in neutral and engage the brake lock.

Allow the engine to run at low idle for approximately five minutes. Turn the engine start switch off and remove the key. Lower the scraper bowl and apron to the ground or on wooden blocks. Release hydraulic pressure by moving the control levers until all movement stops. Turn the disconnect switch off. Close the fuel shut-off valve, if provided. Don't forget to report any malfunctions to your supervisor or the field mechanic.

5.3.0 Basic Maneuvering

Basic maneuvering of the scraper consists of moving forward, moving backward, turning from side to side, and operating the apron, bowl, and ejector.

5.3.1 Moving Forward

Move the bowl and apron to the desired position before moving forward. Apply the service brake and release the parking brake. Move the scraper forward by setting the gear selection lever to the desired gear and depress the engine throttle. Adjust the transmission and apply power to the engines while keeping the front wheels straight. Sometimes it is difficult to hold the scraper in a straight line because of the force of the blade digging into the earth. Adjust the gears and engine speed so that the machine can load without strain.

Sometimes a scraper is pushed by a dozer or other equipment (*Figure 32*). You must make sure everything is lined up properly so the dozer will push in a straight line. If the scraper is not in line, it will cut into the ground on only one side and create an unbalanced load in the bowl. This affects your ability to steer in a straight line and results in an uneven cut.

5.3.2 Moving Backward

Scrapers have a high arched gooseneck as shown in *Figure 33*. This allows the tractor wheels to move under it. The tractor can turn up to 90 degrees to the scraper unit as shown in *Figure 34*. This increases the range of motion and maneuverability of the scraper. At the same time, it also makes it more difficult to handle.

Moving backward in a straight line is not easy. Because the scraper is articulated, it will have a tendency to move from side to side, depending on which way the front wheels are pointed. The process is the same as trying to back up a trailer attached to a car or truck. Once you start moving backwards in one direction, it is hard to straighten the rig out without stopping and pulling forward to get everything in line again.

Before you begin to back the scraper, always look around to make sure there is nothing behind you that you may hit. Also, give yourself enough room on either side of the scraper so you will have enough lateral clearance to correct your direction should you start moving to the right or left. If you start to move at a sharp angle, stop and then pull forward enough to straighten out the scraper. You can then start to back again in a straight line.

Figure 32 ◆ Scraper being pushed by a dozer.

204F32.EPS

Figure 33 ◆ Scraper gooseneck.

204F33.EPS

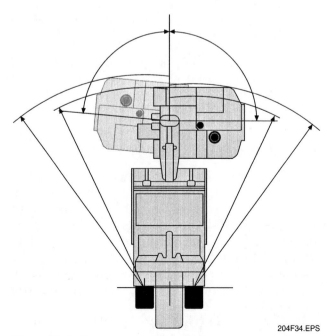

Figure 34 ◆ Scraper range of motion.

204F34.EPS

5.3.3 Turning

Because scrapers are such large pieces of equipment, turning can be tricky. Since they are articulated, they have a pivot point, or hinge, that allows them to turn more sharply than if they were on a rigid frame. However, turning too tightly can jam the neck and the hydraulic lines. This may cause damage to the lines and couplings and put your equipment out of operation.

Steering can be used to walk the tractor out of soft spots, with or without blocking the wheels. This is done by turning the wheels of the tractor first one way and then the other. This will put each wheel on a new footing and may provide enough traction to get out of the soft spot. This technique puts a tremendous strain on the motor and gears. Performing this maneuver makes the motors heat up, so they should be allowed frequent rests in order to cool down.

5.3.4 Operating the Bowl, Apron, and Ejector

When loading, you must be able to operate the bowl lift and apron while moving forward. When unloading or spreading material, you must operate the bowl lift, apron, and ejector. This takes a lot of coordination of your hands and feet. When making a cut, you must lower the bowl to the correct depth using the bowl tilt arm. Raise the apron while moving forward at a speed sufficient to keep the operation moving as shown in *Figure 35*. At the same time, try to keep the wheels from spinning by not applying too much power.

Using the ejector while dumping is a little easier. However, you still have to control the apron and ejector simultaneously so the material is dumped in a smooth even lift.

5.4.0 Working Activities

The activities that a scraper can perform are limited. The machine was designed basically to do one activity well. It can move large quantities of material from one place to another. This involves loading the material, hauling it to another location, and then dumping the material or spreading it over an area to build it up. The variations of this activity and the details of the process are discussed below.

5.4.1 Cut, Load, and Spread

Making cuts and fills with a scraper is a repetitive job that can be organized into a pattern for the scraper operator to follow. The pattern is established based on the number of scrapers available, the terrain, and the required results. Ideally, material scraped from high spots is used to fill in lower spots.

The first cut starts at the top of the first hill (*Figure 36*). The operator cuts as deeply as possible at first. Starting down the hill, the operator raises the bowl slightly. Doing this on each cut will eventually flatten the top of the hill, giving the operator a good flat surface to work from. Then a smooth light cut on a level plain makes loading easier.

Start placing material at the lowest point in the first fill area. Bring the fill up from the lowest point in smooth level lifts until the desired grade is reached. The grade is often marked with stakes as shown in *Figure 37*. The scraper operator is responsible for keeping the fills and cuts level.

The remaining hills would be worked the same way in sequence. When everything has been leveled, there may be a need to do some general smoothing from the beginning to the end.

Try not to load going uphill if at all possible. It is very difficult, and it increases the loading cycle time. Level the top of the hill with several initial passes. Then you can load quickly and smooth out the remaining high and low areas.

204F35.EPS

Figure 35 ◆ Loading the bowl.

204F36.EPS

Figure 36 ◆ Start cut at the top of a hill.

5.4.2 Strip and Stockpile Topsoil

For large construction projects where the topsoil needs to be saved for later use, it can be stripped away with a scraper and stockpiled until it is needed. This can not be done until the clearing and grubbing is completed. Typically, topsoil is considered a **pay material** and therefore keeping it intact is an economic consideration.

Before the **stripping** work begins, the stockpile area needs to be identified and marked. This will depend on how much material has to be moved and what the haul distance will be. To begin removing the topsoil use the following procedures:

Step 1   Move the machine to one end of the work area.

Step 2   Downshift to loading speed.

Step 3   Open the apron to the desired height.

Step 4   Lower the cutting edge to approximately 3 to 6 inches. You should be able to maintain a level cut with little tire spinning (*Figure 38*).

Step 5   Close the apron when loaded and raise the bowl gradually to avoid a shelving effect at the end of the cut.

Step 6   Move to the stockpile area with the bowl just high enough to avoid obstructions (*Figure 39*).

Step 7   Raise or lower the bowl to the desired depth of lift.

Step 8   Raise the apron to the full open position.

Step 9   Engage the ejector lever to push or roll material toward the apron opening, maintaining a smooth, constant flow of material.

Step 10   Lower the apron when the load is spread and return the ejector to full back position.

Step 11   Return to the cut area and repeat the cycle.

NOTE

It is very important to maintain outside limits of the stockpile higher than the middle to avoid the scraper slipping over the edge.

5.4.3 Scraper Operation Using a Bulldozer as Pusher

Assistance is sometimes required to load self-propelled rubber-tired scrapers. While the travel speed of the scraper is increased by the use of wheeled tractors, this is at the expense of traction. For the loading operation where more power is needed, bulldozers are often used as shown in *Figure 40*. The bulldozer blade is reinforced at the center with a heavy steel plate. All scrapers have a pusher block, projecting from the rear, which is engaged by the reinforced area of the blade on the dozer.

204F37.EPS

Figure 37 ◆ Grade stake.

204F38.EPS

Figure 38 ◆ Lower the cutting edge.

204F39.EPS

Figure 39 ◆ Hauling load to stockpile.

Because loading time is of prime importance, it is necessary to organize operations at the loading site so no time will be lost in engaging scrapers as they pull into loading position. There are three ways to organize the work. The space available determines the method used.

- *Back-track loading* – This is used if the loading area is short and wide. After each push, the tractor swings through an arc of 180 degrees and returns to the point adjacent to its original starting point where it turns again before lining up the next scraper.
- *Chain loading* – Use this method where the loading area is long and narrow and may be worked from end to end. Four or five scrapers can be loaded before the dozer needs to return to the starting point. Each of the two turns for the dozer is only 90 degrees.
- *Shuttle loading* – This method would be used in limited areas or in areas that are short but wide. The turning and travel of the pusher are about the same as the chain loading procedure, but access must be provided in both directions. This means that the scrapers should be divided into two groups, with each group moving in opposite directions.

Regardless of the loading arrangement, the push dozer and scraper must work together in a smooth and uniform manner. It is important that the scraper and dozer line up centered on the push block, or the scraper will be pushed sideways instead of straight ahead.

The procedure for doing pushing and loading is as follows:

Step 1 Move the scraper to the end of the cut and set the scraper in position for the next cut close to the dozer. Using hand signals, a good dozer operator can direct the scrapers to the area to be cut next.

Step 2 If at all possible, always move in from the right of the dozer so that you can see the front of the dozer blade and avoid damage to tires or other parts.

Step 3 Continue moving forward while at the same time lowering the cutting edge to the desired depth of cut.

Step 4 Raise the apron to permit material to begin flowing into the bowl.

Step 5 Maintain the cut at the desired depth, using only the throttle to aid the machine and to provide hydraulic power for the attachments (*Figure 41*).

Step 6 Lower the apron when the bowl is full; gradually raise the cutting edge from the cut to avoid shelving the end of the cut.

Step 7 Select the travel gear and move to the fill area as quickly and safely as possible. When you select the travel gear and move away, the dozer will stop pushing and return to the starting point for the next push.

5.4.4 Using Two Scrapers in Tandem

Scrapers are often used in tandem. Two scrapers equipped with a bail and push plate can be linked together so that the engines can provide supplemental power to each other. Two scrapers working in tandem are shown in *Figure 42*. They can be used instead of, or in addition to, a dozer to supply additional power to load the bowl. This can be more cost effective than using a dozer.

Position the first scraper to scrape the area directly in front of it. Move the second scraper directly behind the first. Lower the bail from the second scraper onto the push plate of the scraper in front. Use the power from the rear scraper to load the bowl in the front scraper first. Then load the bowl in the rear scraper.

204F40.EPS

Figure 40 ◆ Bulldozer pushing a scraper to load it.

204F41.EPS

Figure 41 ◆ Let the dozer push the scraper.

Figure 42 ◆ Two scrapers working in tandem.

5.4.5 Picking Up Small Windrows

Self-loading scrapers are sometimes used for fine trimming. The operator must be able to pick up the excess dirt without cutting into the trimmed surface. A self-loading scraper can pick up a small windrow of dirt that has been left after the grader has trimmed the subgrade. The operator must keep the windrow centered with the scraper so one side of the bowl does not fill faster and cause the scraper to lean and dig deeper on one edge.

5.4.6 Finish Grading and Trimming Subgrade

Check the air pressure in the scraper tires before any finish grading or trimming begins. The air pressure in all tires must be the same. If one tire has as little as 10 pounds less pressure than the others, the scraper will lean and dig deeper on the soft tire side as the scraper bowl fills.

Always keep a close watch on the cutting edge of the equipment. It is costly to repair a worn moldboard on the pan of a scraper, and a worn cutting edge will reduce efficiency. For finish work, a good cutting edge is essential. Watch the cutting edge on any paddle wheel scraper that is picking up soil in a trim operation. A scraper with a worn cutting edge or with worn **end shoes** will not make a clean pass, and will leave more work for the grader operator.

Be sure that the end shoes are extended so they are close to the bottom of the cutting edge, but never lower. When they are in good condition, the cutting edge and end shoes should all touch the ground evenly.

Work with a **grade checker** when using a scraper to finish grades. The grade checker stands on the ground and gives directions regarding the depth of cut required or other trimming requirements. Here is how the operation should work:

Step 1 Approach the work area and downshift to a safe working speed.

Step 2 Observe the grade checker's hand signals and lower the bowl to begin cutting finish grade.

> **NOTE**
>
> If a fill is to be made instead of a cut, you must first load the bowl with material either from a stockpile or a previous cut.

Step 3 Continue cutting or filling, observing the grade checker's signals and making necessary corrections, until the bowl is full or empty.

Step 4 Shut the apron when the bowl is loaded and raise the bowl slowly out of cut to avoid a benching effect on the end.

Step 5 Raise the bowl to safe travel height and proceed to the spread area or stockpile.

Step 6 Lower the bowl to the amount of lift to be added following the grade checker's signals. Select the correct speed and raise the apron to the open position. Begin ejecting material from the bowl.

Step 7 Return to the work area when the load is spread, and again observe the grade checker.

Step 8 Line up on the next cut area by putting one wheel on the edge of the previous cut.

Step 9 Select the proper working speed and lower the bowl to begin the cut.

Step 10 Observe the grade checker's hand signals and proceed to make the cut.

Step 11 Continue cutting, making necessary adjustments in depth, until the bowl is full or the job is completed.

5.4.7 Working in Unstable Material

Most of the time a scraper cannot load in a muddy area. A dozer is better suited for this type of work. If you have to use scrapers to move mud, there is a technique that will work safely. Follow these six basic steps:

Step 1 Move into the soft area with the scraper bowl down.

Step 2 Move ahead slowly until the tires start to slip and then stop.

Step 3 Move the dozer into position to push the scraper (*Figure 43*).

Step 4 When contact is made, start to move ahead slowly. Avoid spinning the scraper wheels and don't try to cut too deeply because this will cause the dozer to spin its tracks and get stuck.

Step 5 When the scraper is loaded, apply a little more power, still being careful not to spin the wheels.

Step 6 The dozer operator should keep pushing until the scraper operator can get enough traction to pull away.

In very soft areas you may have to take only half loads until traction improves.

Always watch for soft spots during dumping or loading. If there is a soft area that must eventually be worked through, don't drive right into the center. Drive along the edge first. Move closer to the soft area with each pass. You can decide whether it is getting too soft to hold the scraper on each pass.

CAUTION

If you are working over an area where the mud is at least 3 feet deep and the scraper loses traction, stop and wait to be pulled or pushed out. If you keep spinning the wheels, the scraper will sink deeper into the mud and it will be a major job winching it out, especially if there is a load in the bowl.

5.4.8 Loading a Scraper In Rock

Loading a scraper in rock is a delicate procedure. You must be very alert to keep from spinning the tires or damaging the cutting edge. Spinning the tires may ruin them. Apply little or no power while being pushed. Letting the dozer do the work will help to avoid any tire spin and save the tires.

If you hit a hard spot and the scraper stops or nearly stops, pull the bowl up a little until the boulder that was causing the trouble is passed. Then put the bowl down slowly until the push dozer has to strain slightly to keep going. You can tell when the push dozer's engine is starting to

lug down by watching the exhaust. If you see smoke starting to increase, lift the scraper bowl to reduce the resistance on the dozer.

In loading rock, the scraper operator must raise and lower the scraper bowl constantly. The cutting edge frequently catches on a boulder that will not budge. Don't try to cut through it. Let the dozer perform a **ripping** operation as shown in *Figure 44*. Pick up the ripped material on the next pass. Occasionally the scraper will load a boulder so large that it won't pass under the bowl when the load is dumped, and the bowl can't be raised high enough to pass over the boulder. When this happens, dump the boulder on the ground, back up slightly to give the scraper a little turning room, and then turn sharply so the bowl passes beside the boulder instead of over it.

Always work slowly and carefully in rock. Don't rush. You cannot work fast enough to make up for the loss of a tire or damaged equipment.

5.5.0 Hauling Costs

The scraper is an expensive machine to buy and maintain. Before beginning an operation, use of scrapers to move material should be carefully planned based on the type of material, haul distances, volume of material, and expected results.

Increase of load size decreases hauling cost per yard of material unless the speed and acceleration of the scraper are greatly reduced. In general, a light load can be hauled rapidly, but the cost is divided among fewer yards, making the cost per yard high. Larger loads move more slowly, but costs are spread over more yards, so the cost per yard is lower.

If no push dozer is required for the loading operation, the most efficient loading time would also give the greatest production. When push dozers must be used, their added cost makes the most efficient time shorter than the best production. When scrapers are waiting for a push dozer, it is not cost effective to keep them in the cut any longer than the most efficient time. In this case, it is better to take a lighter load and keep the operation moving.

204F43.EPS

Figure 43 ◆ Move the dozer into position.

204F44.EPS

Figure 44 ◆ Dozer ripping soil.

1. A scraper is also known as a _____.
   a. cutter
   b. grader
   c. pan
   d. load-all

2. On a standard self-propelled scraper, the tractor pulls the _____.
   a. bail
   b. ejector
   c. apron
   d. bowl unit

3. Two scrapers can be connected together with _____.
   a. a bail and hook
   b. a hook and sinker
   c. push plates
   d. connecting rods

4. A tandem scraper has two _____.
   a. bowls
   b. engines
   c. cutting units
   d. cutting blades

5. The tractor and bowl of a scraper are connected by a _____.
   a. push plate
   b. bail
   c. pan
   d. cushioned hitch

6. When starting a dual-engine scraper, you should start _____.
   a. the tractor engine first
   b. the scraper engine first
   c. either engine first
   d. both engines at the same times.

7. The gear selector should be set to N when starting or parking the machine.
   a. True
   b. False

8. The joystick is used to control the _____.
   a. steering
   b. hydraulic elevator
   c. bowl, ejector, and apron
   d. engine speed and transmission setting

9. Increased rimpull is obtained using the _____.
   a. transmission hold
   b. throttle lock
   c. governor override
   d. differential lock

10. If the air pressure decreases below a certain level _____.
    a. the brakes will lock on
    b. the apron will close
    c. the bowl will lower to the ground
    d. the machine will shut down

11. If a warning alarm sounds while an alert indicator is lit, _____.
    a. there is a short in the system
    b. the operator must shut down the machine immediately
    c. the machine should be serviced at the end of the sift
    d. the operator should slow down

12. The type of tire that should be used for rough operation or over rocky terrain is a _____
    a. traction tire
    b. button-type
    c. hard-lug
    d. balloon

13. Before beginning work in a new area, the operator must _____.
    a. reset all the grade stakes
    b. have the engine tuned up
    c. change the blade on the pan
    d. walk around to locate any cliffs, steep banks, or other obstacles

14. On haul roads, give the right of way to _____ machines.
    a. downhill
    b. loaded
    c. empty
    d. returning

15. Conduct your daily walk-around inspection _____.
    a. after the engine is started
    b. in the presence of the field mechanic
    c. before the engine is started
    d. on a hard surface

16. If the scraper is moving too fast when traveling downhill, you should slow the machine by _____.
    a. using the retarder
    b. engaging the differential lock
    c. lifting the bowl
    d. using the transmission hold switch

17. If the pusher bulldozer is not lined up properly with the scraper, it will _____.
    a. not change the operation
    b. create an unbalanced load
    c. cause the bulldozer to flip over
    d. cause the scraper to flip over

18. When beginning a cut with a scraper, the operator should _____.
    a. keep the apron closed
    b. move the ejector forward
    c. speed up and put the scraper in high gear
    d. lower the bowl

19. The removal of overburden or thin layers of pay material is called _____.
    a. ripping
    b. stripping
    c. lug down
    d. winching

20. When cutting an area that is short but wide using a push dozer, use the _____ method.
    a. back-track loading
    b. chain loading
    c. shuttle loading
    d. floating

21. Flat pieces of steel on each side of the scraper bowl that keep the material confined to the front cutting edge are _____.
    a. mold boards
    b. end shoes
    c. winches
    d. lifts

22. Before starting to do finish grading, the operator should _____.
    a. let some air out of the tires
    b. check tire pressure and make sure the pressure is uniform in each tire
    c. arrange for a dozer to push the scraper through the area
    d. take off the end shoes on the bowl

23. Typically, the operator communicates with the grade setter using _____.
    a. a walkie talkie
    b. a piece of paper
    c. hand signals
    d. a string line

24. If a scraper is needed to excavate muddy material, the operator should _____.
    a. always try to fill up the bowl
    b. travel through the area at top speed
    c. try to take a half load at a time
    d. let an excavator or loader load the bowl

25. If a scraper is used in a rocky area, the operator should _____.
    a. sharpen the blade so that it will cut through the rocks
    b. let the bowl float over the ground
    c. apply little or no power to the wheels, letting the dozer do all the work
    d. go as fast as possible to shear any rocks that are in the way

Summary

Scrapers are large machines designed to move large quantities of material from one place to another. They have several different designs, but perform only this one basic function. In doing its work, a scraper sometimes is assisted by a bulldozer that pushes the scraper to provide added power to make deeper cuts into the material.

There are several different scraper designs. These include standard wheel tractor scrapers, elevating wheel scrapers, towed scrapers, and tandem scrapers. The elevating wheel scraper uses a rotating paddle wheel arrangement in the bowl to move the soil to the back of the bowl. The tandem scraper has two engines, one for the tractor in the front, and another at the back which powers the rear wheels on the bowl. Tandem scrapers can be linked together to push and pull each other. Towed scrapers are pulled by a tractor or dozer unit and can be economical on smaller jobs. These designs also come in many different sizes.

To be an effective scraper operator, you must know how to load, haul, and spread material, create a stockpile, and do finish grade work. Scrapers are used to move material from one place to another for the purpose of trimming or leveling uneven terrain. Because of their large size and carrying capacity, they are a good choice for the job, as long as they don't have to work on a side slope.

Finish grading and trimming is one activity where the scraper operator must work with another person, called a grade setter, who controls the cutting or trimming operation from the ground. Scraper operators must learn how to work with grade setters to produce smooth accurate grades.

Notes

Maintenance
Interval Schedule

Maintenance Interval Schedule

Note: All safety information, warnings, and instructions must be read and understood before you perform any operation or any maintenance procedure.

Before each consecutive interval is performed, all of the maintenance requirements from the previous interval must also be performed.

When Required

Automatic Lubrication Grease Tank - Fill
Battery - Recycle
Battery or Battery Cable - Inspect/Replace
Bucket Lift and Bucket Tilt Control - Inspect/Clean
Circuit Breakers - Reset
Engine Air Filter Primary Element - Clean/Replace
Engine Air Filter Secondary Element - Replace
Engine Air Precleaner - Clean
Ether Starting Aid Cylinder - Replace
Fuel System - Prime
Fuses - Replace
Lift Cylinder Pin Oil Level - Check
Loader Boom Pin Oil Level - Check
Oil Filter - Inspect
Radiator Core - Clean
Seat Side Rails - Adjust
Window Washer Reservoir - Fill
Window Wiper - Inspect/Replace

Every 10 Service Hours or Daily

Backup Alarm - Test
Bucket Cutting Edges - Inspect/Replace
Bucket Stops - Inspect/Replace
Bucket Tips - Inspect/Replace
Bucket Wear Plates - Inspect/Replace
Cooling System Coolant Level - Check
Engine Air Filter Service Indicator - Inspect
Engine Oil Level - Check
Hydraulic System Oil Level - Check
Loader Boom Pin and Lift Cylinder Pin - Inspect
Loader Pins and Bearings - Lubricate
Seat Belt - Inspect
Transmission Oil Level - Check
Walk-Around Inspection
Windows - Clean

Every 50 Service Hours or Weekly

Cab Air Filter - Clean/Replace
Fuel System Primary Filter (Water Separator) - Check/Drain
Fuel Tank Water and Sediment - Drain
Tire Inflation - Check

Every 100 Service Hours or 2 Weeks

Axle Oscillation Bearings - Lubricate
Steering Cylinder Bearings - Lubricate

Initial 250 Service Hours

Transmission Oil Filter - Replace

Every 250 Service Hours

Engine Oil Sample - Obtain

Every 250 Service Hours or Monthly

Battery - Clean
Belts - Inspect/Adjust/Replace
Brake Accumulator - Check
Braking System - Test
Differential and Final Drive Oil Level - Check
Engine Air Filter Service Indicator - Inspect/Replace
Engine Oil (High Speed) and Oil Filter - Change
Engine Oil and Filter - Change

Initial 500 Service Hours

Seat Side Rails - Adjust

Initial 500 Hours (for New Systems, Refilled Systems, and Converted Systems)

Cooling System Coolant Sample (Level 2) - Obtain

Every 500 Service Hours

Hydraulic System Oil Sample - Obtain
Transmission Oil Sample - Obtain

Every 500 Service Hours or 3 Months

Axle Oil Cooler Filter - Replace
Cooling System Coolant Sample (Level 1) - Obtain
Differential and Final Drive Oil Sample - Obtain
Engine Oil (High Speed) and Oil Filter - Change
Engine Oil and Filter - Change
Fuel System Primary Filter (Water Separator) Element - Replace
Fuel System Secondary Filter - Replace
Fuel Tank Cap and Strainer - Clean
Hydraulic System Oil Filter - Replace
Transmission Oil Filter - Replace

Every 1000 Service Hours or 6 Months

Articulation Bearings - Lubricate
Battery Hold-Down - Tighten
Case Drain Oil Filters - Replace
Drive Shaft Support Bearing - Lubricate

205A01.EPS

Rollover Protective Structure (ROPS) - Inspect
Transmission Oil - Change

Every 2000 Service Hours or 1 Year

Differential Thrust Pin Clearance - Check
Differential and Final Drive Oil - Change
Engine Crankcase Breather - Clean
Engine Valve Lash - Check
Engine Valve Rotators - Inspect
Hydraulic System Oil - Change
Hydraulic Tank Breaker Relief Valve - Clean
Lift Cylinder Pin Oil Level - Check
Loader Boom Pin Oil - Change
Refrigerant Dryer - Replace

Every 3000 Service Hours or 2 Years

Crankshaft Vibration Damper - Inspect
Engine Mounts - Inspect

Every 3 Years After Date of Installation or Every 5 Years After Date of Manufacture

Seat Belt - Replace

Every 4000 Service Hours or 2 Years

Hydraulic System Oil - Change

Every 4000 Service Hours or 2.5 Years

Electronic Unit Injector - Inspect/Adjust

Every 5000 Service Hours or 3 Years

Alternator - Inspect
Lift Cylinder Pin Oil - Change
Starting Motor - Inspect
Turbocharger - Inspect

Every 6000 Service Hours or 3 Years

Cooling System Coolant Extender (ELC) - Add

Every 6000 Service Hours or 6 Years

Cooling System Water Temperature Regulator - Replace
Engine Water Pump - Inspect

Every 12,000 Service Hours or 6 Years

Cooling System Coolant (ELC) - Change

205A02.EPS

Trade Terms
Introduced in This Module

Apron: A movable metal plate in front of the scraper bowl that can be raised and lowered to control the flow of material out of the bowl.

Bail hitch: Used to attach the front end of a tractor to the rear of a scraper.

Bowl: The main component at the back of a scraper where the material is loaded and hauled.

Ejector: A large metal plate inside the bowl of a scraper that can be activated to push the material forward, causing it to fall out of the bowl.

End shoes: Flat pieces of steel on each side of the scraper bowl that keep the material confined to the front of the cutting edge until the paddles can scoop it up. Sometimes referred to as slobber bits.

Fixed time: Time given to loading, dumping, turning, accelerating, and decelerating.

Grade checker: Person who checks grades and gives signals to the equipment operator.

Lug down: A slowdown in engine speed (rpm) due to increasing the load beyond capacity. This usually occurs when heavy machinery is crossing soft or unstable soil or is pushing or pulling beyond its capability.

Pay material: Deposit of soil valuable enough to stockpile for future use.

Ripping: Loosening hard soil, concrete, asphalt, or soft rock with a ripping attachment.

Stockpile: Material put into a pile and saved for future use.

Stripping: Removal of overburden or thin layers of pay material.

Additional Resources

This module is intended to be a thorough resource for task training. The following reference works are suggested for further study. These are optional materials for continued education rather than for task training.

Excavation and Grading Revised, 1994. Nick Capachi. Carlsbad CA: Craftsman Book Company.

Moving The Earth, Fourth Edition. H.L. Nichols. New York, NY: McGraw-Hill.

Scraper Operation, 1997. Heavy Equipment Training Series Video, Earthwork Productions LLC.

Figure Credits

Terex Construction – Heavy Equipment, 204F01, 204F05

Reprinted courtesy of Caterpillar Inc., 204F02–204F04, 204F06–204F44, Appendix

CONTREN® LEARNING SERIES — USER UPDATE

NCCER makes every effort to keep these textbooks up-to-date and free of technical errors. We appreciate your help in this process. If you have an idea for improving this textbook, or if you find an error, a typographical mistake, or an inaccuracy in NCCER's Contren® textbooks, please write us, using this form or a photocopy. Be sure to include the exact module number, page number, a detailed description, and the correction, if applicable. Your input will be brought to the attention of the Technical Review Committee. Thank you for your assistance.

Instructors – If you found that additional materials were necessary in order to teach this module effectively, please let us know so that we may include them in the Equipment/Materials list in the Annotated Instructor's Guide.

Write: Product Development and Revision
National Center for Construction Education and Research
P.O. Box 141104, Gainesville, FL 32614-1104

Fax: 352-334-0932

E-mail: curriculum@nccer.org

Craft ______________________ Module Name ______________________

Copyright Date __________ Module Number __________ Page Number(s) __________

Description

____

____

____

____

(Optional) Correction

____

____

____

(Optional) Your Name and Address

____

____

____

Heavy Equipment Operations Level Two

22205-06

Loaders

22205-06
Loaders

Topics to be presented in this module include:

Overview

When a lift bucket is mounted on the front of a tractor you have a very versatile machine. Loaders are the perfect complement to a dump truck. Loaders can pick up and dump any type of loose material, including soils, gravel, coal, and woodchips.

Some loaders have crawler tracks, while others have two- or four-wheel drives. In addition, specialty buckets are designed to handle certain materials efficiently. A skilled operator is the key, however, to safe operation. When you understand the different types of loaders and buckets available, you can select the right machine for the job.

A loader operator must understand the instruments and controls and perform daily inspection and maintenance. Once you master basic loading skills, you will find that loaders can perform many tasks on a construction site, including demolition, backfilling, and clearing land. Loaders do have limitations. Safe operators know the limitations of their equipment. For example, working in unstable areas requires special attention to avoid getting stuck.

Objectives

When you have completed this module, you will be able to do the following:

1. Describe the uses of a loader.
2. Identify the components and controls on a typical loader.
3. Explain safety rules for operating a loader.
4. Perform prestart inspection and maintenance procedures.
5. Start, warm up, and shut down a loader.
6. Perform basic maneuvers with a loader.
7. Perform basic earthmoving operations with a loader.
8. Describe the attachments used on loaders.

Trade Terms

Accessories
Blade
Bucket
Debris
Dozing
Grouser
Grubbing
Rubble
Spot
Stockpile
Stripping
Winch

Required Trainee Materials

1. Pencil and paper
2. Appropriate personal protective equipment

Prerequisites

Before you begin this module, it is recommended that you successfully complete *Core Curriculum; Heavy Equipment Operations Level One;* and *Heavy Equipment Operations Level Two,* Modules 22201-06 through 22204-06.

This course map shows all of the modules in the second level of the *Heavy Equipment Operations* curriculum. The suggested training order begins at the bottom and proceeds up. Skill levels increase as you advance on the course map. The local Training Program Sponsor may adjust the training order.

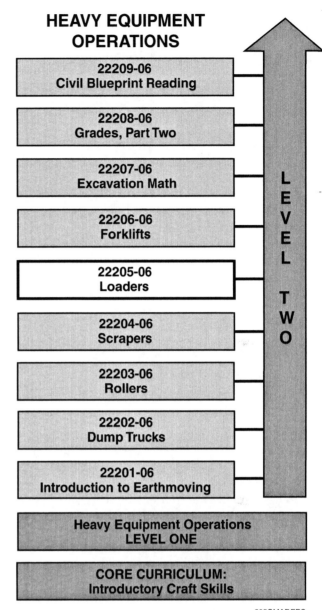

HEAVY EQUIPMENT OPERATIONS

| 22209-06 Civil Blueprint Reading |
| 22208-06 Grades, Part Two |
| 22207-06 Excavation Math |
| 22206-06 Forklifts |
| 22205-06 Loaders |
| 22204-06 Scrapers |
| 22203-06 Rollers |
| 22202-06 Dump Trucks |
| 22201-06 Introduction to Earthmoving |

LEVEL TWO

Heavy Equipment Operations
LEVEL ONE

CORE CURRICULUM:
Introductory Craft Skills

205CMAP.EPS

1.0.0 ◆ INTRODUCTION

Loaders are one of the more popular pieces of equipment. They are characterized by a large **bucket** mounted on two lift arms on the front of the machine. For this reason they are often called front-end loaders. The primary operating components of the loader are the bucket, lift arms, and dump rams.

Loaders are versatile. They are easy to handle and maneuver. They are primarily used for lifting and moving. They also handle various tasks like loading, excavating, demolition, grading, leveling, hauling, and stockpiling. In addition, a loader can carry out other functions with attachments such as brooms, rakes, and forks.

An operator must know how to operate the machine and its **accessories** effectively and must also know the general properties of the materials being worked. Operating a loader is not limited to moving the machine and materials. An operator must perform daily checks and maintenance to keep the machine in good working order. Daily inspections and maintenance are also an important part of safe operations.

1.1.0 Types of Loaders

Loaders are manufactured by several different companies in various sizes and configurations. There are two main categories of loaders, the wheel loader (*Figure 1*) and the crawler loader (*Figure 2*). The wheel loaders have rubber tires, while the crawler loaders have tracks instead of wheels. Additionally, some manufacturers make a small tracked loader similar to a skid steer, which can rotate in place.

Some wheel loaders are articulated; a joint in the frame allows the front and back to move independently. Articulated loaders are easier to reposition because they have a tighter turning radius.

Crawler loaders are usually powered by a diesel engine mounted in back of the operator's cab for greater visibility. The operator sits facing the bucket, which is mounted on the front of the machine. The loader is similar to a bulldozer, except it has a bucket instead of a push **blade**. However, crawler loaders can lift their buckets much higher than bulldozers can lift their blades. Large crawler loaders have a dump height of over nine feet.

Wheel loaders are powered by a gasoline or diesel engine mounted behind the operator's cab. The rigid-frame loader is mounted on four wheels. The bucket and other attachments are mounted on the front. The operator sits facing the bucket. Articulated loaders are also mounted on four wheels, but have a pivot point on the frame

just forward of the operator's cab. The engine is mounted in the rear, behind the operator. This provides added weight over the rear wheels, which provides better traction and helps balance heavy loads in the bucket. *Figure 3* shows an articulated loader hauling soil.

205F01.EPS

Figure 1 ◆ Wheel loader.

205F02.EPS

Figure 2 ◆ Crawler loader.

205F03.EPS

Figure 3 ◆ Articulated loader.

The controls, transmission, and other features will vary with the make and model of the equipment. Many loaders have an electronically controlled hydrostatic transmission. It provides infinitely variable speed within the speed range of the machine. Many have a load sensing feature that automatically adjusts the speed and power to changing load conditions.

Some loaders have specialized designs to be used in a specific industry. For example, loaders are frequently used at transfer stations to move small waste. Several manufacturers make specialized loaders in various sizes for waste handling. They are designed to keep debris out of the machine. Large service doors allow for easy cleaning. Additional guards keep debris out of the engine, hydraulics, and other compartments. The fan reverses to blow debris out of the cores and screens. These features help minimize the amount of waste that gets caught in the cooling system, which helps keep the machine from overheating.

1.2.0 Loader Size and Capacity

Loaders can be generally categorized in three groups; compact, mid-size, and large. The compact models have a bucket capacity under 2 cubic yards. Mid-size or utility loaders have a capacity between 2 and 4 cubic yards. Large or production loaders have a capacity over 4 cubic yards.

Loaders are used to lift and move soils and other materials in many types of job sites from residential landscaping operations to large mining operations. They are designed with a wide range of capacities to meet the demands of various industries.

This module will describe the operations of a wheel loader, specifically the Caterpillar 988. It is a very popular and versatile loader. The features and operations are typical of large wheel loaders. Procedures on other equipment may vary slightly.

NOTE

Always read the operator's manual before operating equipment. Follow all safety and startup procedures.

Some characteristics of the smallest and largest wheel-type loaders are as follows:

- *Power* – 52 hp (smallest); 2,300 hp (largest)
- *Bucket capacity* – 0.78 cubic yards (smallest); 53 cubic yards (largest)
- *Lift capacity* – 3.7 tons (smallest); 80 tons (largest)
- *Vehicle weight* – 9,810 pounds (smallest); 578,000 pounds (largest)
- *Dump height* – 7'-9" (smallest); 24'-0" (largest)

2.0.0 ◆ IDENTIFICATION OF EQUIPMENT

The components of a Caterpillar 988F loader are shown in *Figure 4*. While the components, controls, and indicators will vary with the manufacturer, model, and the options available, they are all similar. Because the information in the following sections is based on the Caterpillar 988 Series loaders, an operator must read and understand the operator's manual for the specific loader before beginning operation.

The engine provides power for loader operations. The power from the engine is transmitted to the wheels via the transmission. The engine and transmission systems on loaders are similar to those in tractors covered in *Heavy Equipment Level One, Tractors*.

The engine also powers the hydraulic system. The engine supplies power to one or more pumps

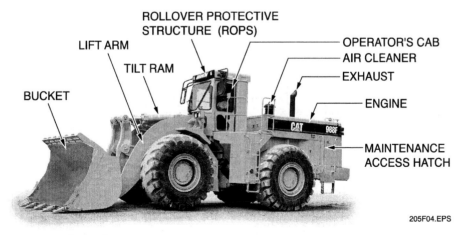

Figure 4 ◆ Typical components of a loader.

that build pressure in the hydraulic reservoir. The controls open and close valves to the hydraulic lines to operate lift and dump rams and other auxiliary hydraulic systems.

The lift arms and dump ram use the hydraulic power to move the bucket. The lift arms are mounted at the front of the frame and are attached to the bottom the bucket. The lift ram raises and lowers the lift arms. The dump ram is attached to the top of the bucket and is used to tilt the bucket forward and backward.

2.1.0 Operator's Cab

The operator's compartment is the central hub for loader operations. An operator must understand the controls and instruments before operating a loader. Study the operator's manual to become familiar with the controls and instruments and their functions.

Figure 5 is an overhead view of the interior of the operator's cab on a Caterpillar 988 wheel loader. Some switches and controls are located outside of the main cab. However, the controls for normal operations are all located within easy reach of the operator's seat. The bucket is controlled with levers located in front of the right armrest. Vehicle movement is controlled with a joystick located above the left armrest. Directly in front of the operator are three instrument display panels. Directly below the instrument panels are foot pedals for the brakes and throttle.

205F05.EPS

Figure 5 ◆ Operator's cab for the Caterpillar 988.

A different cab configuration is shown in *Figure 6*. This is the operator's cab for the Caterpillar 938G wheel loader. On this machine, the bucket is controlled with a joystick in front of the right armrest. The vehicle is maneuvered using the steering wheel. Loaders have different cab configurations. Read the operator's manual before operating any equipment so that you are familiar with the cab configuration and controls on the machine.

The large windows allow the operator to see the area around the machine. Loaders often operate in tandem with one or more dump trucks. It is important that the operator be able to see the movement of other vehicles that are often positioned to the side of the loader. Windshield wipers keep the windshield clean. Lights aid vision at night or in poor lighting conditions.

NOTE

Inspect and clean the windshield and lights daily or whenever they become fouled. An operator can cause equipment damage or injure other workers if he or she cannot see clearly in all directions.

2.2.0 Instruments

An operator must pay attention to the instrument panel. The instrument panel includes the gauges that indicate the vehicle and engine speed. There are also numerous warning lights and gauges that must be monitored for safe operations. An operator can seriously damage the equipment if the instrument panel is not closely monitored.

The instrument panel varies on different makes and models of loaders. Generally they include the instruments and indicators included in the following sections. A typical instrument panel is shown in *Figure 7*. Most of these instruments and indicators are similar to those in other machines and should be familiar from descriptions in previous modules.

The instrument panel on a Caterpillar 988 has three sections. The first section contains four gauges (*Figure 8*). The second section is the speedometer and tachometer display. The third section is the monitoring system display. There are additional switches on a panel beneath the displays.

NOTE

The parenthetical callout numbers or letters in the following sections refer to *Figure 8*.

INSTRUMENT PANEL

STEERING WHEEL

JOYSTICK
BUCKET CONTROL

205F06.EPS

Figure 6 ◆ Operator's cab for the Caterpillar 938G.

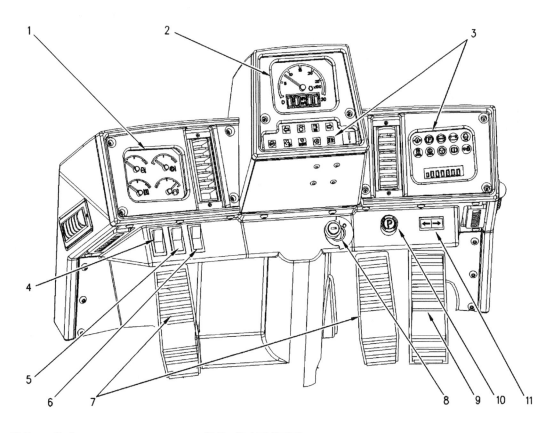

(1) Gauge Display
(2) Speedometer/Tachometer Display
(3) Indicators
(4) Display Modules (Caterpillar Monitoring
System)
(5) Ether Starting Aid

(6) Throttle Lock Control
(7) Service Brake Control
(8) Engine Start Switch
(9) Throttle Control
(10) Parking Brake Control
(11) Light Switches

205F07.EPS

Figure 7 ◆ Instrument panel for the Caterpillar 988 series loader.

2.2.1 Coolant Temperature Gauge

The coolant temperature gauge (A) indicates the temperature of the coolant flowing through the cooling system. Refer to the operator's manual to determine the correct operating range for normal loader operations. Temperature gauges normally read left to right with cold on the left and hot on the right. Most gauges have a section that is red. If the needle is in the red zone, the coolant temperature is excessive. Some machines may also activate warning lights if the engine overheats.

CAUTION

Operating equipment when temperature gauges are in the red zone may severely damage it. Stop operations, determine the cause of problem, and resolve it before continuing operations.

2.2.2 Transmission Oil Temperature Gauge

The transmission oil temperature gauge (B) indicates the temperature of the oil flowing through the transmission. This gauge also reads left to right in increasing temperature. It has a red zone that indicates excessive temperatures.

2.2.3 Hydraulic Oil Temperature Gauge

The hydraulic oil temperature gauge (C) indicates the temperature of the hydraulic oil flowing through the system. This gauge also reads left to right in increasing temperature and has a red zone that indicates excessive temperatures.

CAUTION

Do not operate the loader under a load until the hydraulic oil has reached the correct operating temperature. Operating under a heavy load before the hydraulics have adequately warmed can damage the equipment and may cause failure.

2.2.4 Fuel Level Gauge

This gauge (D) indicates the amount of fuel in the loader's fuel tank. On diesel engine loaders, the gauge may contain a low-fuel warning zone. Some models have a low-fuel warning light. Avoid running out of fuel on diesel engine loaders because the fuel lines and injectors must be bled of air before the engine can be restarted.

NOTE

The parenthetical callout letters in the following section refer to *Figure 9.*

2.2.5 Speedometer/Tachometer

The speedometer and tachometer display is shown in *Figure 9*. The tachometer (A) indicates the engine speed in revolutions per minute (rpm). Most tachometers are marked in hundreds on the meter face and read left to right in an increasing scale. There is a red zone on the high end of the scale that indicates that the engine is overspeeding. On some loaders there is also an indicator light that will warn the operator if the engine speed is too high.

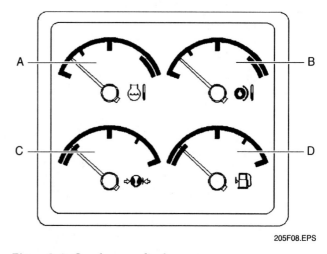

205F08.EPS

Figure 8 ◆ Quad gauge display.

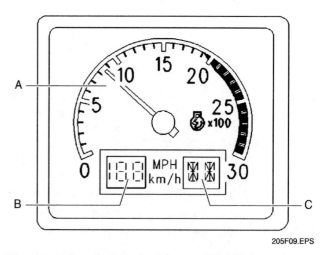

205F09.EPS

Figure 9 ◆ Speedometer/tachometer display.

The speedometer (B) shows the machine's ground speed. It can be set for either miles per hour (mph) or kilometers per hour (kph). The gear/direction readout (C) shows the selected transmission gear and direction.

There are also a series of lights under the tachometer/speedometer. These lights show that various features are activated under normal operating conditions. Generally, they do not indicate that there is something wrong with the machine, merely that a function is active. The exception is the action light, which indicates that there is a fault in the monitoring system.

Typical indicator lights show that the following features are activated:

- Turn signals
- Reduced rimpull
- Autoshift
- Loose material mode
- Automatic ride control
- Torque converter lockout clutch
- Throttle lock

Reduced rimpull allows the operator to control driveline torque while the engine is operating at a high idle. The operator can reduce driveline torque while giving full power to the bucket hydraulics. Activating the reduced rimpull function changes the pressure that must be applied to the foot pedal to activate the brakes. This is useful when inching into a pile to scoop a load. Reduced rimpull reduces wheel slipping.

The reduced rimpull function will provide an impeller clutch pressure that limits rimpull to the desired level when the brakes are fully released. The reduced rimpull function is engaged with a knob that can be set in five positions. The positions represent a maximum allowable percentage of total rimpull. Typically, the intervals are off or 100 percent, 90 percent, 80 percent, 70 percent and 60 percent. The reduced rimpull function can only be activated when the machine is operating in first gear.

The loose material mode is activated via a switch. This provides maximum hydraulic pump flow and provides for faster loading of loose materials. Do not activate this switch if digging conditions are rough. The normal mode provides a reduced hydraulic flow for better digging.

Automatic ride control is used when traveling at higher speeds over rough terrain. The system acts as a shock absorber dampening the forces from the bucket. This minimizes bucket movement and swinging motion and helps stabilize the machine.

The torque converter lockout is also activated with a switch on the side of the operator's cab.

This feature enables the lockup clutch. When the engine is in a certain speed range, the torque converter will lock up. This provides more efficient operations for higher speeds and is used when carrying loads.

The throttle lock is engaged via a switch on the left side of the instrument panel. This switch keeps the engine at the current rpm. Some machines have an indicator light that shows when the throttle lock is active. To disengage the throttle lock, either turn off the switch or depress the brake pedal slightly.

2.2.6 Alert Indicator Display

The loader system has sensors in various parts of the engine and related systems. These sensors are connected to an electronic monitoring system. The monitoring system display (*Figure 10*) has alert indicators that light when the machine is not functioning properly. If the situation needs immediate attention, the indicators may flash, remain lit, or an audible alarm will sound. Each machine has different indicator lights, or they may be arranged differently. Always read the operator's manual so you understand all warning lights before operating the equipment.

> **CAUTION**
> A flashing indicator light may require immediate action by the operator. If you need to stop and look up the meaning of an indicator after it is flashing, you may cause serious damage to the equipment. Know your equipment before you begin operations.

Indicator lights alert the operator to make an adjustment to the machine or how it is being operated. The indicators on a Caterpillar 988 are fairly typical. They include the following:

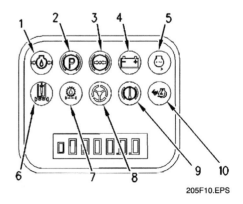

Figure 10 ◆ Electronic monitoring system.

- Engine oil pressure (1) will light when the oil pressure is too low.
- Parking brake (2) will light when the parking brake is engaged.
- Brake oil pressure (3) will light when the axle oil pressure is too low.
- Charging system (4) will light when the battery voltage is outside of the battery charge range. When the engine is running, it will also light when the alternator frequency is low.
- Engine coolant flow (5) will light when there is no coolant flow to the engine.
- Hydraulic oil filter (6) will light when the hydraulic oil filter is plugged and the bypass valve is open.
- Transmission oil filter (7) will light when the transmission oil filter is restricted.
- Secondary steering (8) will light when the primary steering pressure is low.
- Brake oil temperature (9) will light when the axle brake oil temperature is too high.
- Engine overspeed (10) will light when the engine speed is too high.

Below the indicator lights, there is a display window. This display is a digital readout. It can be set to show a digital reading of any of the gauges or set to show other readings. These other readings include: total operating hours, current engine speed, total travel distance, or active diagnostic codes.

The monitoring systems and indicator lights notify the operator of potential problems with the machine. Many monitoring systems have several levels of urgency. The categories are based on the severity of the problem. Different warnings require different actions by the operator. Higher warning levels demand an immediate response by the operator. Some control panels have indicator lights, warning action lights, and an action alarm to show increasing levels of potential danger. The latter two require immediate action by the operator. Typical warning levels are as follows:

- A flashing indicator light means the system needs attention soon.
- A constant action light means the operator must change the machine operation or damage could result.
- When an action alarm sounds, the operator must change machine operations immediately or the machine will be damaged.

- When an action alarm becomes intermittent the operator must immediately perform a safe shutdown. Otherwise, the machine will be damaged or the operator will be injured.

2.3.0 Controls

Depending on the manufacturer and model, a loader may be controlled with a joystick or steering wheel. Foot pedals are often used to steer track loaders. Typically, the vehicle control joystick is located in front of the left armrest. The bucket can be controlled with switches or a joystick. Joystick vehicle movement and lever bucket control will be described in this module. Review the operator's manual to fully understand other types of vehicle movement and bucket controls.

2.3.1 Vehicle Movement Controls

Vehicle movement is controlled with a joystick located in front of the left armrest as shown in *Figure 11*. Moving the lever to the left or right steers the vehicle to the left or right. When the lever is released, it will return to the central position. The machine will maintain the direction it was moving in before the lever was released.

The transmission is controlled with a trigger switch on the side of the joystick. The trigger switch has three positions: forward, neutral, and reverse. Set the trigger switch to select the direction of travel. The selected transmission direction will be displayed on the speedometer.

Some loaders have a lever to select the direction of movement. Typically, these are known as FNR controls, for forward, neutral, and reverse. Often, they are separate from a standard gear selection control. The operator uses this lever to control the direction of the loader.

205F11.EPS

Figure 11 ◆ Steering control joystick.

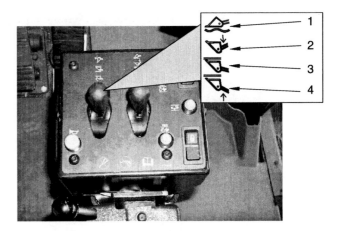

There are two gear selection switches on the top of the joystick. One switch shifts into higher gears, the other downshifts. Typically, loaders have three speed ranges, first, second, and third. Use the upshift switch to move the machine into the next higher speed. Use the downshift to move the machine into the next lower speed.

An automatic shift function is available on some machines. When this function is activated, the control will select the proper gear according to the speed of the machine. An indicator light shows that this feature is activated.

The lever in front of the joystick engages the control lock. This mechanically locks the steering controls. The transmission controls will be electronically locked when this lever is engaged.

2.3.2 Bucket Lift and Tilt Controls

The bucket on a loader is attached to the front frame by two lift arms. This adds stability to the bucket and improves material handling. The arms can be raised and the bucket can be tilted forward and backward. Loaders are equipped with control levers that lift and tilt the bucket. Some models have a joystick that controls both the lift and tilt.

The Caterpillar 988 has a two-switch configuration for bucket control. These switches are located in front of the right armrest. *Figure 12* shows control positions for lifting and tilting the bucket.

The lever positions for the lift controls shown in *Figure 12* are as follows:

- *Float (1)* – Push the lever forward through the detent. This locks the lever into position. The bucket will float with the contour of the ground.
- *Lower (2)* – Push the lever forward in order to lower the bucket. When the lever is released it will return to the hold position and the bucket will stop.
- *Hold (3)* – This is the neutral position. The lever will return to this position unless locked into float position. The bucket will remain in the selected position.
- *Raise (4)* – Pull the lift lever to raise the bucket. Release the lever and the bucket will stop moving. The lever will return to the hold position.

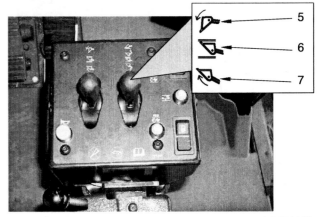

205F12.EPS

Figure 12 ◆ Loader bucket controls for the Caterpillar 988.

The tilt control lever tilts the bucket forward and backward. The lever positions for the tilt control as shown in *Figure 12* are as follows:

- *Dump (5)* – Push the lever forward in order to dump a load from the bucket.
- *Hold (6)* – The lever will return to the hold position when released. The bucket will remain in the selected position.
- *Tilt back (7)* – Pull the bucket tilt lever backward in order to tilt the bucket backward.

Other models may have joystick controls like the Caterpillar 930G shown in *Figure 13A*. Some have two levers to control the bucket like the Caterpillar 994D shown in *Figure 13B*.

Figure 13A ◆ Other bucket control configurations.

Figure 13B ◆ Other bucket control configurations.

2.4.0 Operator Comfort and Other Controls

There are several controls designed to adjust the seat and controls for maximum operator comfort. Adjust the seat, armrests, mirrors, and cab climate controls while the machine is warming up or parked. Do not move or operate the machine before adjusting the seat and mirrors for comfort and visibility, and fastening the seat belt.

There are usually three levers or knobs used to adjust the seat as shown in *Figure 14*. The first lever moves the seat forward and backward. Depress the lever to adjust the seat. Release the lever to lock the seat in place. The second lever changes the angle of the backrest. The third lever is used to adjust the seat height. Adjust the seat so that you can reach all controls comfortably with your back against the back of the seat. You should be able to fully depress all foot pedals and reach all switches.

These adjustments have safety impact in addition to operator comfort. An operator who is adjusting the seat or mirrors after beginning operations is not giving full attention to machine operations. If an operator cannot reach all of the controls, the machine may be hard to control and cause injury or property damage. Properly setting the climate controls will reduce operator fatigue and increase alertness.

If you will be working at night, familiarize yourself with all of the light switches. Once it is dark, it is too late to look for the light switches. Make sure that all lights are functioning properly. Adjust lights so that the work area is properly illuminated.

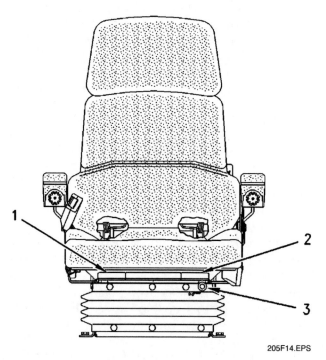

Figure 14 ◆ Seat adjustment levers.

2.5.0 Attachments Used on or with the Loader

The bucket is the main attachment for the loader, although there are several other useful ones. Different buckets are available to maximize productivity under different circumstances. In a quarry setting, the operator may need a stronger bucket with teeth to scrape a wall face. A larger bucket with a higher back plate will aid in handling loose material like dry soils.

In addition to lifting and loading, loaders are sometimes used for other operations such as sweeping and scraping. Brooms, forks, rakes, and snow blades can be attached to a loader. Many loaders are equipped with a hydraulic quick-connect that allows the operator to change attachments from the cab. In other models, the attachments are fixed to the lift arms with pins. The operator must dismount and manually secure the attachment.

 WARNING!
Do not leave the operator's cab while the machine is running. Place the controls in park and shut down the engine before dismounting to adjust attachments.

2.5.1 Buckets

Several types of buckets are made to accommodate different work situations. Working with different materials may require a different bucket design. Some buckets have teeth for scraping or digging, while others have a flat blade for scraping on flat surfaces. Regardless of the design, the bucket width is usually the same as that of the back wheels or tracks of the loader.

General purpose buckets are used for a variety of activities from loading to clearing (*Figure 15*). These buckets are generally larger and sturdier than standard buckets. They have a straight edge, but allow for a toothed edge to be bolted on.

Loose material buckets have a straight edge in front. The back plate is elevated to minimize how much material spills out of the back when the bucket is filled and lifted. These buckets are designed to be used with loose soils, wood chips, light gravel, and similar material.

Heavy-duty, high-abrasion rock buckets (*Figure 16A*) are designed for use in quarries. They are used in bank face loading or in highly abrasive low-to-moderate impact conditions. An upper rock guard aids in load retention. The bucket is available with either a straight or toothed spade

edge. The straight edge offers higher breakout force and increased dump clearance. The toothed spade edge is designed to dig efficiently.

205F15.EPS

Figure 15 ◆ Loader with general purpose bucket.

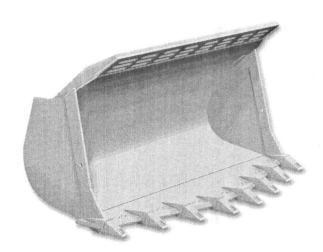

(A) HIGH-ABRASION ROCK BUCKET

(B) MULTIPURPOSE BUCKET

205F16.EPS

Figure 16 ◆ High-abrasion bucket and multipurpose bucket.

A multipurpose bucket is designed to allow the bucket to function in four different ways. It is also known as a four-in-one bucket (*Figure 16B*). It can be used as a standard bucket, dozer blade, clamp, or controlled-discharge bucket. The operator can load material with the bucket, doze with the straight blade, grab items with the hydraulically operated front clamp, or meter out the load with the clamshell. Teeth can be attached to the flat front blade for digging.

The auxiliary hydraulic lines are used to open and close the multipurpose bucket. When the bucket is fully opened, it can function as a dozer blade. The auxiliary control lever also opens and closes the bucket for clamshell and clamp features. When the auxiliary controls are completely closed, the multipurpose bucket functions like a standard bucket.

There are many types of specialty buckets. Several specialty buckets are designed to maximize load capacity for common materials such as coal or wood chips. These buckets have the maximum size given the average density of the material and the capacity of the loader. This ensures maximum operator and machine efficiency.

Some buckets are designed to achieve maximum dump height. They are also called rollout buckets. Finally, there are buckets with a top clamp that can grab and hold loose material. These are often used to move refuse, trash, or brush.

2.5.2 Other Implements

Other types of attachments for a loader include dozer blades, snow plows, brooms, rakes, rippers, and forks. Some of these attachments are designed to meet the needs of specific industries, for example forestry or waste handling. Other attachments such as snow blowers or snow plows enable the loader to be used in the winter when it would otherwise be idle.

There are several loader attachments used in the forestry industry, including log forks and mill yard forks. Log forks (*Figure 17*) are used to load, deck, and sort lumber, logs, or palletized material. The clamp raises to almost 90 degrees to allow the operator to scoop and load the forks. The clamp is then closed to hold the logs in place for transport. Mill yard forks a have a similar design, but the upper tine closes between the lower two tines. This allows the operator to hold a single log or several logs.

Various kinds of dozer-type blades can be attached to the loader. Snow plow blades can be either box shaped, v-shaped, or have the traditional concave blade. Angle blades can be hydraulically adjusted to move the material straight ahead or to either side. This allows the operator to place material without performing multiple passes. The angle blade is used for side-casting, backfilling, cutting ditches, snow removal, pioneering, and maintaining rural or logging roads.

The broom, as shown in *Figure 18*, is another popular attachment. The broom can be angled up to 30 degrees to either side. It is adjusted hydraulically or manually. They are used for clearing parking lots, industrial plants, mill yards, airport runways, streets, driveways, and lanes. The brooms are powered with one or two hydraulic motors. An optional water spray can be added for dust control.

Forks are a common attachment. They can be either attached to the lift arms in place of the bucket or attached to the bucket. *Figure 19* shows a crawler loader with both fork and bucket attachments. With forks, the loader can be used to pick up and move large pieces of rigid material such as lumber, pallets, pipe, or concrete block.

205F17.EPS

Figure 17 ◆ Loader with log fork attachment.

205F18.EPS

Figure 18 ◆ Broom attachment.

Figure 19 ◆ Crawler loader with fork and bucket attachments.

The rake attachment is used for clearing brush. It rakes up brush and other vegetation. It can also lift up and dump it into a haul unit or burn pile.

Some attachments are made for special applications by various manufacturers. The function and operation of an attachment is described in the manufacturer's literature and the operator's manual. Always review the instructions before using any equipment or attachments.

CAUTION
Only use attachments that are compatible with the machine. Using non-standard attachments could damage the equipment or cause injury.

New attachments for specialty applications are always being developed. The Ironwolf Crusher is designed for use in site preparation and excavation projects. The patented cutter drum, housing, and an auxiliary power unit are designed to be mounted on wheel and track loaders. It allows the operator to process rocky soil, grind rocks, and remove roots and stumps.

The cutter drum comes in three sizes: five (5), eight (8), or ten (10) foot wide. The drum diameter options are 40 and 48 inches, enabling cutting depths up to 12 and 16 inches, respectively. The Ironwolf Crusher is used for a variety of construction, reclamation, mining, and forestry applications.

3.0.0 ◆ SAFETY GUIDELINES

Safety can be divided into three areas: safety of the operator, safety of others, and safety of the equipment. You, as the operator, are responsible for performing your work safely, protecting the public and your co-workers from harm, and protecting your equipment from damage. Knowing your equipment and thinking safety are the easiest ways to keep a job site safe.

3.1.0 Operator Safety

Nobody wants to have an accident or be hurt. There are a number of things you can do to protect yourself and those around you from getting hurt on the job. Be alert and avoid accidents.

Know and follow your employer's safety rules. Your employer or supervisor will provide you with the requirements for proper dress and safety equipment. The following are recommended safety procedures for all occasions:

- Clean steps, grab irons, and the operator's compartment.
- Mount and dismount the equipment carefully using three-point contact and facing the machine.
- Wear a hard hat, safety glasses, safety boots, and gloves when operating the equipment (*Figure 20*).
- Do not wear loose clothing or jewelry that could catch on controls or moving parts.
- Keep the windshield, windows, and mirrors clean at all times.
- Never operate equipment under the influence of alcohol or drugs.
- Never smoke while refueling or checking batteries or fluid.
- Never remove protective guards or panels.
- Never attempt to search for leaks with your bare hands. Hydraulic and cooling systems operate at high pressure. Fluids under high pressure can cause serious injury.
- Always lower the bucket or other attachment to the ground before performing any service or when leaving the loader unattended, and turn off the machine.

CAUTION
Getting in and out of equipment can be dangerous. Always face the machine and maintain three points of contact when you are mounting and dismounting. That means you should have three out of four of your hands and feet on the equipment. That can be two hands and one foot or one hand and two feet.

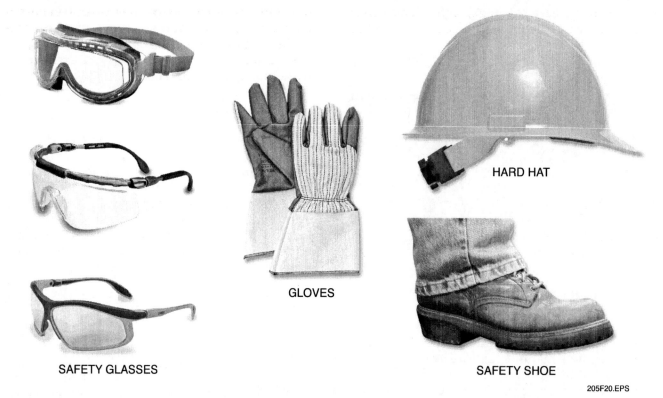

SAFETY GLASSES

GLOVES

HARD HAT

SAFETY SHOE

205F20.EPS

Figure 20 ◆ Wear appropriate personal protection equipment.

3.2.0 Safety of Co-Workers and the Public

You are not only responsible for your personal safety, but also for the safety of other people who may be working around you. Sometimes, you may be working in areas that are very close to pedestrians or motor vehicles. In these areas, take time to be aware of everything going on around you. Remember, it is often difficult to hear when operating a loader. Use a spotter and a radio in crowded conditions.

The main safety points when working around other people include the following:

- Walk around the equipment to make sure that everyone is clear of the equipment before starting and moving it.
- Before beginning work in a new area, take a walk to locate any cliffs, steep banks, holes, power or gas lines, or other obstacles that could cause a hazard to safe operation.
- If you are working in traffic, find out what warning devices are required. Make sure you know the rules and the meaning of all flags, hand signals, signs, and markers.
- Maintain a clear view in all directions. Do not carry any equipment or materials that obstructs your view.
- Always look before changing directions.

- Never allow riders on the loader.
- Keep ground personnel in visual contact.

3.3.0 Equipment Safety

Your loader has been designed with certain safety features to protect you as well as the equipment. For example, it has guards, canopies, shields, rollover protection, and seat belts. Know your equipment's safety devices and be sure they are in working order.

Use to the following guidelines to keep your equipment in good working order:

- Perform prestart inspection and lubrication daily (*Figure 21*).
- Look and listen to make sure the equipment is functioning normally. Stop if it is malfunctioning. Correct or report trouble immediately.
- Always travel with the bucket low to the ground.
- Never exceed the manufacturer's limits for speed, lifting, or operating on inclines.
- Always lower the bucket, engage the parking brake, turn off the engine, and secure the controls before leaving the equipment.
- Never park on an incline.
- Use a rigid-type coupler when towing the loader.

- Make sure clearance flags, lights, and other required warnings are on the equipment when roading or moving.
- When loading, transporting, and unloading equipment, know and follow the manufacturer's recommendations for loading, unloading, and tie-down.

The basic rule is know your equipment. Learn the purpose and use of all gauges and controls as well as your equipment's limitations. Never operate your machine if it is not in good working order. Some basic safety rules of operation include the following:

- Do not operate the loader from any position other than the operator's seat.
- Do not coast. Neutral is for standing only.
- Maintain control when going downhill. Do not shift gears and keep the load low (*Figure 22*).
- Whenever possible, avoid obstacles such as rocks, fallen trees, curbs, or ditches.
- If you must cross over an obstacle, reduce your speed and approach at an angle to reduce the impact on the equipment and yourself.
- Use caution when undercutting high banks, backfilling new walls, and removing trees.
- If you use starting fluid, be sure the engine is cranking before spraying the fluid into the intake.

4.0.0 ◆ BASIC PREVENTIVE MAINTENANCE

Preventive maintenance is an organized effort to regularly perform periodic lubrication and other service work in order to avoid poor performance and breakdowns at critical times. By performing preventive maintenance on the loader, you keep it operating efficiently and safely and avoid the possibility of costly failures in the future.

Preventive maintenance of equipment is essential and is not that difficult if you have the right tools and equipment. The leading cause of premature equipment failure is putting things off. Preventive maintenance should become a habit, performed on a regular basis.

4.1.0 Daily Inspection Checks

Maintenance time intervals for most machines are established by the Society of Automotive Engineers (SAE) and adopted by most equipment manufacturers. Instructions for preventive maintenance are usually in the operator's manual for each piece of equipment. Typical time intervals are: 10 hours (daily); 50 hours (weekly); 100 hours, 250 hours, 500 hours, and 1,000 hours. The operator's manual will also include lists of inspections and servicing activities required for each time interval. *Appendix A* is an example of typical periodic maintenance requirements.

The first thing you must do each day before beginning work is to conduct your daily inspection. This should be done before starting the engine. The daily inspection will identify any potential problems that could cause a breakdown and indicate whether the machine can be operated. The equipment should be inspected before, during, and after operation.

The daily inspection is often called a walk-around. The operator should walk completely around the machine checking various items. Items to be checked and serviced on a daily inspection are as follows:

Figure 21 ◆ Check all fluids daily.

205F21.EPS

Figure 22 ◆ Be careful when moving downhill.

205F22.EPS

NOTE

If a leak is observed, located the source of the leak and fix it. If leaks are suspected, check fluid levels more frequently.

WARNING!

Do not check for leaks with your bare hands. Use cardboard or another device. Pressurized fluids can cause severe injuries to unprotected skin. Long term exposure can cause cancer or other chronic diseases.

- Inspect the cooling system for leaks and faulty hoses.
- Inspect the engine compartment and remove any **debris**. Clean access doors.
- Inspect the engine for obvious damage.
- Check the condition and adjustment of drive belts on the engine.
- Inspect tires for damage and replace any missing valve caps.
- Inspect the axles, differentials, wheel brakes, and transmission for leaks.
- Inspect the hydraulic system for leaks, faulty hoses, or loose clamps.
- Inspect all attachments and the linkage for wear and damage. Check this hardware to make sure there is no damage that would create unsafe operating conditions or cause an equipment breakdown. Make sure the bucket is not cracked or broken. If you have an implement such as a roller attached, make sure the hitch is properly set and that the safety pin is in place.
- Inspect and clean steps, walkways, and handholds.
- Inspect the ROPS for obvious damage.
- Inspect the lights and replace any broken bulbs or lenses.
- Inspect the operator's compartment and remove any trash.
- Inspect the windows for visibility and clean them if they are fouled.
- Adjust the mirrors.
- Test the backup alarm and horn. Put equipment in reverse gear and listen for backup alarm.

Some manufacturers require that daily maintenance be performed on specific parts. These parts are usually those that are the most exposed to dirt or dust and may malfunction if not cleaned or serviced.

For example, your service manual may recommend lubricating specific bearings every 10 hours of operation, or always cleaning the air filter before starting the engine. *Figure 23A* shows the location of points to be inspected on a wheel-type loader. *Figure 23B* shows similar areas to be checked on a crawler-type loader.

Before beginning operation the following fluid levels need to be checked and topped off:

- *Battery* – Check the battery cable connections.
- *Crankcase oil* – Check the crankcase oil level and make sure it is in the safe operating range. (*Figure 24*)
- *Cooling system* – Check the coolant level and make sure it is at the level specified in the operating manual. (*Figure 25*)
- *Fuel level* – Check the fuel level in the fuel tank(s). Do this manually with the aid of the fuel dip stick or marking vial. Do not rely on the fuel gauge at this point. Check the fuel pump sediment bowl if one is fitted on the machine.
- *Hydraulic fluid* – Check the hydraulic fluid level in the reservoir. (*Figure 26*)
- *Transmission fluid* – Measure the level of the transmission fluid to make sure it is in the operating range. (*Figure 27*)
- *Pivot points* – Clean and lubricate all pivot points. (*Figure 28*)

NOTE

In some cold weather it is sometimes preferable to lubricate pivot points at the end of a work shift when the grease is warm. Warm the grease gun before using it for better grease penetration.

For crawler-type loaders there are several other items to check in the same manner as required for bulldozers. These include the following:

- *Idlers* – These components keep tension on the tracks to prevent them from coming off the sprockets. Check for broken or cracked rollers, tension springs, or other damage.
- *Sprockets* – The teeth on a worn sprocket will not be well-defined or may even be broken. This may cause the loader to throw a track and badly damage the track, drive shaft, and bearings.
- *Tracks* – Tracks have a habit of wearing out very fast because they are the one part of the equipment that is constantly in use. Check for broken or missing shoes and bolts. Also, inspect the **grousers** to see if they will need replacing soon. If the grousers are worn down too far, you will not get good traction.

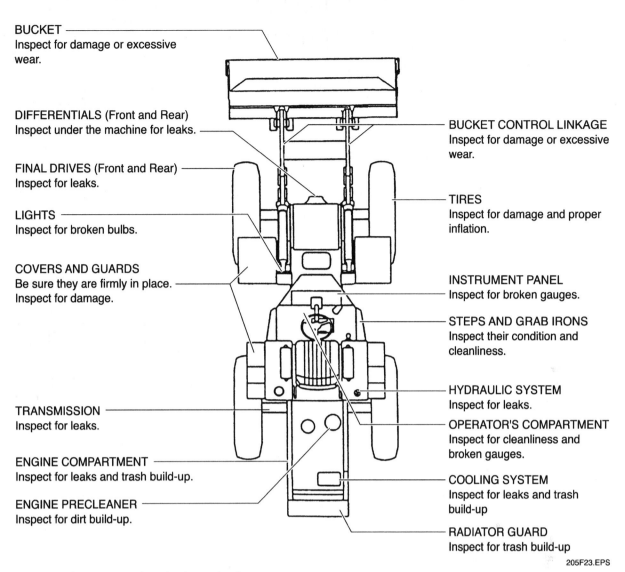

BUCKET —
Inspect for damage or excessive wear.

DIFFERENTIALS (Front and Rear)
Inspect under the machine for leaks.

FINAL DRIVES (Front and Rear)
Inspect for leaks.

LIGHTS —
Inspect for broken bulbs.

COVERS AND GUARDS
Be sure they are firmly in place.
Inspect for damage.

TRANSMISSION —
Inspect for leaks.

ENGINE COMPARTMENT —
Inspect for leaks and trash build-up.

ENGINE PRECLEANER —
Inspect for dirt build-up.

BUCKET CONTROL LINKAGE —
Inspect for damage or excessive wear.

TIRES —
Inspect for damage and proper inflation.

INSTRUMENT PANEL —
Inspect for broken gauges.

STEPS AND GRAB IRONS —
Inspect their condition and cleanliness.

HYDRAULIC SYSTEM —
Inspect for leaks.

OPERATOR'S COMPARTMENT —
Inspect for cleanliness and broken gauges.

COOLING SYSTEM —
Inspect for leaks and trash build-up

RADIATOR GUARD —
Inspect for trash build-up

205F23.EPS

Figure 23 ◆ Daily inspection for wheel-type loaders.

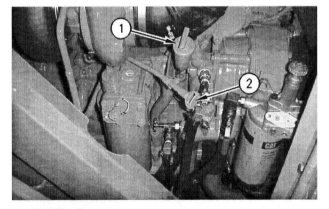

1 - Oil Filler Cap
2 - Engine Oil Dipstick

205F24.EPS

Figure 24 ◆ Check engine oil level.

205F25.EPS

Figure 25 ◆ Check coolant system.

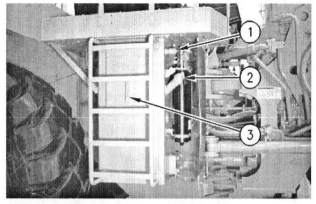

1 - Hydraulic Tank Breaker Relief Valve
2 - Hydraulic Oil Filler Cap
3 - Hydraulic Oil Sight Glass

205F26.EPS

Figure 26 ◆ Check hydraulic fluid.

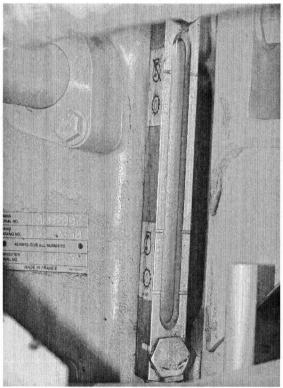

205F27.EPS

Figure 27 ◆ Check transmission fluid.

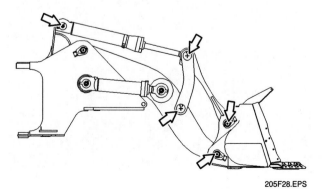

205F28.EPS

Figure 28 ◆ Lubricate pivot points.

The operator's manual usually has detailed instructions for performing periodic maintenance. If you find any problems with your machine that you are not authorized to fix, inform the foreman or field mechanic for correction before beginning operation.

4.2.0 Servicing a Loader

When servicing a loader, follow the manufacturer's recommendations and service chart. Any special servicing for a particular piece of equipment will be highlighted in the manual. Normally, the service chart recommends specific intervals, based on hours of run time, for such things as changing oil, filters, and coolant. Hydraulic fluids should be changed whenever they become dirty or break down due to overheating. Continuous and hard operation of the hydraulic system can heat the hydraulic fluid to the boiling point and cause it to break down. Filters should also be replaced during regular servicing.

Before performing maintenance procedures, always complete the following steps:

Step 1 Park the machine on a level surface to ensure that fluid levels are indicated correctly.

Step 2 Lower all equipment to the ground. Operate the controls to relieve hydraulic pressure.

Step 3 Engage the parking brake.

Step 4 Lock the transmission in neutral.

NOTE

Computers are sometimes used to diagnose problems with the loader. These operations are performed by specially trained technicians.

4.3.0 Preventive Maintenance Records

Accurate, up-to-date maintenance records are essential for knowing the history of your equipment. Each machine should have a record that describes any inspection or service that is to be performed and the corresponding time intervals. Typically, an operator's manual and some sort of inspection sheet are kept with the equipment at all times.

5.0.0 ◆ BASIC OPERATION

Operation of a loader requires constant attention to the controls and the surrounding environment.

Operators must plan their work and movements in advance and be alert to the other operations going on around the equipment. Do not take risks.

If there is doubt concerning the capability of the machine to perform some work, stop the equipment and investigate the situation. Discuss it with the foreman or engineer in charge. Whether it is a slope that may be too steep, or an area that looks too unstable to work, you should know the limitations of the equipment. Decide how to do the job before starting operations. Once you get in the middle of something it may be too late, causing extra work for others or possibly an unsafe condition.

5.1.0 Suggestions for Effective Loader Operation

The following suggestions can help improve operating efficiency:

- Keep equipment clean. Make sure the cab is clean so nothing affects the operation of the controls.
- Calculate and plan operations before starting.
- Do not move the machine until the brake air system reaches 100 pounds per square inch (psi).
- Set up the work cycle so that it will be as short as possible.
- Observe all safety rules and regulations.
- **Spot** trucks properly.
- Level off the work area if necessary.
- Keep transport distances as short as possible.
- Keep the transmission range lever in low range during loading operations and when transporting any load.
- Keep the wind to your back when dumping into a truck.

While traveling, maintain good visibility and loader stability by carrying the bucket low, approximately 15 inches above the ground. When loading trucks from a **stockpile,** as shown in *Figure 29,* use the wait time to clean and level the work area. Cleanup of spillage around the stockpile will smooth loader cycles and lessen operator fatigue. Maintain traction when loading the bucket by not putting excessive down pressure on the bucket. Excessive down pressure forces the front wheels or the front portion of the tracks to raise up off the ground.

To control dumping, move the bucket tilt control lever to the dump position. Repeat this operation until the bucket is empty. When handling dusty material, try to dump the material with the wind to your back. This will keep dust from entering the engine compartment and the operator's cab.

Use the appropriate size bucket. Check the work to be done and choose the right bucket for the job. Using the wrong bucket increases the wear on the bucket. Using the wrong bucket also increases the potential to exceed the machine's operating limits, which reduces its service life. *Appendix B* shows a typical chart for the rated volume and loads for various types of buckets.

5.2.0 Preparing to Work

Preparing to work involves getting yourself organized in the cab and starting your machine.

Mount your equipment using the grab rails and foot rests. Adjust the seat to a comfortable operating position. The seat should be adjusted to allow full brake pedal travel with your back against the seat back. This will permit the application of maximum force on the brake pedals. Make sure you can see clearly and reach all the controls.

NOTE

Always maintain three points of contact when mounting equipment. Keep grab rails and foot rests clear of dirt, mud, grease, ice, and snow.

WARNING!

OSHA requires that approved seat belts and Rollover Protective Structure (ROPS) be installed on virtually all heavy equipment. Old equipment must be retrofitted. Do not use heavy equipment that is not equipped with these safety devices.

205F29.EPS

Figure 29 ◆ Loading a truck from a stockpile.

Operator stations vary depending on the manufacturer, size, and age of the equipment. However, all stations have gauges, indicators and switches, levers, and pedals. Gauges tell you the status of critical items such as water temperature, oil pressure, battery voltage, and fuel level. Indicators alert the operator to low oil pressure, engine overheating, clogged air and oil filters, and electrical system malfunctions. Switches are for activating the glow plugs, starting the engine, and turning accessories, such as lights, on and off. Typical instruments and controls were described previously. Review the operator's manual so that you know the specifics of the machine you will be operating.

The startup and shutdown of an engine is very important. Proper startup lengthens the life of the engine and other components. A slow warm-up is essential for proper operation of the machine under load. Similarly, the shutdown of the machine is critical because of all the hot fluids circulating through the system. These fluids must cool so that they can cool the metal parts before the engine is switched off.

5.2.1 Startup

There may be specific startup procedures for the piece of equipment you are operating, but in general, the startup procedure should follow this sequence:

Step 1   Be sure all controls are in neutral.

Step 2   Engage the parking brake (*Figure 30*). This is done with either a lever or a knob, depending on the loader make and model.

Step 3   Place the ignition switch in the ON position.

 NOTE
When the parking brake is engaged an indicator light on the dash will light up or flash. If it does not, stop and correct the problem before operating the equipment.

Step 4   Depress the throttle control (*Figure 31*) approximately one-third the total distance.

Step 5   Press the starter button.

 NOTE
Never operate the starter for more than 30 seconds at a time. If the engine fails to start, wait two to five minutes before cranking again.

Step 6   Warm up the engine for at least five minutes.

 NOTE
Warm up the machine for a longer period in colder temperatures.

Step 7   Check all the gauges and instruments to make sure they are working properly.

Step 8   Shift the gears to low range.

Step 9   Release the parking brake and depress the service brakes.

Step 10   Check all the controls for proper operation.

Step 11   Check service brakes for proper operation.

Step 12   Check the steering for proper operation.

Step 13   Manipulate the controls to be sure all components are operating properly.

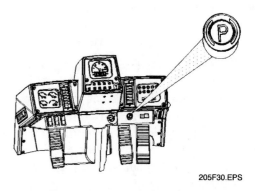

205F30.EPS

Figure 30 ◆ Parking brake knob on Caterpillar 988.

205F31.EPS

Figure 31 ◆ Throttle control pedal on Caterpillar 988.

Step 14 Shift the gears to neutral and lock.

Step 15 Reset the brake.

Step 16 Make a final visual check for leaks, unusual noises, or vibrations.

If the machine you are using has a diesel engine, there are special procedures for starting the engine in cold temperatures. Many diesel engines have glow plugs that heat up the engine for ignition. See *Table 1* for recommended glow plug heating times.

To use the glow plug heater, push in and turn the heat start switch or depress the glow plug button for the indicated time. After holding that position for the indicated time, push in and turn the start switch to start. Some units have a small sight glass to observe the glow plugs. When they stop glowing, the engine is ready to START. Some units are also equipped with ether starting aids. Review the operator's manual so that you fully understand the procedures for using these aids.

As soon as the engine starts, release the starter switch and adjust the engine speed to approximately half throttle. Let the engine warm up to operating temperature before moving the loader.

Let the machine warm up for a longer period of time when it is cold. If the temperature is at or above freezing, 32°F (0°C), let the engine warm up for 15 minutes. If the temperature is between 32°F (0°C) and 0°F (–18°C), warm the engine for 30 minutes. If the temperature is less than 0°F (–18°C) or hydraulic operations are sluggish, additional time is needed. Follow the manufacturer's procedure for cold starting.

5.2.2 Checking Gauges and Indicators

Keep the engine speed low until the oil pressure registers. The oil pressure light should come on briefly and then go out. If the oil pressure light does not turn off within 10 seconds, stop the engine, investigate, and correct the problem.

Check the other gauges and indicators to see that the engine is operating normally. Check that the water temperature, ammeter, and oil pressure indicators are in the normal range. If there are any problems, shut down the machine and investigate or get a mechanic to look at the problem.

Table 1 Recommended Glow Plug Heating Times

Starting Aid Chart	
Starting Temperature	**Glow Plug Heat Time**
Above 60°F (16°C)	None
60°F (16°C) to 32°F (0°C)	1 Minute
32°F (0°C) to 0°F (-18°C)	2 Minutes
Below 0°F (-18°C)	3 Minutes

205T01.EPS

5.2.3 Shutdown

Shutdown should also follow a specific procedure. Proper shutdown will reduce engine wear and possible damage to the machine.

Step 1 Find a dry, level spot to park the loader. Stop the loader by decreasing the engine speed, depressing the clutch, and placing the direction lever in neutral. Depress the service brakes and bring the machine to a full stop.

NOTE

If you must park on an incline, block the tires.

Step 2 Place the transmission in neutral and engage the brake lock. Lock out the controls if the machine has a control lock feature (*Figure 32*).

Step 3 Lower the bucket so that it rests on the ground. If you have any other attachment, be sure it is also lowered.

Step 4 Place the speed control in low idle and let the engine run for approximately five minutes.

205F32.EPS

Figure 32 ◆ Engage steering and transmission control lock.

Step 5 Turn the start switch off.

Step 6 Release hydraulic pressure by moving the control levers until all movement stops.

Step 7 Turn the disconnect switch to OFF and remove the key.

Some machines have additional disconnect switches for added security. The Caterpillar 988 has a battery disconnect. Other machines have a fuel shutoff switch. These controls provide an additional safety feature and deter unauthorized users. Always engage any additional security systems when leaving the loader unattended.

5.3.0 Basic Maneuvering

To maneuver the loader you must be able to move forward, backward, and turn. Basic maneuvering was covered in detail in *Heavy Equipment Level One*. This section serves as a review.

On wheel-type loaders, direction is controlled using a steering wheel or a joystick. Crawler-type loaders run on tracks that are controlled by foot pedals and levers or joysticks. For loaders with joystick steering controls, it will take some practice to coordinate the control of the hand levers and foot pedals to steer the machine, while at the same time operating the bucket or other attachment. This section will highlight joystick steering as it may be unfamiliar.

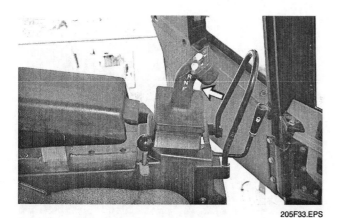

205F33.EPS

Figure 33 ◆ Transmission controls used with joystick steering.

5.3.1 Moving Forward

The first basic maneuver is learning to drive forward. To move forward, follow these steps:

Step 1 Before starting to move, raise the bucket assembly by pulling the boom and bucket control lever. Raise the bucket to about 15 inches above the ground. This is the travel position.

Step 2 Put the shift lever in low forward (*Figure 33*). Release the parking brake, and press the accelerator pedal to start the loader moving.

Step 3 Steer the machine using a steering wheel or joystick for wheel-type loaders or levers and pedals or joystick for crawler-type loaders.

Step 4 Once underway, shift to a higher gear to drive on the road. To shift from a lower to a higher gear, move the shift lever forward. Remember, high gear is used only for traveling on the road.

NOTE

Always travel with the bucket tilted back and low to the ground (12 to 18 inches). This will provide better visibility.

5.3.2 Moving Backward

To back up or reverse direction, always come to a complete stop. Then move the shift lever (*Figure 34*) to reverse. Once in reverse gear, you can apply some acceleration and begin to move backwards.

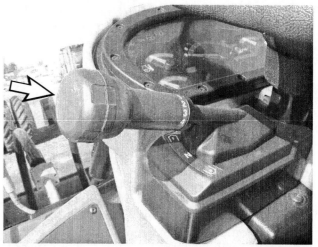

205F34.EPS

Figure 34 ◆ Transmission controls used with steering wheel.

For articulated loaders, it may take additional practice to be able to back in a straight line. Steering one of these machines is about the same as trying to back up a trailer attached to truck. If you begin to turn too sharply, stop and pull forward to straighten out before continuing to move backward.

5.3.3 Steering and Turning

How you steer a loader depends on whether it has a steering wheel or a joystick. Some wheel-type loaders will have a joystick for one-hand use. This allows the operator to keep the other hand on the bucket controls for simultaneous operations.

Steering wheels on wheel-type loaders operate in the same manner as steering wheels on cars and trucks. Moving the wheel to the right turns the loader to the right. Turning the wheel to the left moves the wheels to the left.

Some loaders use a joystick for steering instead of a wheel. The joystick may also incorporate a gear change lever. Moving the steering lever to the left steers the machine to the left (*Figure 35*). The further the steering lever is moved, the faster the machine steers to the left. Moving the lever to the right accomplishes the same action to the other side.

The turning radius of a rigid-frame wheel-type loader is greater than that of an articulated loader. Therefore, articulated equipment can be used in tighter work areas without having to pull forward and backward to be repositioned. *Figure 36* shows an example of an articulated loader with a turning capability of 40 degrees left and right.

Crawler-type loaders require the use of foot pedals and levers or a joystick to steer the machine. This allows the operator to maintain at least one hand on the bucket controls at all times. Crawler-type loaders are steered in the same manner as bulldozers.

5.3.4 Operating the Bucket

Generally, the bucket on a loader is attached to the front frame by two lift arms. This adds stability to the bucket and improves material handling. Loaders may be equipped with one or two control levers. These controls operate the bucket lift and bucket tilt. Joysticks have replaced the lever operation in some models. All movements for both bucket tilt and lift are incorporated into one joystick. Bucket controls were covered previously in this module.

A loader with one control lever can accomplish all the functions of a two-lever unit. This operation would be similar to the use of a joystick, except that the joystick may have additional controls in the form of buttons or switches. Additional controls perform functions such as load metering or operation of special attachments.

All loaders have some type of bucket level indicator. The purpose of the indicator is to show the operator the position of the bucket as it is being raised or lowered. This helps keep the bucket from being rolled back during a high lift. A roll-back can cause the contents of the bucket to spill out, causing injury to the operator and damage to the machine.

Mechanical bucket level indicators have several different designs. One type uses two pointers on the bucket links. The bucket is level to the ground when the two pointers are opposite each other. Another type has an indicator rod that travels back and forth inside a tube attached to the dump

205F35.EPS

Figure 35 ◆ Joystick steering.

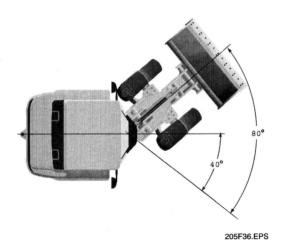

205F36.EPS

Figure 36 ◆ Articulated loader turn radius.

ram. When the end of the indicator rod is flush with the end of the tube, the bucket is level in any boom position.

Newer models of loaders may have an automatic self-leveling feature. The bucket is leveled throughout the hoist cycle. This reduces spillback and maintains better load control.

6.0.0 ◆ WORK ACTIVITIES

Operation of the loader is not as complex as some machines, but it does require constant attention and planning. By thinking ahead, you will have no trouble operating the loader. The basic work activities performed with either the wheel-type or crawler-type loader are described in this section.

> **NOTE**
>
> The controls on specific loaders may be different than those described in the procedures. Check your operator's manual for information about the controls and limitations of your equipment.

6.1.0 Basic Activities Performed By the Loader

The basic activities performed by a loader are all accomplished using one of the bucket attachments. Most loaders can perform several different activities with the same bucket. However, the bucket may not be the most effective or efficient for the job. Always check your operator's manual to make sure the bucket is designed for the intended purpose. Exceeding the model's design limits will reduce the service life of the equipment.

205F37.EPS

Figure 37 ◆ Position bucket before loading.

6.1.1 Loading Operations

Loaders are frequently used to load trucks, bins, and other containers. Usually, this loading is done by taking material from a stockpile. The procedure for carrying out a loading operation from a stockpile is as follows:

Step 1 Travel to the work area with the bucket in the travel position.

Step 2 Position the bucket parallel to and just skimming the ground, as shown in *Figure 37*.

Step 3 Drive the bucket straight into the stockpile.

Step 4 Adjust the controls and raise the bucket to fill it, as shown in *Figure 38*.

Step 5 Place the bucket and boom control levers on hold when the bucket is filled.

Step 6 Work the tilt control lever back and forth to move material to the back of the bucket. This is referred to as bumping. When the bucket is full, move the tilt control lever to the tilt back position.

Step 7 Shift the gears to reverse and back the loader away from the stockpile.

Step 8 Place the bucket in the travel position and move the loader to the truck.

Step 9 Center the loader with the truck bed and raise the bucket high enough to clear the side of the truck (*Figure 39*).

Step 10 Move the bucket over the truck bed and shift the bucket control lever forward to dump the bucket. At the same time, pull the boom control lever to the rear in order

205F38.EPS

Figure 38 ◆ Filling loader bucket.

to retract the bucket. *Figure 40* shows the proper position for dumping from a standard bucket.

Step 11 Pull the bucket control lever to retract the empty bucket and back the loader away from the truck as soon as the bucket is empty.

Step 12 Lower the bucket to the travel position and return to the stockpile.

Step 13 Repeat the cycle until the truck is loaded.

As the truck fills, the material will need to be pushed across the truck bed to even the load. As the leading edge of the bucket passes the sideboard of the truck, roll the bucket down quickly. Dump the material in the middle of the bed. Back up and push the load across the truck as the bucket is raised. By raising the bucket and backing up slowly, the material will be distributed evenly across the bed.

205F39.EPS

Figure 39 ◆ Center the bucket over the truck bed.

205F40.EPS

Figure 40 ◆ Proper dumping position.

Loading material into a truck with a loader requires that the operator have good reflexes and distance judgment. The loader must be placed close to the side of the truck in order to get the bucket positioned to dump properly.

There are two main points where accidental contact is the most common. The first area of contact is between the bucket and the side of the truck as indicated by arrow A in *Figure 41*. Either the operator has misjudged the height of the truck bed or he has approached too quickly, not allowing sufficient time to raise the bucket before getting to the truck. The second contact point is between the front of the loader and the side of the truck body as indicated by arrow B in *Figure 41*. This is due to the operator misjudging the distance between the front of the loader and the side of the truck. Contact from these situations can cause severe damage to both pieces of equipment.

When loading from a stockpile or bank, the placement of both truck and loader are variable and must be adjusted to local operating conditions. Conditions to be considered are the weight of material, the gradient of the loading area, traction, and the turning capability of the loader. If the loader works too close to the truck, it will be necessary to pause during each cycle for the bucket to clear the side of the truck. If the loader works too far from the truck, the cycle will be excessively long, with a resulting waste of time. The operator, by experience, must determine the most efficient arrangement for the particular operation and direct the trucks accordingly.

While there are many ways to maneuver a loader, the two most common patterns for a truck loading operation are the I-pattern and the Y-pattern.

For the I-pattern, both the loader and the dump truck move in only a straight line, backward and forward (*Figure 42*). This is a good method for

205F41.EPS

Figure 41 ◆ Loader contact with a truck.

small, cramped areas. The loader fills the bucket and backs approximately 20 feet away from the pile. The dump truck backs up between the loader and the pile. The loader dumps the bucket into the truck. The truck moves out of the way and the cycle repeats.

To perform this maneuver, position the loader so you are loading from the driver's side of the truck. That way, you have eye contact with the driver. Fill the bucket, as shown in *Figure 42A*. Back far enough away from the pile. Signal the truck driver with the horn. The truck will back to a predetermined position, as shown in *Figure 42B*. Move the loader forward and center it on the truck bed. Raise the bucket to clear the side of the truck and place it over the truck bed. Move the boom and bucket control lever to the left to dump the bucket. At the same time, raise the boom to make sure the bucket clears the truck bed. When the bucket is empty, move the boom and bucket control from side to side to shake out the last of the material. Back the loader away from the truck and signal the truck driver to move. When the truck is out of the way, lower the bucket and position the loader to return for another bucket of material.

The other loading pattern is the Y-pattern. This method is used when larger open areas are available. The dump truck remains stationary and as close as possible to the pile. The loader does all the moving in a Y-shaped pattern as shown in *Figure 43*.

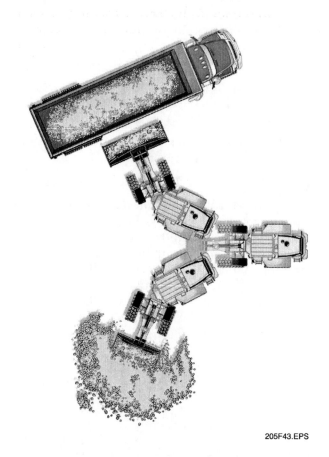

205F43.EPS

Figure 43 ◆ Y-pattern for loading.

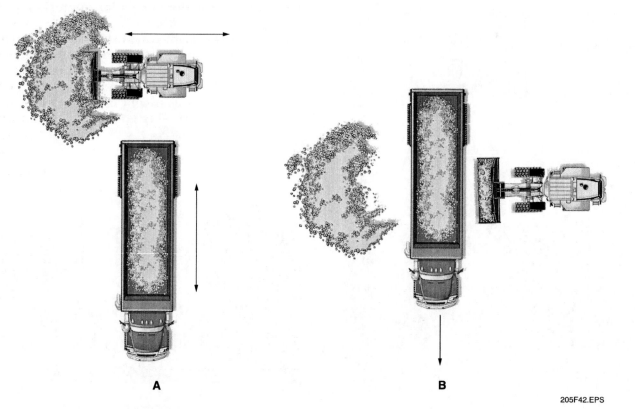

A

B

205F42.EPS

Figure 42 ◆ I-pattern for loading.

Position the dump truck or trucks so you have eye contact with the driver. Fill the bucket with material. While backing up, turn the loader to the right or left, depending on the position of the truck. Shift forward and turn the loader while approaching the truck slowly. Stop when you are lined up with the truck bed. Dump the bucket in the same way you would for the I-pattern. When the bucket is empty, back away from the truck while turning toward the pile. Drive forward into the pile to repeat the pattern. Repeat the cycle until the truck is full.

Some loaders have different configurations that change the dump height. Note the difference in dump clearance on the two loaders shown in *Figure 44*. Check the operator's manual for different configurations and dump height. Make sure that you have enough clearance for the project.

6.1.2 Leveling and Grading

Leveling and grading operations can be performed with the loader under most conditions. The multipurpose bucket is well suited for grading (*Figure 45*). The general purpose bucket is also used for this operation. Leveling can be done by tilting the bucket down and placing the cutting edge on the ground surface. Backing up with the bucket in this position will smooth out loose material.

For grading operations, the bottom of the bucket should be parallel to the ground surface. While maintaining this position, load material into the bucket and use the loaded bucket as the main **dozing** blade.

To perform a grading operation with the loader, follow these steps:

Step 1 Line up bucket and loader on the area to be graded.

Step 2 Move the boom control lever forward and position the bucket on the ground or at the desired height.

Step 3 Shift to low gear and press the accelerator to begin moving forward.

Step 4 Maintain a low steady speed and keep the bucket at a constant height. As high spots are encountered, they will be trimmed by the blade and loaded into the bucket (*Figure 46*).

205F44.EPS

Figure 44 ◆ Loader dump height.

205F45.EPS

Figure 45 ◆ Grading with the multipurpose bucket.

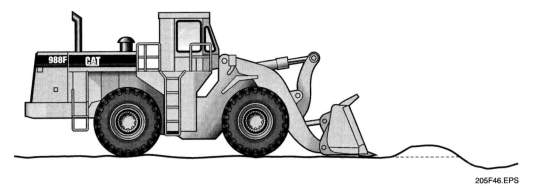

205F46.EPS

Figure 46 ◆ Trimming a high spot with a loader.

Step 5 When low spots are encountered, tilt the bucket forward and dump the material. Back up over the dumped material and smooth it out with the back of the blade.

Step 6 Reposition the loader and make another pass to the left or right of the area just graded.

6.1.3 Demolition

Demolition work is a good activity for a crawler-type loader. These loaders have the power and size to demolish structures such as walls and small buildings while at the same time being able to load and haul the **rubble** for removal from the area.

Use the following procedure to perform a demolition activity with a crawler-type loader:

Step 1 Travel to the building site with the bucket raised in the travel mode.

Step 2 Downshift into first gear and raise the boom to an overhead position.

Step 3 Approach the building cautiously and place the bucket on the building wall.

Step 4 Very slowly, move forward to push the top of the wall in.

Step 5 As the building begins to fall, select the reverse gear and back away.

Step 6 Continue placing the bucket on the wall until all of the building is down.

Step 7 Load the bucket and approach the dump area or haul unit.

Step 8 Raise the bucket to clear the side of the haul unit or truck.

Step 9 Dump the material slowly in order to avoid jarring or damaging the truck.

Step 10 Continue loading until all the debris is removed from the site.

Demolition work sometimes requires the breaking of large concrete slabs. If the slab is on a footing, you may have to excavate until you can slide the edge of the bucket under the corner of the slab. Do not attempt to lift slabs that may be too large for the loader to handle.

Begin lifting the slab until it breaks or the section is high enough to drop and break. Repeat this process around the edge of the slab. It will break it up into pieces small enough to load for hauling.

6.1.4 Excavating Operation

Using a loader to excavate a ground area has to be done with shallow cuts to keep traction on the machine and maintain production. This is different from loading the bucket from a bank or stockpile where the loader remains stationary while the bucket is raised against the material.

The procedure for excavating includes the following steps:

Step 1 Start the cut along the outer edge of the excavation by lowering the bucket to the ground and positioning for a straight digging angle.

Step 2 Align the bucket edge approximately 12 inches from the string lines or stakes when excavating foundations.

Step 3 Lower the bucket to the ground and move forward. Tilt the cutting edge down slightly until the bucket begins to dig into the soil.

Step 4 Move forward and load the bucket. Manipulate the bucket until it is full of material.

Step 5 Place the bucket and boom control levers on "hold" when the bucket is filled.

Step 6 Shift the gears to reverse and back the loader out of the material.

Step 7 Position the bucket in the travel position and proceed to the dump site. The travel position is about 12 to 18 inches above ground with the bucket tilted back, as shown in *Figure 47*.

Step 8 Move the bucket control lever forward to dump the bucket. At the same time, pull the boom control lever to the rear to raise the bucket.

Step 9 Pull the bucket control lever to retract the empty bucket. At same time, back the loader away from the pile.

Step 10 Position the bucket in the travel position and return to the excavation.

Step 11 Repeat the cycle until the excavation is completed.

Remember that you can not make deep cuts with the loader's bucket. Check the operator's manual for the recommended maximum allowable cut that can be made with your equipment.

6.1.5 Stockpiling

Stockpiling materials and loading from a stockpile are probably the most frequent tasks a loader operator will perform. There are different stockpile configurations; these include standard, ramp, and bin. Each configuration requires a slightly different method to build and maintain. The loading operation was described earlier. The procedures for maintaining the stockpiles and working the material are described in this section.

To make a standard stockpile as shown in *Figure 48*, use the basic bucket loading technique. Pick up the material from the bottom of the pile while moving forward and raising the bucket to the top of the pile. Raise the bucket high enough to clear the pile of material. Move forward so the raised bucket is over the top of the pile (*Figure 49*). Dump the bucket and allow the material to spread from the top of the pile downward.

Put the loader in reverse, back away from the pile, and start over. Work this way in a pattern all around the stockpile, moving the material continually toward the center. Always start at the point furthest from the center and work the area smooth.

A ramp stockpile is used to store large quantities of material in a small area with a larger loader. The shape of the stockpile will be a high, long, and narrow ramp.

To make a ramp stockpile, start the ramp close to the work area. Dump the material (*Figure 50*), then lower the bucket to approximately 15 inches above the ground. Use it to push the highest area of the pile forward. Dump the next load of material beside the first dump area and spread this material out the same way. Repeat the steps of dumping onto the pile and dragging the bucket back to build the ramp with a gradual slope. The base of the stockpile should be twice as wide as the loader so that when the ramp is complete, it will still be wide enough to support the loader.

205F47.EPS

Figure 47 ◆ Travel position with loaded bucket.

205F49.EPS

Figure 49 ◆ Lift materials over the top of the pile.

205F48.EPS

Figure 48 ◆ Standard stockpile.

205F50.EPS

Figure 50 ◆ Ramp stockpile.

Follow these tips to be more efficient in maintaining a ramp stockpile:

- After each dump on the ramp, push only the top half of the material off the end. The other half will add to the ramp.
- Do not run the loader in the same track all the time. Move the loader from one side of the ramp to the other to help compact the material.
- As the ramp grows, level it off and continue dumping until you reach the limit of your stockpile area.
- To remove material from a ramp stockpile, reverse the process used to make the ramp. As you get more material from the ramp, you reduce the size layer by layer.
- Begin loading the bucket at the point you last placed material.

A bin stockpile is contained in some sort of three-sided enclosure (*Figure 51*). The bin's walls are usually made of wood, concrete, or metal. The bin is usually rectangular in shape, with a floor of concrete or asphalt. Bin-type stockpiles provide good storage and keep different types of materials separated. They are usually found at materials manufacturing plants and maintenance facilities. Use extreme care when working with material in a bin. Bumping the side of the bin or pushing material into the wall can cause damage to the structure as well as the loader.

To fill a bin, place the material close to the wall at the farthest point from the entrance. Dump the material along this wall first. Next, dump material on top of this first row, allowing it to fall forward. As the back of the bin begins to fill, place the additional material in front of the last pile of material. Do not let any material run over the sides of the bin. Always enter the bin from the center of the opening and work out from the back of the bin.

To consolidate material or clean up the floor of the bin, pick up the material and place it on the top of the pile. Do not push material forward. This will put pressure on the walls and damage the bin.

6.1.6 Clearing Land

Clearing land involves the removal of vegetation, trees, and other obstructions above the surface. It may also include **stripping** away the top soil and doing some rough grading. Wheel-type loaders normally do not do this type of work because the rubber tires do not perform well under the conditions surrounding clearing and **grubbing**.

Crawler-type loaders are better suited for this activity because of their tracks. The tracks give them the stability and firmness to clear brush and small trees and move small boulders. Crawler-type loaders are able to maneuver the rough terrain and are free of problems associated with rubber-tired equipment.

6.1.7 Backfilling

Backfilling can be accomplished with either the wheel-type loader or crawler-type loader. The first method requires that materials be loaded from a stockpile, carried to the site, and dumped. A second method involves using spoils material located close to the area to be backfilled (*Figure 52*). If you are backfilling a trench, have someone observe the operation from a safe vantage point and direct the operation so that you do not come too close to the edge of the trench or collapse the side of the trench from the weight of the loader.

When backfilling from the spoils pile, use the bucket to push the material toward the area. Do not overfill the bucket and try to push too much material up against a structure. This may put too much pressure on the structure, causing it to fall over.

205F51.EPS

Figure 51 ◆ Bin stockpile.

205F52.EPS

Figure 52 ◆ Backfilling.

6.1.8 Excavating Work in Confined Areas

Sometimes, you will be required to work in small spaces or confined areas. This requires careful planning and execution.

When working in a confined area, the type of loader used will have a great impact on both the time it takes to do the job and the quality of the work. Different types of confined areas will require different approaches. The first type of confinement is below grade, where the space is limited and enclosed by walls or other vertical restrictions. The second type is at grade, where space is limited by some obstruction such as a trench, wall, building, or other equipment.

When working below grade, you must exercise extreme caution and not run into or damage any wall supports or shoring. This may cause a collapse of the embankment, burying people and equipment. Usually, loading out of a deep excavation such as a foundation or pit will require workers on the ground to direct truck traffic and spot equipment for loading. Always watch your spotter for signals when loading trucks in a confined area.

Loading trucks from within an excavation is basically the same as the standard loading operation. Loading will normally be from a stockpile of spoils placed in an open area by other equipment such as hydraulic excavators or bulldozers that are digging out the excavation (*Figure 53*). The loader is used for loading from the stockpile into the truck. When the excavation is complete, the loader can drive out on the remaining ramp. After all equipment has been removed, the ramp can be removed by equipment from outside the excavation.

Working in limited space between vertical structures may restrict the bucket height and turning ability of the loader. Articulated loaders are a better choice under these conditions because they have sharper turning and better maneuverability.

205F53.EPS

Figure 53 ◆ Excavation.

Make sure the view in all directions is unobstructed. If raising the bucket limits the view, have someone stand off to the side and assist by spotting your loading operation.

6.2.0 Working in Unstable Soils

Working in mud or unstable soils that will not support the loader can be aggravating and dangerous. This is a problem even for experienced operators.

When you see that you are about to enter a soft or wet area, go very slowly. If you feel the front of the loader start to settle, stop and back out immediately. That settling is the first indication that the ground is too soft to support the equipment. The engine will lug slightly and the front end of the loader will start to settle.

After backing out, examine how deep the wheels or tracks sank into the ground. If they sink deep enough that the material hits the bottom of the loader, the ground is too soft to work in a normal way.

To work in soft or unstable material, follow this procedure:

Step 1 Start from the edge and work forward slowly.

Step 2 Push the mud ahead of the bucket and be sure the ground below is firm.

Step 3 Don't try to move too much material on any one pass.

Keep the wheels or tracks from slipping and digging in.

Partially stable material can also be a hazard because you may drive in and out over relatively firm ground many times, while it slowly gets softer because the weight of the wheels or tracks pumps more water to the surface. If this happens, the wheels or tracks will sink a little more each time until the loader finally gets high-centered. Then, the machine must be pulled out with a **winch**.

To keep this from happening, do not run in the same track each time you enter or leave an area. Move over slightly in one direction or the other so the same tracks are not pushed deeper into the unstable material each time.

6.3.0 Using Special Attachments

In addition to the various types of buckets used with a loader, there are several special attachments that expand the loader's operational capability. The three main attachments are the multipurpose bucket, ripper, and forklift.

6.3.1 Multipurpose Bucket

A multipurpose bucket (*Figure 54*) has several movable parts that are hydraulically operated. The bucket can perform four basic functions. It can be used as a clamshell bucket, scraper, dozer, and regular loader bucket.

The clamshell can be used for removing stumps and large rocks, as well as picking up debris and brush. To do this, the operator must open the bowl and position the bucket over the material to be loaded, then lower the bucket and close the bowl to fill the bucket. Material can then be transported to the truck or stockpile for dumping. Use of the clamshell configuration gives the loader added height for dumping and better handling of sticky material such as wet clay soils.

Using the bucket as a scraper requires the operator to open the bowl and use the back side of the bucket to cut material. When the material is filling the back of the bucket, close the bowl over the material and raise the bucket for transporting to the dump site.

For use as a dozer or pusher, the operator must open the bowl fully and use the back of the bucket as a blade. Level cutting is maintained by the bucket lift control. The material can then be pushed to an area and stockpiled.

With the added height of the multipurpose bucket, loading trucks becomes easier because the boom can remain higher and stay away from the side of the truck. The clamshell configuration also makes it easier to dump sticky material because the bucket does not compact the material.

6.3.2 Ripper Operation

Ripping can be done with a crawler-type loader because of the good traction and heavier weight of the machine. Ripping aids in breaking up hard material for ease of loading. After the material is broken loose, it can be removed to the haul unit or stockpiled. Good traction is required to pull the ripper teeth through the material without spinning the tracks.

When ripping, the loader should be operated at a low travel speed and only one or two ripper teeth should be used. If the material breaks up easily, teeth can be added. As the ripping operation continues, keep some of the loose material on the ground to cushion the loader and improve traction. The basic procedure for using a ripper with a crawler-type loader is the same as with a bulldozer. Refer to the operator's manual for specific instructions on location and use of controls.

6.3.3 Forklift Attachment

Use a forklift attachment on the loader (*Figure 55*) to move palletized materials. Some models have forks that attach to a general purpose bucket, while other models attach directly to the lift arms. They come in several different sizes depending on their lifting weight and reach. Operating the loader as a forklift is basically the same procedure for all models.

The following steps are used to unload material from a flatbed truck:

Step 1   Position the loader at either side of the truck bed.

Step 2   Manipulate the control levers in order to obtain the appropriate fork height and angle.

Step 3   Drive the forks into the opening of the pallet or under the loose material. Use care not to damage any material.

205F54.EPS

Figure 54 ◆ Multipurpose bucket.

205F55.EPS

Figure 55 ◆ Forklift attachment for a loader.

Step 4 Manipulate the controls and brake pedal as required to lift the material slightly off the bed.

Step 5 Tilt the forks back to keep the pallet or other material from sliding off the front of the forks (*Figure 56*).

Step 6 Shift the gear to reverse and back the loader away from the truck.

Step 7 Lower the forks to the travel position and travel to the stockpile area.

> **CAUTION**
> Do not lower the forks with the boom and bucket control lever in the float position. This could cause equipment damage.

Step 8 Position the loader so that the material can be placed in the desired area.

Step 9 Lower the boom until the material is set on the required surface.

Step 10 Adjust the forks with the boom lever in order to relieve the pressure under the pallet.

Step 11 Back the loader away from the pallet.

Step 12 Repeat the cycle until the truck is unloaded.

Safe operation of the loader as a forklift requires extreme caution. Make sure you follow these safety requirements:

• Do not swing loads over the heads of workers. Make sure you have a clear area to maneuver.

• Do not allow workers to ride on the forks or the bucket.

• Using the forklift attachment usually requires operation in a confined area. Always know what clearance is available for maneuvering.

6.4.0 Transporting the Loader

If the loader needs to be transported from one job site to another, it may either be driven if it is a short distance, or loaded and hauled on a transporter.

When roading the loader from one site to another, make sure the necessary permits for traveling on a public road have been obtained. Flags must be mounted on the left and right corners of the machine. Lights and flashers should be switched on. Depending on the location, a scout vehicle may also be required. Because the wheel-type loader is top heavy and prone to bouncing, drive at a slow rate of speed, especially around corners and over rough terrain. Keep the bucket in the travel position of 12 to 18 inches above the ground.

If the equipment must be moved a long distance, it should be transported on a properly equipped trailer or other transport vehicle. Before beginning to load the equipment for transport, make sure the following tasks have been completed:

• Check the operator's manual to determine if the loaded equipment complies with height, width, and weight limitations for over-the-road hauling.

• Check the operator's manual to identify the correct tie-down points on the equipment (*Figure 57*).

• Be sure to get the proper permits, if required.

• Plan the loading operation so that the loading angle is at a minimum.

205F56.EPS

Figure 56 ◆ Tilt forks back to secure load.

205F57.EPS

Figure 57 ◆ Loader tie-down point.

Once these tasks are complete and the loading plan has been determined, carry out the following procedures:

Step 1 Position the trailer or transporting vehicle. Always block the wheels of the transporter after it is in position but before loading is started (*Figure 58*).

Step 2 Place the loader bucket in the travel position and drive the loader onto the transporter. Whether the loader is facing forward or backward will depend on the recommendation of the manufacturer. Most manufacturers recommend backing the loader onto the transporter.

Step 3 If the loader is articulated, connect the steering frame lock link to hold the front and rear frames together. The steering frame lock is secured by a pin when not in use (*Figure 59*). Remove the cotter pin and holding pin. Swing the lock to the transport position and secure it with a pin and cotter pin (*Figure 60*).

Step 4 Lower the bucket to the floor of the transporter.

Step 5 Move the transmission lever to neutral, engage the parking brake, and turn off the engine.

Step 6 Manipulate the bucket raise and tilt controls to remove any remaining hydraulic pressure.

Step 7 Remove the engine start key and place the fuel valve in the OFF position.

Step 8 Lock the door to the cab as well as any access covers. Attach any vandalism protection.

Step 9 Secure the machine with the proper tie-down equipment as specified by the manufacturer. Place chocks at the front and back of all four tires.

Step 10 Cover the exhaust and air intake openings with tape or a plastic cover.

Step 11 Place appropriate flags or markers on the equipment if needed for height and width restrictions.

Unloading the equipment from the transporter would be the reverse of the loading operation.

205F59.EPS

Figure 59 ◆ Steering frame lock in storage position.

205F60.EPS

Figure 60 ◆ Secure steering frame lock for transport.

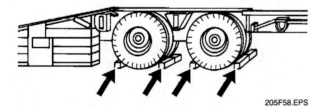

205F58.EPS

Figure 58 ◆ Block trailer wheels.

1. Loaders are primarily used for demolition.
   a. True
   b. False

2. The engine on a wheel-type loader is mounted _____.
   a. in the front
   b. on each wheel
   c. in the back
   d. underneath the bucket

3. An articulated loader _____.
   a. is mounted on a rigid frame
   b. is mounted on an articulator
   c. is mounted on a hinged frame
   d. has tracks instead of wheels

4. Large production loaders used in construction work have a capacity of _____.
   a. 2 to 4 cubic yards
   b. over 4 cubic yards
   c. over 10 cubic yards
   d. over 25 cubic yards

5. The lift and dump ram move the basket using _____ power.
   a. electrical
   b. hydraulic
   c. pneumatic
   d. internal combustion

6. The reduced rimpull function can only be activated when the machine is _____.
   a. switched off
   b. in first gear
   c. operating at high speeds
   d. in reverse gear

7. To lower the bucket, move the loader bucket controls to the _____ position.
   a. float
   b. tilt
   c. lower
   d. dump

8. Attachments are fixed to the lift arms with hydraulic quick-connects or _____.
   a. pins
   b. bolts
   c. locks
   d. steel plates

9. You can remove protective guards or plates on a loader if they are in your way.
   a. True
   b. False

10. If you cross over an obstacle with a loader _____.
    a. increase your speed so you will clear it
    b. shift to a higher gear for better traction
    c. approach at an angle to reduce the impact
    d. raise the load to balance the machine

11. On crawler-type loaders, grousers are attached to the _____.
    a. bucket
    b. tracks
    c. exhaust stack
    d. winch

12. When loading trucks from a stockpile, use the wait time to _____.
    a. do your daily inspections
    b. clean and level the work area
    c. check fluid levels
    d. read load charts

13. When loading a dump truck, keep the wind _____.
    a. at your back
    b. at your front
    c. at your right side
    d. in your face

14. When shutting down the loader, perform all of the following operations *except* _____.
    a. keep the hydraulics at operating pressure
    b. turn the disconnect switch to off
    c. lower the bucket to the ground
    d. decrease the engine speed to idle

15. When moving the loader, you should keep the bucket _____.
    a. in the float position
    b. as high as possible
    c. approximately 15 inches off the ground
    d. in any position that is good for you

16. Compared to an articulated loader, the turning radius of a rigid-frame loader is _____.
    a. less
    b. wider
    c. the same
    d. tighter

17. When using the loader to load from a stockpile it is best to _____.
    a. push the bucket into the top of the pile
    b. push the bucket into the center of the pile
    c. push the bucket into the bottom of the pile
    d. scoop from different areas

18. When loading a dump truck you should dump the load _____.
    a. on the closest side first
    b. in the middle of the bed
    c. on the far side of the bed
    d. as close to the back as possible

19. The Y-pattern for loading trucks is better for _____.
    a. areas where there is more room
    b. areas where there is less room
    c. confined areas
    d. muddy areas

20. When using a loader to smooth and level areas, you should _____.
    a. raise the bucket off the ground
    b. let the bucket float
    c. tilt the bucket down
    d. tilt the bucket up

21. When using the loader for demolition you should always _____.
    a. get a running start at the wall so you will have enough force to knock it over
    b. use the bucket to lift everything off the ground
    c. slowly move forward to push the top of the wall in
    d. slowly move forward and push the bottom of the wall in

22. Crawler loaders are better suited for clearing land than wheel loaders.
    a. True
    b. False

23. In unstable soils, if you feel the loader start to settle you should _____.
    a. go forward slowly
    b. go forward quickly
    c. lower the load
    d. back out immediately

24. The multipurpose bucket is operated _____.
    a. manually
    b. hydraulically
    c. electrically
    d. periodically

25. The loader is secured to the trailer by attaching tie-downs to the loader's _____.
    a. wheels
    b. drive shaft
    c. axle
    d. tie-down points

Summary

Loaders are used primarily for loading material from the ground or from a stockpile. They can also be used for digging, grading, hauling, and light clearing work. There are two basic types of loaders: the wheel-type loader and the crawler-type loader. There are two basic configurations for the wheel-type loader: a rigid-frame machine and an articulated machine.

All loaders have a large steel bucket mounted on two rigid arms extending from the frame. The bucket can dig, scoop, and curl upward in order to pick up large quantities of material. A loader can be equipped with a standard bucket, a multipurpose bucket, or other specialty buckets. Other attachments include forks, plows, rakes, and brooms.

Safety considerations when operating a loader include keeping the loader in good working condition, obeying all safety rules, being aware of other people and equipment in the same area where you are operating, and not taking chances. If you don't know the terrain or operation, stop the machine, get down, look around, and discuss the work with your supervisor or the resident engineer. When you start to dig, be sure you are in the right place and are clear of any underground utilities or other structures.

Use caution when loading into dump trucks. Do not use excessive speed when making turns with a loaded bucket. Always carry the bucket low to the ground.

The two common loading patterns are the I- and Y- patterns. They are named after the configuration the loader makes when performing the loading operation. With the I-pattern, the loader moves up to fill the bucket and back to allow the dump truck to move in between the loader and the stockpile. The Y-pattern is used when there is more room to maneuver. In this operation, the loader moves up to load the bucket, then moves back and turns right or left approximately 45 to 90 degrees to approach the truck waiting to the side of the loader. The loader creates a Y-pattern when making one complete loading cycle.

Use caution when working in unstable soils. If you feel the front of the loader start to settle, stop and back out immediately. This is an indication that the ground is too soft. When working in soft material, work from the edge and move forward slowly. Push any soft material ahead of the bucket and be sure the ground below is firm. Keep the wheels or tracks from slipping and digging in.

Notes

Trade Terms Introduced in This Module

Accessories: Attachments used to expand the use of a loader.

Blade: An attachment on the front end of a loader for scraping and pushing material.

Bucket: A U-shaped closed-end scoop that is attached to the front of the loader.

Debris: Rough broken bits of material such as stone, wood, glass, rubbish, and litter after demolition.

Dozing: Using a blade to scrape or excavate material and move it to another place.

Grouser: A ridge or cleat across a track that improves the track's grip on the ground.

Grubbing: Digging out roots and other buried material.

Rubble: Fragments of stone, brick, or rock that have broken apart from larger pieces.

Spot: To line up the haul unit so that it is in the proper position.

Stockpile: Material put into a pile and saved for future use.

Stripping: Removal of overburden or thin layers of pay material.

Winch: A power control unit that is used with wire rope to pull equipment and remove stumps.

Typical Periodic
Maintenance Requirements

Maintenance Interval Schedule

Note: All safety information, warnings, and instructions must be read and understood before you perform any operation or any maintenance procedure.

Before each consecutive interval is performed, all of the maintenance requirements from the previous interval must also be performed.

When Required

Automatic Lubrication Grease Tank - Fill
Battery - Recycle
Battery or Battery Cable - Inspect/Replace
Bucket Lift and Bucket Tilt Control - Inspect/Clean
Circuit Breakers - Reset
Engine Air Filter Primary Element - Clean/Replace
Engine Air Filter Secondary Element - Replace
Engine Air Precleaner - Clean
Ether Starting Aid Cylinder - Replace
Fuel System - Prime
Fuses - Replace
Lift Cylinder Pin Oil Level - Check
Loader Boom Pin Oil Level - Check
Oil Filter - Inspect
Radiator Core - Clean
Seat Side Rails - Adjust
Window Washer Reservoir - Fill
Window Wiper - Inspect/Replace

Every 10 Service Hours or Daily

Backup Alarm - Test
Bucket Cutting Edges - Inspect/Replace
Bucket Stops - Inspect/Replace
Bucket Tips - Inspect/Replace
Bucket Wear Plates - Inspect/Replace
Cooling System Coolant Level - Check
Engine Air Filter Service Indicator - Inspect
Engine Oil Level - Check
Hydraulic System Oil Level - Check
Loader Boom Pin and Lift Cylinder Pin - Inspect
Loader Pins and Bearings - Lubricate
Seat Belt - Inspect
Transmission Oil Level - Check
Walk-Around Inspection
Windows - Clean

Every 50 Service Hours or Weekly

Cab Air Filter - Clean/Replace
Fuel System Primary Filter (Water Separator) - Check/Drain
Fuel Tank Water and Sediment - Drain
Tire Inflation - Check

Every 100 Service Hours or 2 Weeks

Axle Oscillation Bearings - Lubricate
Steering Cylinder Bearings - Lubricate

Initial 250 Service Hours

Transmission Oil Filter - Replace

Every 250 Service Hours

Engine Oil Sample - Obtain

Every 250 Service Hours or Monthly

Battery - Clean
Belts - Inspect/Adjust/Replace
Brake Accumulator - Check
Braking System - Test
Differential and Final Drive Oil Level - Check
Engine Air Filter Service Indicator - Inspect/Replace
Engine Oil (High Speed) and Oil Filter - Change
Engine Oil and Filter - Change

Initial 500 Service Hours

Seat Side Rails - Adjust

Initial 500 Hours (for New Systems, Refilled Systems, and Converted Systems)

Cooling System Coolant Sample (Level 2) - Obtain

Every 500 Service Hours

Hydraulic System Oil Sample - Obtain
Transmission Oil Sample - Obtain

Every 500 Service Hours or 3 Months

Axle Oil Cooler Filter - Replace
Cooling System Coolant Sample (Level 1) - Obtain
Differential and Final Drive Oil Sample - Obtain
Engine Oil (High Speed) and Oil Filter - Change
Engine Oil and Filter - Change
Fuel System Primary Filter (Water Separator) Element - Replace
Fuel System Secondary Filter - Replace
Fuel Tank Cap and Strainer - Clean
Hydraulic System Oil Filter - Replace
Transmission Oil Filter - Replace

Every 1000 Service Hours or 6 Months

Articulation Bearings - Lubricate
Battery Hold-Down - Tighten
Case Drain Oil Filters - Replace
Drive Shaft Support Bearing - Lubricate

205A01.EPS

Rollover Protective Structure (ROPS) - Inspect
Transmission Oil - Change

Every 2000 Service Hours or 1 Year

Differential Thrust Pin Clearance - Check
Differential and Final Drive Oil - Change
Engine Crankcase Breather - Clean
Engine Valve Lash - Check
Engine Valve Rotators - Inspect
Hydraulic System Oil - Change
Hydraulic Tank Breaker Relief Valve - Clean
Lift Cylinder Pin Oil Level - Check
Loader Boom Pin Oil - Change
Refrigerant Dryer - Replace

Every 3000 Service Hours or 2 Years

Crankshaft Vibration Damper - Inspect
Engine Mounts - Inspect

Every 3 Years After Date of Installation or Every 5 Years After Date of Manufacture

Seat Belt - Replace

Every 4000 Service Hours or 2 Years

Hydraulic System Oil - Change

Every 4000 Service Hours or 2.5 Years

Electronic Unit Injector - Inspect/Adjust

Every 5000 Service Hours or 3 Years

Alternator - Inspect
Lift Cylinder Pin Oil - Change
Starting Motor - Inspect
Turbocharger - Inspect

Every 6000 Service Hours or 3 Years

Cooling System Coolant Extender (ELC) - Add

Every 6000 Service Hours or 6 Years

Cooling System Water Temperature Regulator - Replace
Engine Water Pump - Inspect

Every 12,000 Service Hours or 6 Years

Cooling System Coolant (ELC) - Change

205A02.EPS

Rated Loads for Various Types of Buckets

Rated Load				
Ground Engaging Tools	**Rated Volume**	**Rated Operating Load**	**Dump Clearance A**	**Reach B**
Spade Rock with Teeth and Segments (Standard Bucket)	6.4 m³ (8.4 yd³)	11340 kg (25000 lb)	4243 mm (167 inch)	1625 mm (64 inch)
Wide Spade Rock with Teeth and Segments	6.9 m³ (9 yd³)	11340 kg (25000 lb)	4246 mm (167 inch)	1659 mm (65 inch)
General Purpose with Bolt-On Cutting Edges	7 m³ (9.2 yd³)	11340 kg (25000 lb)	4300 mm (169 inch)	1849 mm (73 inch)
Straight Rock with Teeth and Segments	6.3 m³ (8.2 yd³)	11340 kg (25000 lb)	4449 mm (175 inch)	1680 mm (66 inch)
Rock Quarry (Heavy Duty) with Teeth and Segments	6.4 m³ (8.4 yd³)	11340 kg (25000 lb)	4252 mm (167 inch)	1627 mm (64 inch)
Mining (Heavy Duty) with Teeth and Segments	6.4 m³ (8.4 yd³)	11340 kg (25000 lb)	4242 mm (167 inch)	1623 mm (64 inch)
Spade Rock with Teeth and Segments	6.6 m³ (8.6 yd³)	11340 kg (25000 lb)	4246 mm (167 inch)	1666 mm (66 inch)

205A03.EPS

Additional Resources

This module is intended to be a thorough resource for task training. The following reference works are suggested for further study. These are optional materials for continued education rather than for task training.

Excavation and Grading Revised, 1994. Nick Capachi. Carlsbad, CA: Craftsman Book Company.

Moving the Earth, 1999. Fourth Edition. H.L. Nichols. New York, NY: McGraw-Hill.

Figure Credits

Reprinted courtesy of Caterpillar Inc., 205F01, 205F02, 205F04–205F19, 205F21, 205F24–205F26, 205F28–205F33, 205F35–205F41, 205F43–205F45, 205F47–205F56, 205F58, Appendices A and B

Deere & Company, 205F03, 205F22

Bacou-Dalloz, 205F20 (safety glasses)

North Safety Products USA, 205F20 (gloves)

Bullard, 205F20 (hard hat)

Topaz Publications, Inc., 205F20 (safety shoe), 205F27, 205F34, 205F57, 205F59, 205F60

Reprinted with permission of the Texas Engineering Extension Service, 205F23

The NCCER makes every effort to keep these textbooks up-to-date and free of technical errors. We appreciate your help in this process. If you have an idea for improving this textbook, or if you find an error, a typographical mistake, or an inaccuracy in NCCER's Contren® textbooks, please write us, using this form or a photocopy. Be sure to include the exact module number, page number, a detailed description, and the correction, if applicable. Your input will be brought to the attention of the Technical Review Committee. Thank you for your assistance.

Instructors – If you found that additional materials were necessary in order to teach this module effectively, please let us know so that we may include them in the Equipment/Materials list in the Annotated Instructor's Guide.

Write: Product Development and Revision
National Center for Construction Education and Research
P.O. Box 141104, Gainesville, FL 32614-1104

Fax: 352-334-0932

E-mail: curriculum@nccer.org

Craft _____ Module Name _____

Copyright Date _____ Module Number _____ Page Number(s) _____

Description _____

_____

_____

_____

(Optional) Correction _____

_____

_____

(Optional) Your Name and Address _____

_____

_____

22206-06

Forklifts

22206-06
Forklifts

Topics to be presented in this module include:

Overview

Forklifts were originally designed to lift and place palletized material. Today, forklifts have several attachments that expand their use, including booms and hooks, buckets, personnel lifts, and specialty forks. With the right attachments, forklifts can handle many aboveground tasks.

The key to safe operation is to understand the instruments and controls and the lifting capacity of the machine. Lifting heavy loads is dangerous. The operator must know how to select the right attachment, rig the load securely, and lift and place the load safely. Losing a load can cause significant property damage and personal injury. Plan your work and do not exceed the capacity of the forklift. When operated safely, a forklift can save many back-breaking hours of material movement.

Objectives

When you have completed this module, you will be able to do the following:

1. Describe the uses of a forklift.
2. Identify the components and controls on a typical forklift.
3. Explain safety rules for operating a forklift.
4. Perform prestart inspection and maintenance procedures.
5. Start, warm up, and shut down a forklift.
6. Perform basic maneuvers with a forklift.
7. Perform basic lifting operations with a forklift.
8. Describe the attachments used on forklifts.

Trade Terms

Articulated steering
Circle steering
Crab steering
Four-wheel steering
Oblique steering
Powered industrial trucks
Telehandler

Required Trainee Materials

1. Pencil and paper
2. Appropriate personal protective equipment

Prerequisites

Before you begin this module, it is recommended that you successfully complete *Core Curriculum; Heavy Equipment Operations Level One; Heavy Equipment Operations Level Two*, Modules 22201-06 through 22205-06.

This course map shows all of the modules in the second level of the *Heavy Equipment Operations* curriculum. The suggested training order begins at the bottom and proceeds up. Skill levels increase as you advance on the course map. The local Training Program Sponsor may adjust the training order.

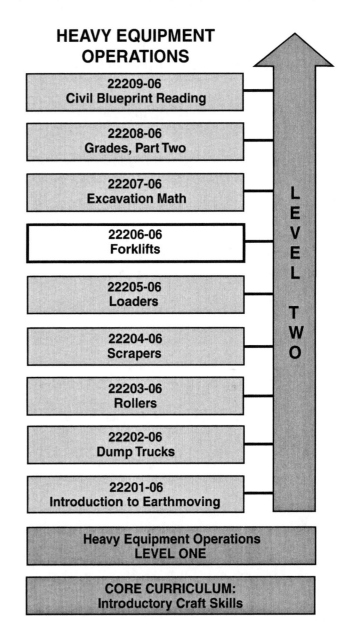

HEAVY EQUIPMENT OPERATIONS

22209-06
Civil Blueprint Reading

22208-06
Grades, Part Two

22207-06
Excavation Math

22206-06
Forklifts

22205-06
Loaders

22204-06
Scrapers

22203-06
Rollers

22202-06
Dump Trucks

22201-06
Introduction to Earthmoving

LEVEL TWO

Heavy Equipment Operations
LEVEL ONE

CORE CURRICULUM:
Introductory Craft Skills

206CMAP.EPS

1.0.0 ◆ INTRODUCTION

Forklifts are characterized by a pair of flat tines that can be raised and lowered to lift and move materials. Also called **powered industrial trucks**, they are manufactured in many configurations to meet the needs of different industries. This module will use the term forklift.

Like other equipment, some forklifts can be fitted with different attachments to increase their versatility. Operators need to know how to operate the machine and its accessories. They also need to understand a load chart and the dynamics of lifting loads. Operating a forklift is not simply a matter of moving the machine and materials. Daily inspections and maintenance are also an important part of safe operations.

1.1.0 Types of Forklifts

Forklifts are used in various industries. They are used indoors in warehouses and outdoors in materials yards and construction sites. Different types of forklifts are available for these purposes. Generally, forklifts are divided into two broad categories: fixed-mast and telescoping-boom. Within each of these categories, forklifts are further differentiated by their drivetrain, steering, and capacity.

There are two types of fixed-mast forklifts: rough terrain and warehouse. Rough terrain forklifts (*Figure 1*) are designed for outdoor use and are used in construction work. They have higher ground clearance, larger tires, and leveling devices. Warehouse forklifts (*Figure 2*) are designed for indoor use. They have hard rubber tires and cleaner emissions. This module will only cover rough terrain forklifts.

Forklifts have either two-wheel or four-wheel drive. The distinction and controls are similar to those of other vehicles.

A two-wheel-drive forklift has a drivetrain that transmits power to the front or rear wheels. The front wheels are often larger than the rear wheels. Two-wheel-drive forklifts typically have rear-wheel steering. The rear wheels move when the steering wheel is turned. Rear-wheel steering provides greater maneuverability. This is especially useful when backing up with a load. Some rear-wheel-drive forklifts may have only one rear wheel. These forklifts are good for general use, but may lack power in rough terrain.

A four-wheel-drive forklift has a drivetrain that transmits power to all four wheels. These forklifts are well suited for rough terrain and other conditions that require additional traction. When four-wheel drive is engaged, power is transmitted to all four wheels.

Rough terrain forklifts may also have different steering modes, as illustrated in *Figure 3*. Different steering modes enable the forklift to maneuver in smaller spaces. Some models can be operated in the following steering modes:

206F01.EPS

Figure 1 ◆ Fixed-mast rough terrain forklift.

206F02.EPS

Figure 2 ◆ Fixed-mast warehouse forklift.

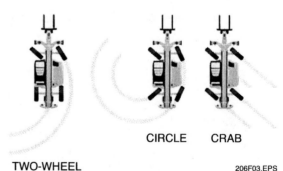

206F03.EPS

Figure 3 ◆ Steering modes for forklifts.

- The rear wheels can be locked so only the front wheels steer. This is called two-wheel steering.
- The front and rear wheels may move in opposite directions. This is called **articulated steering**, **four-wheel steering**, or **circle steering**.
- All wheels may move in the same direction. This is called **crab steering** or **oblique steering**.

Forklifts can be powered with several different types of fuel, including electricity, liquid propane (LP), diesel, or gasoline. Warehouse forklifts commonly use electricity or LP, which do not produce harmful emissions. Indoors, emissions from a combustion engine could build to toxic levels. Construction forklifts typically have diesel engines.

NOTE

Forklifts that are used indoors are designed so that they have little, if any, emissions. Otherwise toxic gases from the exhaust would pose a threat to indoor air quality. Without proper ventilation, carbon monoxide from the exhaust could build to deadly levels. Indoor forklifts use electricity or LP for power. Electric forklifts do not produce toxic gases. The only emissions from LP forklifts are carbon dioxide and water vapor.

1.1.1 Fixed-Mast Forklifts

The mast is the upright structure that the forks travel along. A typical fixed-mast forklift has a mast that tilts 20 degrees forward and 10 degrees rearward. These types of forklifts are suitable for placing loads vertically and traveling with the load. However, their reach is limited to how close the machine can be driven to the pickup or landing point. For example, a fixed-mast forklift cannot place a load beyond the leading edge of a roof because of its limited reach. For this reason, telescoping-boom forklifts are frequently used in construction.

1.1.2 Telescoping-Boom Forklifts

Telescoping-boom forklifts (*Figure 4*) have more versatility in placing loads. They are also called shooting-boom forklifts or telehandlers. They are a combination of a crane and a forklift. They can easily place loads on upper floors, well beyond the leading edge.

Some models of telescoping-boom forklifts have a level-reach fork carriage, sometimes called a squirt-boom. A squirt boom allows the fork carriage to be moved forward while the boom remains in a stationary position.

This module will describe the operations of a telescoping-boom forklift similar to the Skytrack or Lull telescoping-boom forklift or the Caterpillar TH83 off-road **telehandler**. These machines are typically used on a construction site for light and medium work, including staging of materials and equipment.

NOTE

Always read the operator's manual before operating equipment. Follow all safety and startup procedures.

2.0.0 ◆ IDENTIFICATION OF EQUIPMENT

The main features of a Caterpillar TH83 telehandler are shown in *Figure 5*. The components, controls, and indicators will vary with the manufacturer, model, and options available, but they are all similar. Because the information in the following sections is based on the Caterpillar TH83 series forklift, an operator must read and understand the operator's manual for the specific forklift before beginning operation.

Most rough-terrain forklifts are powered by a diesel engine. Air is drawn in through the air cleaner into the cylinder. The power from the engine is

206F04.EPS

Figure 4 ◆ Telescoping-boom forklift.

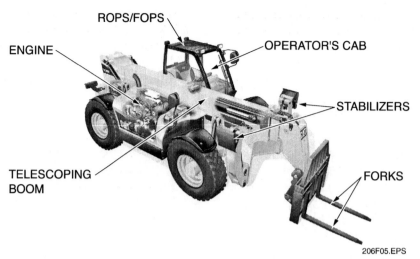

ROPS/FOPS

ENGINE

OPERATOR'S CAB

STABILIZERS

TELESCOPING
BOOM

FORKS

206F05.EPS

Figure 5 ◆ Typical components of the Caterpillar TH83.

transmitted to the wheels via the transmission. Some forklifts have a manual transmission which is actuated by a clutch. However, many forklifts have a hydrostatic transmission that does not require a clutch.

The hydraulic system is used to operate the boom and forks. The hydraulic fluid is pumped from the main tank through a series of hoses and pumps to actuate the ram. The lift ram raises and lowers the boom. The boom is retracted and extended with a hydraulic belt. The hydraulic system is also used to move auxiliary features and attachments.

2.1.0 Operator's Cab

The operator's compartment is the central hub for forklift operations. An operator must understand controls and instruments before operating a forklift. Study the operator's manual to become familiar with the controls and instruments and their functions.

Figure 6 is an overhead view of the interior of the operator's cab on a Caterpillar TH83. Some switches and controls are located outside of the main cab. However, the controls for normal operations are all located within easy reach of the operator's seat. A steering wheel, foot throttle, and brakes are used for maneuvering. The instrument panel is to the right of the steering wheel. The boom and fork are controlled with a joystick located on right side of the cab. There are foot pedals for the brakes and throttle.

The engine is started with a key switch. Various switches to the left and right of the steering wheel control the stabilizers and other machine functions, which will be covered in detail in the following sections. A controller on the right side of the steering wheel operates the directional signals and the windshield wipers. A lever on the left side of the steering wheel controls the transmission.

206F06.EPS

Figure 6 ◆ Operator's cab.

2.2.0 Instruments

An operator must pay attention to the instrument panel. The instrument panel includes the gauges that indicate engine and transmission temperature. There are several warning lights and indicators that must also be monitored. An operator can seriously damage the equipment by ignoring the instrument panel.

The instrument panel varies on different makes and models of forklifts. Generally they include the instruments and indicators covered in the following sections. A typical instrument panel is shown in *Figure 7*. Most of these instruments and indicators are similar to those in other machines and should be familiar from descriptions in previous modules.

The instrument panel on a Caterpillar TH83 has four gauges. They include fuel level, engine coolant temperature, transmission oil temperature, and service hour meter. Other types of forklifts may have different gauges. Refer to the operator's manual for the specific gauges on the machine you are operating.

NOTE

The parenthetical callout numbers in the following section refer to *Figure 7*.

2.2.1 Fuel Level Gauge

This gauge (1) indicates the amount of fuel in the forklift's fuel tank. On diesel engine forklifts, the gauge may contain a low fuel warning zone. On this model, the warning zone is yellow. Some models have a low fuel warning light. Avoid running out of fuel on diesel engine forklifts because the fuel lines and injectors must be bled of air before the engine can be restarted.

2.2.2 Engine Coolant Temperature Gauge

The engine coolant temperature gauge (2) indicates the temperature of the coolant flowing through the cooling system. Refer to the operator's manual to determine the correct operating range for normal forklift operations. Temperature gauges normally read left to right, with cold on the left and hot on the right. If the gauge is in the white zone, the coolant temperature is in the normal range. Most gauges have a section that is red. If the needle is in the red zone, the coolant temperature is excessive. Some machines may also activate warning lights if the engine overheats.

CAUTION

Operating equipment when temperature gauges are in the red zone may severely damage it. Stop operations, determine the cause of the problem, and resolve it before continuing operations.

If the engine temperature gets too high, stop operations immediately. Get out of the machine and investigate the problem. There are several checks that the operator can perform. First, check the engine coolant level. Add more fluid if it is too low. Check that the fan belt is loose or broken. Replace it if necessary. Check that the radiator fins are not fouled and clean them if necessary. These are the three primary causes of the engine overheating. If initial troubleshooting fails to resolve the problem, stop operations and take the machine out of service.

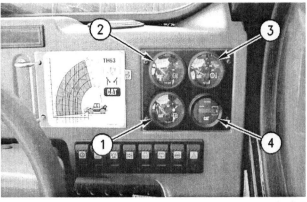

Figure 7 ◆ Instrument panel for the Caterpillar TH83 series forklift.

WARNING!

Engine coolant is extremely hot and under pressure. Check the operator's manual and follow the procedure to safely check and fill engine coolant.

2.2.3 Transmission Oil Temperature Gauge

The transmission oil temperature gauge (3) indicates the temperature of the oil flowing through the transmission. This gauge also reads left to right in increasing temperature. It has a red zone that indicates excessive temperatures. When the weather is colder, allow the transmission oil to warm up sufficiently before operating the machine.

2.2.4 Service Hour Meter

This meter (4) indicates the total hours of operation. It indicates the period of time the machine has been running over the life of the machine. Typically, it cannot be reset.

Periodic maintenance is scheduled based on hours of service. The operator must keep track of the hours since the last service using an hour meter and a service log. Periodic maintenance will be covered in more detail later in this module.

2.2.5 Speedometer/Tachometer

Some machines also have a tachometer and speedometer. This machine does not have these gauges. The tachometer indicates the engine speed in revolutions per minute (rpm). The speedometer shows the machine's ground speed. Typically they can be set for either miles per hour (mph) or kilometers per hour (kph).

2.2.6 Indicators

There are a series of indicator lights above the steering wheel as shown in *Figure 8*. These lights show when various features are activated under normal operating conditions. Generally, they do not indicate that there is something wrong with the machine, merely that a function is active. The exceptions are the alternator light and oil pressure light. When either is lit, the system is not functioning properly. If the alternator light is lit, the battery is not receiving a charge. If the oil pressure light is lit, the oil pressure is too low. Stop the machine and resolve the problem before continuing operations.

Typical indicator lights show that the following features are activated:

- Left turn signal (1)
- Alternator (2)
- Oil pressure (3)
- High beam (4)
- Circular steer (5)
- Crab steer (6)
- Right turn signal (10)

NOTE

In this model, lights in positions 7, 8, and 9 are not used.

CAUTION

A flashing indicator light may require immediate action by the operator. If you need to stop and look up the meaning of an indicator after it is flashing, you may cause serious damage to the equipment. Know your equipment before you begin operations.

2.2.7 Longitudinal Stability Indicator

Load stability is an important consideration in forklift operations. The longitudinal stability indicator is located on the right side of the front windshield, as shown in *Figure 9A*. This way, the operator can easily monitor the indicator and the load at the same time. This indicator determines the forward stability of the load. It produces visual and audible signals to indicate the limits of forward stability of the machine. The indicator alerts the operator to make adjustments to the machine or how it is being operated. Carefully monitor this indicator as the load is being raised to maintain load stability.

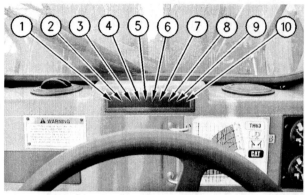

206F08.EPS

Figure 8 ◆ Indicators lights on the Caterpillar TH83 series forklift.

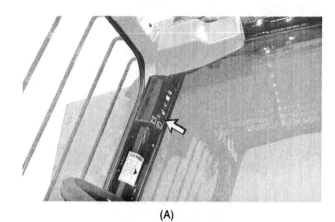

(A)

(B)   206F09.EPS

Figure 9 ◆ Load stability indicator.

The visual display (*Figure 9B*) has a meter that shows the percent of the load. Green lights will illuminate as the load on the machine increases from 20 to 80 percent of the forward stability. When the load reaches 85 percent of forward stability, amber lights will light and an alarm will sound every two seconds. At this point the operator must not increase the outreach of the load. If the machine reaches 100 percent, red lights will illuminate and the alarm will sound continuously. The operator must immediately lower or retract the load to avoid an accident.

Test the system daily. The load stability indicator can be adjusted for operations on tires or with the stabilizers lowered. The visual and audible alarms may operate momentarily if the machine is carrying a load close to the maximum, especially when traveling over rough terrain.

2.3.0 Controls

Controls and their locations can vary between makes and models of forklifts. Vehicle movement controls on forklifts can be similar to those of cars and trucks. A steering wheel is used in combination with foot pedals for throttle and brakes to control vehicle movement. The boom and forks are controlled with a joystick and switches or levers. Some models have alternative joystick settings for boom and fork controls. Switches and levers activate ancillary and specialty machine functions. The controls for a Caterpillar TH83 will be described in this module. Review the operator's manual to fully understand other types of vehicle movement, boom, and fork controls.

2.3.1 Disconnect Switches

Some models are equipped with disconnect switches located outside of the cab. These switches disconnect critical functions so that the machine cannot be operated. These switches must be activated before the machine can be started. Turn them on before mounting the machine. They offer an additional level of safety and security. Typically, unauthorized users will not activate these switches and will not be able to operate the equipment.

Some models have a fuel shutoff valve or switch that must be turned to the ON position before the machine can be operated. The fuel shutoff valve or switch physically prevents fuel from flowing from the fuel tank into the supply lines. This prevents unwanted fuel flow during idle periods or when transporting the machine, significantly reducing the potential for fuel leaks.

Alternatively, some machines have a battery disconnect switch as shown in *Figure 10*. When the battery disconnect switch is turned off, the entire electrical system is disabled. This switch should be turned off when the machine is idle overnight or

longer to prevent a short circuit or active components from draining the battery. Before mounting the machine, check that the switch is in the ON position.

2.3.2 Seat and Steering Wheel Adjustment

Upon first entering the cab, the operator should adjust the seat and steering wheel. While seat (*Figure 11*) and steering wheel (*Figure 12*) adjustments are not directly involved in forklift operation, correct positioning of these items can aid in safe operation. Before operating the forklift, the operator should adjust the seat and steering wheel position and then fasten their seat belt.

206F10.EPS

Figure 10 ◆ Battery disconnect switch.

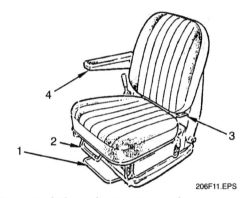

206F11.EPS

Figure 11 ◆ Seat adjustment controls.

206F12.EPS

Figure 12 ◆ Steering wheel adjustment controls.

Most seats can be moved up or down and forward or backward. Some provide an adjustable shock absorber function using springs or hydraulics. Lift the fore/aft lever (1) and slide the seat to the desired position. Release the lever to lock the seat into position. Adjust the armrests by rotating a knob on the underside of the armrest (4). The angle of the back of the seat can be set by depressing the seat recline lever (3). The cushion height and angle can be set in different positions using the second lever (2) below the front of the seat. The seat should be adjusted so that the operator's legs are almost straight when the clutch or brake pedal is fully depressed and the operator's back is flat against the back of the seat.

The steering wheel on some forklifts can be adjusted up or down and in or out as desired after the seat is correctly positioned. In *Figure 12*, rotating the knob counterclockwise unlocks the steering column. The steering column can then be tilted and positioned correctly. Turn the knob clockwise to lock the steering column into position. Because seat and steering wheel adjustment devices vary widely for various makes and models, refer to the operator manual for specific instructions.

2.3.3 Engine Start Switch

The ignition switch functions can vary widely between makes and models of forklifts. Some only activate the starter and ignition system. Others may activate fuel pumps, fuel valves, and starting aids. In some cases, the starter and starting aids are engaged by other manual controls. The multipurpose ignition switch for the Caterpillar TH83 series forklift is shown in *Figure 13*. This switch is located to the right of the steering wheel on the console.

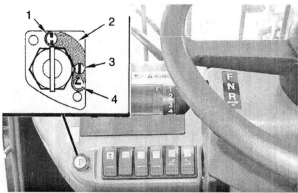

Figure 13 ◆ Engine start switch.

206F13.EPS

The engine start switch has the following four positions:

- *Off (1)* – Turning the key to this position stops the engine. It will also disconnect power to electrical circuits in the cab. However, several lights remain active when the key is in the OFF position, including the hazard warning light, the interior light, and the parking lights.
- *Auxiliary (2)* – If available, turning the key to this position only applies power to accessory power outlets located on top of the front console, or to devices such as radios that may be connected to accessory power. Power is not supplied to the instrument panel or other electrical controls, other than a battery gauge or indicator.

- *On (3)* – Turning the key to this position activates all of the electrical circuits except the starter motor circuit. When the key is first turned to the ON position, it may initiate a momentary instrument panel and indicator bulb check.
- *Start (4)* – The key is turned to this position to activate the starter, which starts the engine. This position is spring-loaded to return to the ON position when the key is released. If the engine fails to start, the key must be returned to the OFF position before the starter can be activated again. To reduce battery load during starting, the ignition switch of some forklifts may be configured to shut off power to accessories and lights when the key is in the start position.

2.3.4 Vehicle Movement Controls

Forklift movement is controlled in a manner similar to that of cars and trucks. The throttle and brakes are operated by foot pedals. A steering wheel (*Figure 14*) is used to turn the vehicle. The transmission is controlled with a lever on the right of the steering column. Moving the steering wheel to the left or right steers the vehicle to the left or right.

Some steering wheels are equipped with a knob (A, *Figure 14*). This gives the operator greater control of the wheel when steering with one hand. This is important when the operator is using one hand to steer and the other to operate the boom and folks with the joystick.

The transmission is controlled via a lever on the left side of the steering column (*Figure 15*). The lever has three positions; forward, neutral, and reverse. Move the lever to the indicated position to set the lever to select the direction of travel (1). There are four transmission speeds indicated on the collar (2). Rotate the transmission control lever to select the transmission speed. The first three speeds can be used for either forward or reverse. Fourth gear is reserved for forward motion only. Do not skip gears when downshifting.

> **CAUTION**
>
> Always come to a complete stop before changing travel direction. Changing travel direction while the machine is moving can damage some machines. Although it is possible to change gears while in motion on machines with a hydrostatic transmission, it is not recommended. Changing directions suddenly can dislodge the load.

2.3.5 Forklift Controls

The boom and forks on a forklift can be controlled with levers or with a joystick. The forks can be moved in several directions. The forks can be tilted up and down and the boom can be raised and lowered. Some forks can be moved side to side and others can be tilted or angled from side to side. Telescopic booms can be extended and retracted.

Older forklifts are controlled with a series of levers. Each lever controls one aspect of the motion of the forks. For example, some forklifts have three levers. One lever will move the forks up and down, while another tilts them forward and backward. The third lever moves the forks from side to side. Because different makes and models have different controls, review the operator's manual to familiarize yourself with the controls before operating any forklift. This module will describe joystick controls for fork and boom movement.

The joystick is used in conjunction with switches, triggers, or buttons to control the boom and fork movement. The six ways to move the fork and boom are as follows:

- *Boom raise* – Raises the boom
- *Boom lower* – Lowers the boom
- *Boom extend* – Extends the boom
- *Boom retract* – Retracts the boom

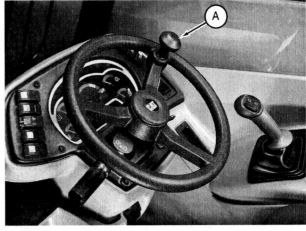

206F14.EPS

Figure 14 ◆ Steering wheel.

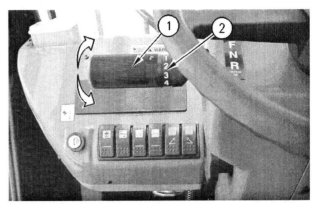

206F15.EPS

Figure 15 ◆ Transmission control.

- *Tilt forward* – Tilts the forks forward or downward toward the ground
- *Tilt back* – Tilts the forks backward or upward toward the sky

The speed of the movement is controlled by how far the joystick is moved and/or the engine speed. Increase the engine speed and then move the joystick slowly until the forks are moving at the desired speed. When the joystick is released it will return the central or neutral position and movement will cease. Avoid sudden movement with the forks as that can dislodge the load or cause accidents.

A typical control arrangement is shown in *Figure 16*. The right and left movement of the joystick tilts the forks forward and back. Backward and forward movement of the joystick raises and lowers the boom. There are switches on top of the joystick. The switch on the left retracts and extends the telescopic boom. The switches on the right raise and lower the stabilizers. In this configuration, the speed at which the boom extends and retracts depends only on engine speed.

Moving the joystick diagonally will tilt the forks while raising or lowering the boom. The switch can

Figure 16 ◆ Forklift controls.

206F16.EPS

be activated to retract or extend the boom. The joystick is moved diagonally to tilt the forks while raising or lowering the boom. The forks can be moved in several directions at the same time by activating multiple controls. With practice, you will be able to move the boom and forks smoothly.

Some joysticks have a trigger control, as shown in Figure 17. In this configuration, the joystick (1) only controls the boom movement. Moving the joystick to the right or left retracts or extends the boom. Moving the joystick forward and back lowers and raises the boom.

The trigger (3) must be depressed to tilt the forks. Activating the trigger changes the joystick controls so that right and left joystick movement controls the forks. Depress the trigger (3) and move the joystick to the right to tilt the forks up. Depress the trigger and move the joystick to the left to tilt the forks

down. When the trigger is activated, the boom cannot be retracted or extended. The switch on top of the joystick (2) controls auxiliary features.

The carriage on some forklifts can be shifted from side to side or tilted up to 10 degrees to either side. The tilt feature allows the operator to position the forks to pick up a load that is not level and is particularly useful on rough terrain. The side-shift feature allows the forks to move side to side horizontally while the boom remains stationary. The operator can precisely position a load without repositioning the forklift.

2.3.6 Switches

There are two sets of switches to either side of the steering wheel. They activate various features and functions of the forklift. Some switches are used in conjunction with other controls, including the steering mode select and the quick coupler.

Figure 18 shows the controls on the left side of the steering wheel. These switches operate the following functions:

- Quick coupler (1)
- Steering mode select (2)
- Fog lights (3)
- Reading lights (4) (main head lights)
- Left stabilizer (5)
- Right stabilizer (6)

Figure 19 shows the switches on the right side of the steering wheel. Some of these switches operate the rotating beacon, roof window wipers,

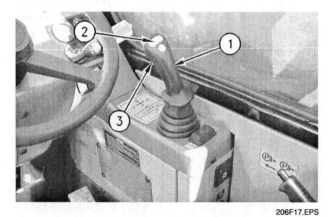

206F17.EPS

Figure 17 ◆ Alternate forklift controls.

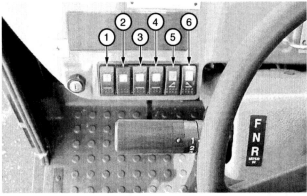

Figure 18 ◆ Lights and other switches.

206F18.EPS

Figure 19 ◆ Switches to the right of the steering wheel.

206F19.EPS

starting aids/ and hazard and working lights. Several activate specialized features which are described in the following section. These switches operate the following functions:

- Transmission neutralizer (1)
- Work lights (2)
- Rotating beacon (3)
- Master steering select (4)
- Starting aid (5)
- Roof window wiper switch (6)
- Differential lock control (7)
- Hazard flashers (8)

2.3.7 Specialized Features

There are several features that override the normal functions of a forklift. These include the transmission neutralizer and the differential lock. These features are an important part of operating the forklift. They offer added control of the machine in specialized situations.

The transmission neutralizer switch is a two-position switch. With the switch activated, the transmission will be neutralized when the service brake is applied. With the switch turned off, the transmission will remain engaged when the service brakes are applied.

The differential lock control overrides the normal operation of the front axle differential. Torque is transmitted to both wheels, even though one wheel may not have traction. This helps maintain traction when ground conditions are soft or slippery.

If the wheels start to spin, release pressure on the accelerator until the wheels stop spinning. Engage the differential lock and increase pressure on the accelerator. Once clear of the area, release pressure on the accelerator and release the switch to disengage the differential lock.

CAUTION

Limit steering maneuvers while the differential lock is engaged. Steering maneuvers with the differential lock engaged can damage the machine.

2.4.0 Operator Comfort and Other Controls

There are several controls designed to adjust the seat and controls for maximum operator comfort. Adjust the seat, armrests, mirrors, and cab climate controls while the machine is warming up or parked. Do not move or operate the machine before adjusting the seat and mirrors for comfort and visibility.

These adjustments affect safety in addition to operator comfort. Operators who are adjusting the seat or mirrors after beginning operations are not giving their full attention to machine operations. If an operator cannot reach all of the controls, the machine may be hard to control and cause injury or property damage. Properly setting the climate controls will reduce operator fatigue and increase alertness. Failure to properly set the climate controls can reduce visibility if the windshield is fogged.

If you will be working at night, familiarize yourself with all of the light switches. Once it is dark, it is too late to look for the light switches. Make sure that all lights are functioning properly. Adjust the lights so that the work area is properly illuminated.

2.5.0 Attachments Used on or with the Forklift

The main attachment on a forklift is the forks. However, on many types of forklifts, the forks can be detached and replaced with different forks or other attachments. The length and configuration of the forks can be changed to maximize productivity when handling certain materials. Some forks are designed to handle specific materials like pallets, cubes of brick and block, or sheet metal.

In addition to lifting and moving, forklifts can be used for other operations. Different attachments, such as sweepers, buckets, or hoppers, can be fitted onto some forklifts. A forklift can be used as a raised work station when an access platform is attached. These attachments greatly expand the capabilities of the forklift and increase its usefulness on the construction site. Use the proper attachment to lift loads safely. Although a sling can be attached to the forks, it is safer to use the boom and hook attachment to lift loads from the top. Make sure that the attachment is certified by the manufacturer for the intended lift.

> **CAUTION**
>
> Only use attachments that are designed for the machine. Using attachments that were not designed for the specific model forklift can cause the machine to fail.

Some forklift attachments are designed for use in specific industries such as farming or waste handling. A bale handler is attached to the forklift shown in *Figure 20*. The forklift can easily move bales of scrap even though they are not palletized. The attachment is hooked up to the forklift's hydraulic system. The bale grabber is operated using auxiliary switches on the forklift controls.

2.5.1 Forks

Forks are available in various sizes to handle different types of materials, including standard forks, cube forks, and lumber forks. Standard forks are 48 inches long, 2 inches thick, and 4 inches wide. Cube forks have narrower, 2-inch-wide tines. They are used in a set of four rather than two. Lumber forks are normally longer, 60 inches, and wider, 7 inches.

Some carriages have fixed forks. On other models, the space between the forks can be adjusted. To manually adjust the forks, lift the tip of the fork and tilt it upward. Slide it along the carriage until it reaches the desired position. Lower the tip of the fork to lock it into place.

> **WARNING!**
>
> Manually adjusting the forks poses a severe pinch hazard. Read the operator's manual and follow all safety precautions. Do not drop forks into place. Lower them carefully to avoid crushing your fingers or hands.

206F20.EPS

Figure 20 ◆ Waste handling forklift.

The forklift's hydraulic system can be used to power the attachment. For example, some forks are designed to swing from side to side. The carriage is connected to the forklift's hydraulic system. The auxiliary hydraulic switches are used to rotate the carriage up to 45 degrees to either side. This provides better maneuverability and allows loads to be lifted or placed when the forklift cannot be directly perpendicular to the load.

2.5.2 Booms and Hooks

Some loads are not palletized or easily lifted from the bottom. These loads must be lifted from the top using slings and a lifting hook. The lifting hook and boom extension allow the forklift to operate like a crane. This arrangement is often used in construction to place trusses or to set pipe. The operator should always remain at the controls while a load is suspended from either the boom extension or the lifting hook.

The boom extension is used when additional reach or height is needed (*Figure 21*). It is frequently used to place lighter loads such as trusses. The triangular boom extension is attached to the end of the boom. The load is rigged to the lifting hook on the end of the boom extension.

The lifting hook is used for heavier loads as shown in *Figure 22*. The hook is mounted on a carriage that is attached to the end of the boom. Only use the boom extension and lifting hook with loads that can be rigged using chains or slings.

Figure 21 ◆ Boom extension.

Figure 22 ◆ Lifting hook.

 WARNING!

Always use the appropriate rigging methods to secure a load to the hook. Rigging can fail, causing the load to fall and cause injury or property damage. Use tag lines to prevent the load from swinging.

2.5.3 Personnel Platform

Some forklifts can be equipped with an access platform (*Figure 23*). This provides an elevated work station for two workers and their tools and materials. The platform uses the forklift's hydraulic system and is operated with controls located on the platform. Once the machine is positioned and the stabilizers lowered, the operator controls the movement of the platform from the platform.

As an additional safeguard, other machine controls are locked out with a key switch. The lockout switch disables the stabilizers, frame level, transmission control, and quick coupler. This prevents the machine from being moved while the platform is raised. Moving the machine while workers are on the platform is dangerous and must be prevented.

Figure 23 ◆ Access platform.

 WARNING!

Always use a fall arrest system when working on an elevated work station. Follow all safety procedures for operating the access platform. Improper operation of the platform could result in injury or death.

Review the operator's manual or other literature before operating the access platform. Follow all safety precautions to minimize hazards from slips and falls. Keep the platform clear of power lines, and do not allow the platform to contract any structure. Do not move the machine unless the access platform is lowered to 18 inches or less off the ground.

2.5.4 Other Attachments

Other types of attachments for a forklift include concrete buckets, hoppers, utility buckets, multipurpose buckets, augers, and sweepers. These attachments enable the forklift to be used for a wide variety of tasks on the construction site.

Several types of buckets can be used with the forklift. General-purpose buckets are used for a variety of activities from digging to loading loose material such as dirt or gravel. This bucket is shown in *Figure 24*. There are several size buckets for different types of materials. Light material buckets are generally larger than the standard bucket. They are designed to be used for low-density materials such as wood chips.

A multipurpose bucket allows the bucket to function in four different ways. It is also known as a 4-in-l bucket or clamshell (*Figure 25*). It can be used as a standard bucket, dozer blade, clamp, or controlled-discharge bucket. It is connected to the machine's hydraulic system and is operated with the auxiliary switches on the joystick controls.

Some attachments are made for special applications. The sweeper is a powered broom. It is used for clearing parking lots, industrial plants, mill yards, airport runways, streets, driveways, and

Figure 24 ◆ Forklift with general-purpose bucket.

Figure 25 ◆ Multipurpose bucket.

lanes. The brooms are powered with one or two hydraulic motors. An optional water spray can be added for dust control.

Hoppers and concrete buckets allow the forklift to carry loose or wet materials. The hopper can be manually dumped as shown in *Figure 26*. The concrete bucket can lift ½ or 1 cubic yard of concrete over columns or other forms. The concrete bucket has a clamshell gate on the bottom which is opened to place the concrete.

Figure 26 ◆ Forklift with hopper.

The function and operation of an attachment is described in the manufacturer's literature and the operator's manual. Always review the instructions before using any equipment or attachments. Improper operation can result in personal injury, machine damage, or property damage.

 WARNING!
Only use attachments that are compatible with the machine. Using non-standard attachments could damage the equipment or cause injury.

3.0.0 ◆ SAFETY GUIDELINES

Safe operation is the responsibility of the operator. Operators must develop safe working habits and recognize hazardous conditions to protect themselves and others from injury or death. Always be aware of unsafe conditions to protect the load and the forklift from damage. Become familiar with the operation and function of all controls and instruments before operating the equipment. Read and fully understand the operator's manual.

Forklift Fatality

A 39-year-old supply motorman was helping a forklift operator place crib blocks on a mine flat car. The motorman was using a shovel to remove ice and snow from the bottom of a crib block while it was raised on the forklift. The load slipped and crushed the motorman. He later died from the injury.

The Bottom Line: Do not work beneath or around a raised forklift load.

Source: National Institute for Occupational Safety and Health.

3.1.0 Operator Safety

Nobody wants to have an accident or be hurt. There are a number of things you can do to protect yourself and those around you from getting hurt on the job. Be alert and avoid accidents.

Each year in the United States nearly 100 workers are killed and another 20,000 are seriously injured in forklift-related incidents. The most frequent type of accident is a forklift striking a pedestrian. This accounts for 25 percent of all forklift accidents.

Know and follow your employer's safety rules. Your employer or supervisor will provide you with the requirements for proper dress and safety equipment. The following are recommended safety procedures for all occasions:

- Only operate the machine from the operator's cab.
- Mount and dismount the equipment carefully.
- Wear a hard hat, safety glasses, safety boots, and gloves when operating the equipment.
- Do not wear loose clothing or jewelry that could catch on controls or moving parts.
- Keep the windshield, windows, and mirrors clean at all times.
- Never operate equipment under the influence of alcohol or drugs.
- Never smoke while refueling (*Figure 27*).
- Do not use a cell phone and avoid other sources of static electricity while refueling.
- Never remove protective guards or panels.
- Never attempt to search for leaks with your bare hands. Hydraulic and cooling systems operate at high pressure. Fluids under high pressure can cause serious injury.
- Always lower the forks or other attachments to the ground before performing any service or when leaving the forklift unattended.

206F27.EPS

Figure 27 ◆ Refuel safely.

3.2.0 Safety of Co-Workers and the Public

You are not only responsible for your personal safety, but also for the safety of other people who may be working around you. Sometimes, you may be working in areas that are very close to pedestrians or motor vehicles. In these areas, take time to be aware of what is going on around you. Create a safe work zone using cones, tape, or other barriers. Remember, it is often difficult to hear when operating a forklift. Use a spotter and a radio in crowded conditions.

The main safety points when working around other people include the following:

- Walk around the equipment to make sure that everyone is clear of the equipment before starting and moving it.
- Always look in the direction of travel.
- Do not drive the forklift up to anyone standing in front of an object or load.
- Make sure that personnel are clear of the rear area before turning.
- Know and understand the traffic rules for the area you will be operating in.
- Use a spotter when landing an elevated load or when you do not have a clear view of the landing area.
- Exercise particular care at blind spots, crossings, and other locations where there is other traffic or where pedestrians may step into the travel path.
- Do not swing loads over the heads of workers. Make sure you have a clear area to maneuver.
- Do not allow workers to ride in the cab or on the forks or the bucket.

When working around pedestrians or in other public areas, create a safe work zone for forklift operations. Use barrels, cones, tape, or barricades to keep others out of your work area. This protects both you and the community.

3.3.0 Equipment Safety

Your forklift has been designed with certain safety features to protect you as well as the equipment. For example, it has guards, canopies, shields, rollover protection, and seat belts. Know your equipment's safety devices and be sure they are in working order.

Forklift overturns are the leading cause of deadly forklift accidents. They represent 25 percent of all forklift-related deaths. Know the weight of the load and the limits of your machine. Don't take risks and overload the machine. Be especially careful moving suspended loads in windy conditions.

Use the following guidelines to keep your equipment in good working order:

- Perform pre-start inspection and lubrication daily (*Figure 28*).
- Look and listen to make sure the equipment is functioning normally. Stop if it is malfunctioning. Correct or report trouble immediately.
- Always travel with the forks or load low to the ground.
- Never exceed the manufacturer's limits for speed, lifting, or operating on inclines.
- Know the weight of all loads before attempting to lift them. Review the appropriate load chart. Do not exceed the rated capacity of the forklift.
- Always lower the forks, engage the parking brake, turn off the engine, and secure the controls before leaving the equipment.
- Never park on an incline.
- Maintain a safe distance between your forklift and other equipment that may be on the job site.

Know your equipment. Learn the purpose and use of all gauges and controls as well as your equipment's limitations. Never operate your machine if it is not in good working order.

4.0.0 ◆ BASIC PREVENTIVE MAINTENANCE

Preventive maintenance is an organized effort to regularly lubricate and service the machine in order to avoid poor performance and breakdowns at critical times. By performing preventive maintenance on the forklift, you keep it operating efficiently and safely and avoid costly failures in the future.

Preventive maintenance of equipment is essential and is not that difficult, if you have the right tools and equipment. Preventive maintenance

206F28.EPS

Figure 28 ◆ Check all fluids daily.

should become a habit, performed on a regular basis. Inspect and lubricate the machine on a daily basis. Be aware of hours of service and have the machine serviced at appropriate intervals.

CAUTION

Forklift service is normally based on hours of service. A service schedule is contained in the operator's manual. Failure to perform scheduled maintenance could damage the machine.

4.1.0 Daily Inspection Checks

Maintenance time intervals for most machines are established by the Society of Automotive Engineers (SAE) and adopted by most equipment manufacturers. Instructions for preventive maintenance are usually in the operator's manual for each piece of equipment. Typical time intervals are: 10 hours (daily); 50 hours (weekly); 100 hours, 250 hours, 500 hours, and 1,000 hours. The operator's manual will also include lists of inspections and servicing activities required for each time interval. *Appendix A* is an example of typical periodic maintenance requirements.

The first thing you must do each day before beginning work is to conduct your daily inspection. Some companies have an inspection checklist that must be completed daily. A model inspection checklist for forklifts that was developed by OSHA is included in *Appendix B*. This should be done before starting the engine. It will identify any potential problems that could cause a breakdown and indicate whether the machine can be operated. The equipment should be inspected before, during, and after operation.

The daily inspection is often called a walk-around. The operator should walk completely around the machine checking various items. The following items should be checked and serviced on a daily inspection:

WARNING!

Do not check for hydraulic leaks with your bare hands. Use cardboard or another device. Pressurized fluids can cause severe injuries to unprotected skin. Long term exposure to these fluids can cause cancer or other chronic diseases.

- Look around and under the machine for leaks, damaged components, or missing bolts or pins.
- Inspect the cooling system for leaks or faulty hoses. Remove any debris from the radiator (*Figure 29*).

- Inspect all attachments and implements for wear and damage. Make sure there is no damage that would create unsafe operating conditions or cause an equipment breakdown.
- Inspect the boom cylinders.
- Inspect and clean steps, walkways, and handholds.
- Inspect the engine compartment and remove any debris. Clean access doors.
- Inspect the rollover protective structure (ROPS) for obvious damage.
- Inspect the hydraulic system for leaks, faulty hoses, or loose clamps.
- Inspect the lights and replace any broken bulbs or lenses.
- Inspect the axles, differentials, wheel brakes, and transmission for leaks.
- Inspect tires for damage and replace any missing valve caps. Check that the tires are inflated to the correct pressure.
- Check the condition and adjustment of drive belts on the engine.
- Inspect the operator's compartment and remove any trash.
- Inspect the windows for visibility and clean them if needed.
- Adjust the mirrors.
- Test the backup alarm and horn. Put equipment in reverse gear and listen for the backup alarm.

Some manufacturers require that daily maintenance be performed on specific parts. These parts are usually those that are the most exposed to dirt or dust and may malfunction if not cleaned or serviced. For example, your service manual may recommend lubricating specific bearings every 10 hours of operation, or always cleaning the air filter before starting the engine.

Before beginning operation the following fluid levels need to be checked and topped off:

- *Engine oil* – Check the crankcase oil level and make sure it is in the safe operating range. The dipstick is located at arrow A in *Figure 30*.
- *Cooling system* – Check the coolant level and make sure it is at the level specified in the operating manual. The coolant level can be seen at arrow B in *Figure 30*.
- *Fuel level* – Check the fuel level in the fuel tank(s). Refuel if necessary.
- *Hydraulic fluid* – Check the hydraulic fluid level in the reservoir (*Figure 31*).
- *Transmission fluid* – Measure the level of the transmission fluid to make sure it is in the operating range (*Figure 32*).
- *Pivot points* – Clean and lubricate all pivot points (*Figure 33*).

The operator's manual usually has detailed instructions for performing periodic maintenance. If you find any problems with your machine that you are not authorized to fix, inform the foreman or field mechanic before operating the machine.

Forklift tires can be inflated with one of several substances including air or a liquid mixture. Tires inflated with liquid ballast provide added machine stability. The liquid is a mixture of water and calcium chloride; the latter provides antifreeze protection. No matter what is inside the tires, check the pressures daily to make sure they meet the recommended pressures in the operator's manual.

206F29.EPS

Figure 29 ◆ Clean radiator core.

4.2.0 Servicing a Forklift

When servicing a forklift, follow the manufacturer's recommendations and service chart. Any special servicing for a particular piece of equipment will be highlighted in the manual. Normally, the service chart recommends specific intervals, based on hours of run time, for such things as changing oil, filters, and coolant.

Hydraulic fluids should be changed whenever they become dirty or break down due to overheating. Continuous and hard operation of the hydraulic system can heat the hydraulic fluid to the boiling point and cause it to break down. Filters should also be replaced during regular servicing.

Before performing maintenance procedures, always complete the following steps:

Step 1 Park the machine on a level surface to ensure that fluid levels are indicated correctly.

Step 2 Lower all attachments to the ground. Operate the controls to relieve hydraulic pressure.

Step 3 Engage the parking brake.

Step 4 Lock the transmission in neutral.

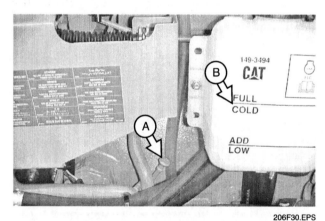

206F30.EPS

Figure 30 ◆ Check the engine oil and coolant levels.

206F31.EPS

Figure 31 ◆ Check hydraulic fluid.

4.3.0 Preventive Maintenance Records

Accurate, up-to-date maintenance records are essential for knowing the history of your equipment. Each machine should have a record that describes any inspection or service that is to be performed and the corresponding time intervals. Typically, an operator's manual and some sort of inspection sheet are kept with the equipment at all times.

5.0.0 ◆ BASIC OPERATION

Operation of a forklift requires constant attention to the controls and instruments, the load, and the surrounding environment. Operators must plan

206F32.EPS

Figure 32 ◆ Check transmission fluid.

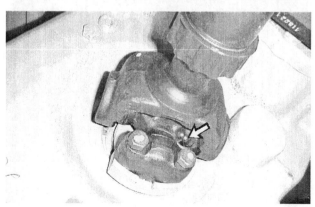

206F33.EPS

Figure 33 ◆ Lubricate all pivot points.

their work and movements in advance and be alert to the other operations going on around the equipment. Do not take risks.

The most important factor to consider when operating any forklift is its capacity. Each forklift is designed with an intended capacity that must never be exceeded. Exceeding the capacity jeopardizes the equipment, the load, and the safety of everyone near the equipment. Capacity information is provided in the load charts, which are included in the operator's manual or posted in the forklift. Be sure to read and follow the load chart.

The ability of a forklift to lift a load without tipping is called the rated capacity. The capacity of a forklift varies depending on the angle and height of the boom. Load charts, like the one shown in *Figure 34*, provide data on the maximum capacity of the forklift. A forklift operator must be able to read load charts and make sure that the load does not exceed the capacity.

Different load charts are used for different machine configurations. For example, the capacity changes if the stabilizers are lowered or raised. Using different attachments also requires that you use a different load chart. Make sure that you are using the correct load chart for the lift you are performing.

5.1.0 Suggestions for Effective Forklift Operation

Before starting forklift operations, make sure that you are familiar with the load and the area of operations. Check the area for both vertical and horizontal clearances. Make sure that the path is clear of electrical power lines and other obstacles.

The following suggestions can help improve operating efficiency:

• Observe all safety rules and regulations.
• Determine the weight of the load, review the load charts, and plan operations before starting.
• Use a spotter if you cannot see the area where the load will be placed.

5.2.0 Preparing to Work

Preparing to work involves getting yourself organized in the cab, fastening your seat belt, and starting your machine. Mount your equipment using the grab rails and foot rests. Adjust the seat to a comfortable operating position. The seat should be adjusted to allow full pedal travel with your back against the seat back. This will permit the application of maximum force on the brake pedals. Make sure you can see clearly and reach all the controls.

NOTE

Always maintain three points of contact when mounting equipment. Keep grab rails and foot rests clear of dirt, mud, grease, ice, and snow.

WARNING!

OSHA requires that approved seat belts and a rollover protective structure (ROPS) be installed on virtually all heavy equipment. Old equipment must be retrofitted. Do not use heavy equipment that is not equipped with these safety devices.

Operator stations vary, depending on the manufacturer, size, and age of the equipment. However, all stations have gauges, indicators, switches, levers, and pedals. Gauges tell you the status of critical items such as water temperature, oil pressure, battery voltage, and fuel level. Indicators alert the operator to low oil pressure, engine overheating, clogged air and oil filters, and electrical system malfunctions. Switches exist for activating the glow plugs, starting the engine, and operating accessories, such as lights. Typical instruments and controls were described previously. Review the operator's manual so that you know the specifics of the machine you will be operating.

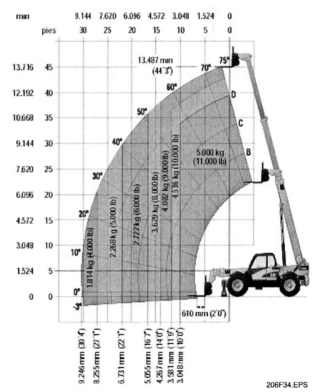

Figure 34 ◆ Load chart.

The startup and shutdown of an engine is very important. Proper startup lengthens the life of the engine and other components. A slow warm up is essential for proper operation of the machine under load. Similarly, the machine must be shut down properly to cool the hot fluids circulating through the system. These fluids must cool so that they can cool the metal parts of the engine before it is switched off.

5.2.1 Startup

There may be specific startup procedures for the piece of equipment you are operating, but in general, the startup procedure should follow this sequence:

Step 1 Be sure the transmission control is in neutral.

Step 2 Engage the parking brake (*Figure 35*). This is done with either a lever or a knob, depending on the forklift make and model.

 NOTE
When the parking brake is engaged an indicator light on the dash will light up or flash. If it does not, stop and correct the problem before operating the equipment.

Step 3 Depress the throttle control slightly.

Step 4 Turn the ignition switch to the start position. The engine should turn over. Never operate the starter for more than 30 seconds at a time. If the engine fails to start, wait two to five minutes before cranking again.

Step 5 Warm up the engine for at least 5 minutes. Warm up the machine for a longer period in colder temperatures.

Step 6 Check all the gauges and instruments to make sure they are working properly.

Step 7 Shift the transmission into forward and rotate the gear control to low range.

Step 8 Release the parking brake and depress the service brakes.

Step 9 Check all the controls for proper operation.

Step 10 Check service brakes for proper operation.

Step 11 Check the steering for proper operation.

Step 12 Manipulate the controls to be sure all components are operating properly.

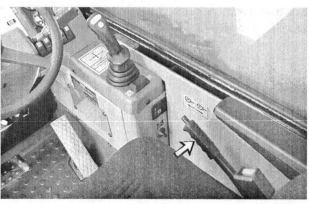

Figure 35 ◆ Parking brake.

206F35.EPS

 CAUTION
Do not pick up a load if the hydraulics are sluggish. Allow the machine to warm up until the hydraulics function normally. The hydraulics can fail if not warmed up completely.

Step 13 Shift the gears to neutral and lock.

Step 14 Reset the parking brake.

Step 15 Make a final visual check for leaks, unusual noises, or vibrations.

If the machine you are using has a diesel engine, there are special procedures for starting the engine in cold temperatures. Many diesel engines have glow plugs that heat up the engine for ignition. Some units are also equipped with ether starting aids. Review the operator's manual so that you fully understand the procedures for using these aids. Follow the manufacturer's instructions for starting the machine in cold weather.

As soon as the engine starts, release the key; it should return to the ON position. Adjust the engine speed to approximately half throttle. Let the engine warm up to operating temperature before moving the forklift.

 WARNING!
Ether is a highly flammable gas. It mixes with the air in the carburetor. As the pistons compress the gases, ignition takes place. The flame from the ether produces enough heat to overcome the compression heat lost through the cold metal of the engine. Ether should only be used when temperatures are below 0°F and when the engine is cold.

5.2.2 Checking Gauges and Indicators

Keep the engine speed low until the oil pressure registers. The oil pressure light should initially light and then go out. If the oil pressure light does not turn off within 10 seconds, stop the engine, investigate, and correct the problem.

Check the other gauges and indicators to see that the engine is operating normally. Check that the water temperature indicator, ammeter, and oil pressure indicator are in the normal range. If there are any problems, shut down the machine and investigate or get a mechanic to look at the problem.

5.2.3 Shutdown

Shutdown should also follow a specific procedure. Proper shutdown will reduce engine wear and possible damage to the machine.

Step 1 Find a dry, level spot to park the forklift. Stop the forklift by decreasing the engine speed, depressing the clutch, and placing the direction lever in neutral. Depress the service brakes and bring the machine to a full stop. If you must park on an incline, chock the wheels

Step 2 Place the transmission in neutral and engage the parking brakes.

Step 3 Release the service brake and make sure that the parking brake is holding the machine.

Step 4 Make sure that the boom is fully retracted. Lower the forks so that they are resting on the ground. Lower any other attachments.

Step 5 Place the speed control in low idle and let the engine run for approximately five minutes.

CAUTION
Failure to allow the machine to cool down can cause excessive temperatures in the engine. This can cause the oil to overheat or boil. This damages the oil and it must be replaced.

Step 6 Turn the engine start switch to the OFF position.

Step 7 Release hydraulic pressure by moving the control levers until all movement stops.

Step 8 Turn the engine switch to OFF and remove the key.

Some machines have security panels or vandalism caps for added security. The panels cover the controls and can be locked when the machine is not in use. Lock the cab door, and secure and lock the engine enclosure. Always engage any security systems when leaving the forklift unattended.

5.3.0 Basic Maneuvering

To maneuver the forklift, you must be able to move forward and backward and turn. Basic maneuvering was covered in detail in *Heavy Equipment Level One*. The first part of this section serves as a review.

5.3.1 Moving Forward

The first basic maneuver is learning to drive forward. To move forward, follow these steps:

Step 1 Before starting to move, use the joystick to raise the forks to about 15 inches above the ground. This is the travel position.

Step 2 Put the shift lever in low forward. Release the parking brake, and press the accelerator pedal to start the forklift moving.

Step 3 Steer the machine using a steering wheel or joystick.

Step 4 Once underway, shift to a higher gear to drive on the road. To shift from a lower to a higher gear, rotate the collar of the shift lever.

CAUTION
Always travel with the forks low to the ground (12 to 18 inches). Be aware of the tips of the forks and avoid hitting things with them.

5.3.2 Moving Backward

To back up or reverse direction, always come to a complete stop. Then move the shift lever to reverse. Once in reverse gear, you can apply some acceleration and begin to move backwards.

NOTE
Although it is possible to change directions while moving if the machine has a hydrostatic transmission, it is not recommended. Changing direction while in motion can cause a sudden jolt to the operator and the load.

5.3.3 Steering and Turning

How you steer a forklift depends on the make and model; however, most have a steering wheel. Some forklift steering wheels can be operated with one hand. This allows the operator to use the other hand to control the forks and boom. The steering wheels on a forklift operate in the same manner as steering wheels on cars and trucks. Moving the wheel to the right turns the forklift to the right. Turning the wheel to the left moves the wheels to the left.

Some forklifts have different steering modes, including two-wheel, circle, and crab steering. When the machine is in two-wheel steering, only the front wheels move. Use two-wheel steering when you road the machine. Use circle steering for normal operations. When the machine is in circle steer mode, the front and back wheels turn in opposite directions. This allows the machine to make tight turns. When working in a confined area, use crab steering. In crab steering, the front and back wheels turn in the same direction. The machine will travel forward and to one side or backward and to one side.

CAUTION

Always straighten the wheels before switching modes. You can damage the steering if the wheels are not centered before operating the machine in two-wheel steer.

NOTE

It is possible for the steering to go out of synchronization if the correct procedures are not followed when changing steering modes. Review the operator's manual before changing steering modes. Follow the procedure in the operator's manual to synchronize the wheels if they go out of sync.

5.3.4 Leveling the Forklift

Keeping the forklift level is an vital part of safe operations. The forklift must be level and balanced before loads are lifted. If the forklift is not level when a load is lifted, it could tip over. Forklifts overturning are the leading cause of death in forklift operations.

Many forklifts have some type of level indicator. In the Caterpillar TH83 the level indicator is located on the frame above the front windshield as shown in *Figure 36*. An air bubble or small bead is contained within a small arced sightglass. When the bubble or bead is centered in the middle of the

sightglass, the forklift frame is level (*Figure 37*). A scale on the side of the glass is marked zero in the center and increases to either side. The numbers represent the number of degrees off level.

There are two ways to level the forklift. One is to lower and adjust the stabilizers until the machine is level. When the forklift must be leveled while the machine is on its wheels, use the frame leveling control to level the machine. This feature rotates the frame on the wheels to level the machine. The frame level controls cannot be used after the stabilizers are lowered.

The frame leveling controls are located on the console near the joystick (*Figure 38*). The default position for the three-position switch is center or

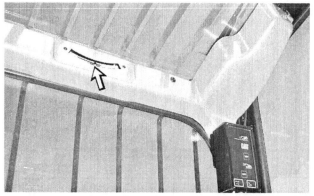

206F36.EPS

Figure 36 ◆ Forklift level indicator.

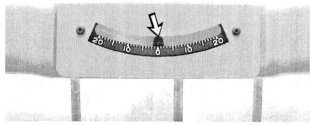

206F37.EPS

Figure 37 ◆ Reading the level indicator.

206F38.EPS

Figure 38 ◆ Frame leveling controls.

hold. Depressing the switch to either side rotates the frame. Pressing the right side of the switch will lower the right side of the machine. Pressing the left side of the switch will lower the left side.

Use the frame leveling controls when a load must be lifted from an uneven surface. The controls will tilt the frame 10 degrees to the left or right. Lower the boom before using the level controls. The boom should be close to, but not on, the ground. Depress the appropriate switch until the machine is level. When you release the switch, it will return to the hold position and the machine will stop rotating.

 CAUTION

Do not operate the frame leveling controls when the boom is raised. Doing so could damage the machine.

The stabilizers can also be used to level the machine. The stabilizer controls are used to lower and raise the left and right stabilizer arms. The stabilizers are controlled with two three-position switches located to the left of the steering wheel. Lowering the stabilizer arms firmly to the ground provides a firm and level base for lifting. When the stabilizers are lowered, the forklift can handle heavier loads.

 WARNING!

Check that the ground will support the stabilizers before you lower them. If the soil collapses from the weight of the machine, the forklift could overturn and cause injury or death. Make sure that all personnel stand clear when you are operating the stabilizers. Moving stabilizers poses a significant crushing hazard.

Before lowering the stabilizers, check that the area is free from obstruction and that the ground will support the stabilizers. Lower and retract the boom. The boom should be close to, but not on, the ground. Use the frame leveling controls to level the machine. Run the engine at sufficient speed to supply power to the stabilizers. Check that all personnel are clear of the area. Depress the bottom of the right stabilizer switch to lower it. Release the switch when the stabilizer has reached the desired position. Repeat the procedure for the left stabilizer. When both stabilizers are lowered, the front wheels of the forklift should be off the ground (*Figure 39*). Adjust the positions of the stabilizers until the frame is level as indicated in the sightglass.

206F39.EPS

Figure 39 ◆ Forklift with lowered stabilizers.

Before raising the stabilizers, fully retract and lower the boom. Check that all personnel are clear of the area before raising the stabilizers. Make sure that both stabilizers are fully raised before moving the forklift.

6.0.0 ◆ WORK ACTIVITIES

Operation of the forklift is fairly straightforward, but it requires an attention to detail. With proper planning, you will have no trouble operating the forklift. The basic work activities performed with a forklift are described in this section.

 NOTE

The controls on specific forklifts may be different from those described in the procedures. Check your operator's manual for information about the controls and limitations of your equipment.

6.1.0 Basic Operational Movement

The most important factor to consider when using a forklift is its capacity. Each forklift is designed with an intended capacity which must not be exceeded. Exceeding the capacity jeopardizes the equipment, the operator, and anyone nearby. Each manufacturer supplies a capacity chart for each forklift. Read and follow the capacity chart.

6.1.1 Picking Up a Load

To pick up a load, first check the position of the forks. They should be centered on the carriage. If they must be adjusted, check the operator's manual for the proper procedure. Usually, there is a pin at the top of each fork. When the fork is lifted, you can slide it along the upper backing plate until it reaches the desired position. Lock them into place.

Travel to the area at a safe rate of speed. Use the foot throttle and the steering wheel to maneuver. Always keep the forks low and tilted slightly back.

Before picking up the load, make sure it is stable. If it looks like the load might shift when picked up, secure the load with strapping or other tie-downs. The center of gravity is critical to load stability. If necessary, make a trial lift and adjust the load.

Approach the load so that the forks straddle the load evenly. It is important that the weight of the load is distributed evenly on the forks. Overloading one fork can damage it or cause the forklift to overturn. In some cases, it may be advisable to measure and mark the center of gravity.

Drive up to the load with the forks straight and level (*Figure 40*). Approach the load so that the forklift is square and level to the intended load. If necessary, level the forklift using the frame level controls or the stabilizers. If the load is on a pallet, make sure that the forks will clear the pallet board. If the load is on blocks, you may need to get out and check if the forks will clear.

WARNING!

Set the parking brake and turn off the engine before exiting the cab. Do not leave a machine running unattended.

Move forward slowly until the leading edge of the load rests against the back of both forks. If you cannot see the forks engage the load, ask someone to signal for you. This avoids expensive damage and injury. Raise the carriage slowly until the forks contact the load. Continue raising the carriage until the load safely clears the ground (*Figure 41*). Tilt the mast fully rearward to cradle the load. This minimizes the possibility of the load slipping in transport.

CAUTION

Ice, mud, grease, rain, or snow can cause the forks to become very slippery. Clean the forks whenever they become wet or fouled. Wet metal on wet metal is extremely slippery. Be extremely careful when picking up metal loads with wet forks.

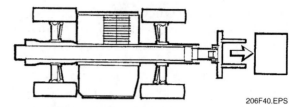

206F40.EPS

Figure 40 ◆ Approach the load squarely.

206F41.EPS

Figure 41 ◆ Raise the load slowly.

6.1.2 Picking Up an Elevated Load

Picking up a load from an elevated position is more difficult than picking a load off the ground. First, it is more difficult to see an elevated load. Use a spotter if necessary. Second, the load can fall farther and cause other materials to fall. Use extreme caution when picking up elevated loads. Remember that the capacity is diminished as load height increases. Use the load chart to make sure that you can safely lift and place the load.

Follow the procedure above to approach the elevated load square and level. Raise and extend the boom so that the forks are in line with the load (*Figure 42*). Extend and lower the boom to move the forks under the load. The forks will remain level. Be careful not to hit the stack.

Carefully raise the boom to lift the load, then tilt the forks slightly backward to cradle the load (*Figure 43*) before continuing to raise the load. Once the load is cradled, retract the boom slowly until the load is clear. Lower the boom to the travel position before moving the forklift. Be careful not to hit the stack as you retract the load.

6.1.3 Traveling with a Load

Always travel at a safe rate of speed. Never travel with a raised load. Keep the load as low as possible (*Figure 44*). Be sure that the carriage is level and tilted slightly rearward so the load does not slip off.

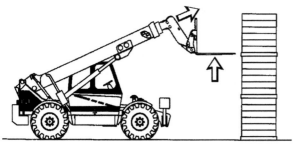

Figure 42 ◆ Raise the boom to align the forks with the load.

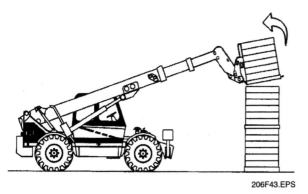

Figure 43 ◆ Tilt the forks to cradle the load.

As you travel, keep your eyes open and stay alert. Watch the load and conditions ahead of you. Alert others to your presence. Avoid sudden stops and abrupt changes in direction. Be careful when downshifting because sudden deceleration can cause the load to shift. Be aware of front and rear swing when turning.

If you have to drive on a steep slope, keep the load as low as possible. Do not drive across steep slopes, because the forklift could overturn. Drive up and down the slope, but make sure the load is pointed uphill with a load and downhill without. If you have to turn on an incline, make the turn wide and slow. This minimizes the risk of tipping over.

 WARNING!
Do not carry passengers on the forks or in the cab. Never lift personnel on the forks; use only equipment specifically designed for this purpose to elevate personnel.

Driving over bumps and holes with a load can cause you to lose a load. If the machine bumps, the load can bounce off. Avoid driving over bumps and holes. If impossible to avoid, drive over them slowly and keep the load low.

Figure 44 ◆ Keep the load low.

6.1.4 Placing a Load

Position the forklift at the landing point so the load can be placed where you want it. Remember that it is easier to reposition the forklift than to reposition the load after it is placed. If the load is not palletized, place blocking material where the load is to be placed. The blocking will create space under the load so that you can remove the forks. Make sure that the blocking will adequately support the load and be sure that everyone is clear of the area.

The area under the load must be clear of obstruction and must be able to support the weight of the load. If you cannot see the placement, use a spotter to guide you.

When the forklift is in position, tilt the forks to the horizontal position. Lower the load slowly. You can usually feel when the load is resting on the ground or blocking. When the load has been placed, lower the forks a little more to clear the underside of the load. Back carefully away from the load to disengage the forks.

 NOTE
Do not lower the forks too much after placing the load. You can lower the forks too far and dig up the ground under the load when backing up. Lower the forks just enough to clear the bottom of the load but still remain off the ground.

6.1.5 Placing an Elevated Load

You must take extra precautions when placing elevated loads. It is extremely important to level the machine before lifting the load. Failure to do so can cause the machine, the load, or the existing stack to tip over.

One of the biggest safety hazards for elevated load placement is poor visibility (*Figure 45*). There may be workers in the area who cannot be seen. The landing point itself may not be visible. Your depth perception decreases as the height of the load increases. To be safe, use a signal person to help you spot the load. Use tag lines for long loads.

Drive the forklift as close as possible to the landing point with the load kept low. Set the parking brake. Lower the stabilizers if necessary. Raise the load slowly and carefully while maintaining a slight rearward tilt to keep the load cradled. Do not tilt the load forward until the load is over the landing point and ready to be set down.

If the forklift's rear wheels start to lift off the ground, stop immediately but not abruptly. Lower the load. Slowly reposition it, break it down into smaller loads, or lower the stabilizers (*Figure 46*). If the surface conditions are bad at the unloading site, it may be necessary to reinforce the surface to provide more stability.

6.1.6 Unloading a Flatbed Truck

The following steps are used to unload material from a flatbed truck:

Step 1 Position the forklift at either side of the truck bed.

Step 2 Manipulate the control levers in order to obtain the appropriate fork height and angle.

Step 3 Drive forks into the opening of the pallet or under the loose material. Use care not to damage any material.

Step 4 Adjust the controls as required to lift the material slightly off the bed.

Step 5 Tilt the forks back to keep the pallet or other material from sliding off the front of the forks (*Figure 47*).

Step 6 Retract the boom or back the forklift away from the truck.

Step 7 Lower the forks to the travel position and move to the stockpile area.

Step 8 Position the forklift so that the material can be placed in the desired area.

Step 9 Lower the boom until the material is set on the required surface.

Step 10 Adjust the forks with the boom lever in order to relieve the pressure under the pallet.

Step 11 Back the forklift away from the pallet.

Step 12 Repeat the cycle until the truck is unloaded.

206F46.EPS

Figure 46 ◆ Use the stabilizers to place heavy loads.

206F47.EPS

Figure 47 ◆ Tilt forks back to secure load.

206F45.EPS

Figure 45 ◆ Place elevated loads carefully.

6.2.0 Using Special Attachments

In addition to the various types of forks, there are several special attachments that expand the forklift's operational capability. The three main attachments are the hook, the boom extension, and the bucket. Always read the operator's manual to make sure you follow the proper procedure for securing attachments to the forklift.

6.2.1 Changing Attachments

Some forklifts have a coupler system that allows the operator to easily change attachments. *Figure 48* shows the main features of the coupler. The coupler is activated with switches located next to the steering wheel. On some models the coupler is controlled with the joystick after the switch is activated.

Before detaching the forks from the forklift, you need to disengage the hydraulics. Typically, there is a diverter valve on the hydraulic hoses which must be closed. Check the operator's manual for the correct procedures. If the attachments are not secured properly, they could fail, causing property damage and significant injury.

 WARNING!
The hydraulic system is under pressure. Follow the safety procedures listed in the operator's manual for relieving pressure before disconnecting hydraulic hoses. The release of fluids under pressure can cause significant injury.

To attach the coupler to the forks or other attachments, position the coupler in line with the attachment. Tilt the coupler forward so that it is below the levels of the hooks. Move the forklift forward or extend the boom until the coupler contacts the carriage. Tilt the coupler back until the lower part of the carriage contacts the coupler, then secure the attachment to the coupler.

6.2.2 Hook and Boom Extension

The boom extension and lifting hook allow the forklift to lift objects from above rather than from below. These attachments feature a sturdy hook mounted on the carriage or the end of a boom extension (*Figure 49*). Loads must be securely rigged using approved lifting equipment. Each attachment has different lifting capacities. Be sure to use the correct load chart when planning the lift. Do not exceed the lifting capacity of the equipment and the attachment.

 WARNING!
Only use approved chains, slings, hooks, and other rigging. Nonstandard rigging can fail and cause property damage and personal injury.

Position the hook or lifting point directly above the load. If it is not, the load could swing when it is lifted. Secure the load to the hook. Using shorter slings also reduces swinging (*Figure 50*). Use tag lines to control load swing and placement. Once the load is secured, use the boom controls to lift and position the load.

 CAUTION
In extremely cold temperatures, the load can freeze to the ground. Free the load before attempting to lift it. Lifting a frozen load will cause a jolt that could affect the stability of the machine or dislodge the load.

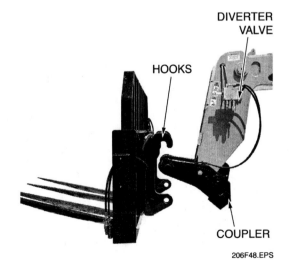

Figure 48 ◆ Quick coupler.

Figure 49 ◆ Forklift with boom extension.

6.2.3 Using a Bucket

Loading trucks, bins, and other containers can be done using a forklift with a bucket attachment. Usually, this loading is done by taking material from a stockpile. The procedure for carrying out a loading operation from a stockpile is as follows:

Step 1 Travel to the work area with the bucket in the travel position.

Step 2 Position the bucket parallel to and just skimming the ground.

Step 3 Drive the bucket straight into the stockpile, as shown in *Figure 51*.

Step 4 Tilt the bucket backwards to fill it, as shown in *Figure 52*.

Step 5 Work the tilt control lever back and forth to move material to the back of the bucket. This is called bumping. When the bucket is full, move the tilt control lever to the tilt back position.

Step 6 Shift the gears to reverse and back the forklift away from the stockpile.

Step 7 Place the bucket in the travel position and move the forklift to the truck.

Step 8 Center the forklift with the truck bed and raise the bucket high enough to clear the side of the truck.

Step 9 Move the bucket over the truck bed and shift the bucket control lever forward to dump the bucket (*Figure 53*).

Step 10 Pull the bucket control lever to retract the empty bucket, and back the forklift away from the truck as soon as the bucket is empty.

Step 11 Lower the bucket to the travel position and return to the stockpile.

Step 12 Repeat the cycle until the truck is loaded.

206F50.EPS

Figure 50 ◆ Lift heavy loads with the hook attachment.

206F52.EPS

Figure 52 ◆ Filling forklift bucket.

206F51.EPS

Figure 51 ◆ Drive the bucket into the stockpile.

206F53.EPS

Figure 53 ◆ Center the bucket over the truck bed.

As the truck fills, the material will need to be pushed across the truck bed to even the load. As the leading edge of the bucket passes the sideboard of the truck, roll the bucket down quickly. Dump the material in the middle of the bed. The load is then pushed across the truck as the bucket is raised. By raising the bucket and backing up slowly, the material will be distributed evenly across the bed.

CAUTION

Avoid hitting the side of the truck with the boom or bucket when you are unloading.

While there are many ways to maneuver a forklift, the two most common patterns for a truck loading operation are the I-pattern and the Y-pattern. These maneuvers were covered in the *Heavy Equipment Operations Level Two* module *Loaders*. Review that module to fully understand various loading patterns.

6.2.4 Multipurpose Bucket

A multipurpose bucket has a hydraulically operated clamshell design. The bucket can perform four basic functions. It can be used as a clamshell bucket, scraper, dozer, and regular forklift bucket.

The clamshell can be used for removing stumps and large rocks, as well as picking up debris and brush. To do this, the operator must open the bowl and position the bucket over the material to be loaded, then lower the bucket and close the bowl to fill the bucket. Material can then be transported to the truck or stockpile for dumping. Use of the clamshell configuration gives the forklift added height for dumping and better handling of sticky material such as wet clay soils.

Using the bucket as a scraper requires the operator to open the bowl and use the back side of the bucket to cut material. When the material is filling the back of the bucket, close the bowl over the material, and raise the bucket for transporting to the dump site.

For use as a dozer or pusher, the operator must open the bowl fully and use the back of the bucket as a blade. Level cutting is maintained by the bucket lift control. The material can then be pushed into an area to create a stockpile. The multipurpose bucket is useful for roughing-in access roads as shown in *Figure 54*.

With the added height of the multipurpose bucket, loading trucks becomes easier because the boom can remain higher and stay away from the

206F54.EPS

Figure 54 ◆ Multipurpose bucket.

side of the truck. The clamshell configuration also makes it easier to dump sticky material because the bucket does not compact the material.

6.3.0 Transporting the Forklift

If the forklift needs to be transported from one job site to another, it may either be driven if it is a short distance, or loaded and hauled on a transporter.

Before roading the forklift from one site to another, make sure the necessary permits for traveling on a public road have been obtained. Inflate the tires to the correct air pressure for road travel. Switch on lights and flashers. Drive at a moderate rate of speed, especially around corners and over rough terrain. Travel with the boom fully retracted and the forks as low as possible while maintaining adequate ground clearance.

If the forklift must be moved a long distance, it should be transported on a properly equipped trailer or other transport vehicle. Before beginning to load the equipment for transport, make sure the following tasks have been completed:

- Check the operator's manual to determine if the loaded equipment complies with height, width, and weight limitations for over-the-road hauling.
- Check the operator's manual to identify the correct tie-down points on the equipment (*Figure 55*).
- Be sure to get the proper permits, if required.

Once the above tasks are completed and the loading plan determined, carry out the following procedures:

Step 1   Position the trailer or transporting vehicle. Always block the wheels of the transporter after it is in position but before loading is started.

Step 2   Place the forks in the travel position and drive the forklift onto the transporter. Whether the forklift is facing forward or backward will depend on the recommendation of the manufacturer. Most manufacturers recommend backing the forklift onto the transporter.

Step 3 Lower the forks or other attachments to the floor of the transporter.

Step 4 Ensure that the boom is fully retracted and lowered.

Step 5 Move the transmission lever to neutral, engage the parking brake, turn off the engine, and remove the key.

Step 6 Manipulate the hydraulic controls to relieve any remaining hydraulic pressure.

TIE DOWN POINTS (FRONT)

TIE DOWN POINTS (REAR)   206F55.EPS

Figure 55 ◆ Forklift tie-down points.

Step 7 Lock the door to the cab and any access covers. Attach any vandalism protection.

Step 8 Secure the machine with the proper tie-down equipment as specified by the manufacturer. Place chocks at the front and back of all four tires.

Step 9 Cover the exhaust and air intake openings with tape or a plastic cover.

Step 10 Place appropriate flags or markers on the equipment if needed for height and width restrictions.

> **WARNING!**
> The machine may shift while in transit if it is not properly tied down. If the machine shifts in transport it could cause personal injury or death. Follow the manufacturer's safety procedures.

Unloading the equipment from the transporter would be the reverse of the loading operation.

Vehicles on public roads that are traveling at speeds of less than 25 miles per hour must have a sign like the one shown in *Figure 56*. The triangular sign is fluorescent yellow-orange with a dark red reflective border.

206F56.EPS

Figure 56 ◆ Slow vehicle sign.

1. Warehouse forklifts commonly use electricity or liquid propane because they _____.
   a. are quieter
   b. do not produce harmful emissions
   c. are smaller
   d. can move heavier loads

2. Circle steering is also called _____.
   a. crab steering
   b. two-wheel drive
   c. two-wheel steering
   d. four-wheel steering

3. Fixed-mast forklifts are limited to ____.
   a. indoor use only
   b. operation on smooth, hard, or paved surfaces
   c. loads under 5,000 pounds
   d. how close the machine can get to the landing point

4. A telehandler is another name for a(n) _____.
   a. fixed-mast forklift
   b. rough-terrain forklift
   c. telescoping-boom forklift
   d. electric forklift

5. Most rough terrain forklifts have a(n) _____ engine.
   a. diesel
   b. LP
   c. electric
   d. propane

6. If the load stability indicator reaches 100 percent, the operator must immediately _____.
   a. raise the load
   b. retract the boom
   c. extend the boom
   d. shut down the machine

7. On some forklifts, the joystick operates the _____.
   a. boom and forks
   b. instrument panel
   c. load stability indicator
   d. transmission

8. To change the travel direction, _____.
   a. release the parking brakes
   b. switch steering modes
   c. rotate the transmission control lever
   d. move the joystick forward

9. How fast the boom is extended depends on _____.
   a. engine speed
   b. air pressure
   c. hydraulic pressure
   d. oil pressure

10. On all forklifts, the forks and boom are controlled with a joystick.
    a. True
    b. False

11. Cube forks are normally used in a set of _____.
    a. 2
    b. 3
    c. 4
    d. 6

12. Standard forks are _____ inches long.
    a. 24
    b. 48
    c. 54
    d. 60

13. When using the access platform, lock out other machine functions with _____.
    a. a key switch
    b. the differential lockout
    c. the transmission lock
    d. the joystick

14. For safety, only operate the forks _____.
    a. with the stabilizers lowered
    b. when the engine is not running
    c. when the machine is stationary
    d. from the operator's cab

15. If necessary, riders are allowed in the cab of the forklift.
    a. True
    b. False

16. In a daily inspection you should look under the machine for _____.
    a. leaks
    b. debris
    c. missing inventory
    d. the hydraulic reservoir

17. The most important factor to consider when operating a forklift is the _____.
    a. machine's capacity
    b. travel distance
    c. position of the forks
    d. height of the lift

18. If you cannot see the area where the load will be placed use a _____.
    a. boom extension
    b. squirt-boom
    c. spotter
    d. load stability indicator

19. While the engine is warming up you should _____.
    a. get out and adjust the mirrors
    b. test-lift the load
    c. check if the components are working properly
    d. refuel

20. If one of the gauges is *not* working, _____.
    a. don't worry, you can use other gauges
    b. keep working
    c. have it checked at your next break
    d. shut down the machine immediately

21. When working in a confined area, use _____.
    a. crab steering
    b. two-wheel steering
    c. circle steering
    d. overdrive

22. When the forklift is in crab steering mode _____.
    a. only the front wheels move
    b. only the rear wheels move
    c. the front and back wheels turn in the same direction
    d. the front and back wheels turn in opposite directions

23. For normal operations use the _____ steering mode.
    a. circle
    b. crab
    c. two-wheel
    d. oblique

24. The leading cause of death in forklift operations is _____.
    a. forklifts tipping over
    b. being crushed by a load
    c. being run over
    d. electrocution from overhead power lines

25. The frame level controls will tilt the frame up to _____ degrees to the left or right.
    a. 10
    b. 15
    c. 20
    d. 25

Summary

Forklifts are used primarily for lifting and loading material. The forks are used to lift palletized loads and other bundled material. The forklift can be fitted with several attachments that enable it to lift and place loose or bulky material. These attachments include a lifting hook, a boom extension, a hopper, and a bucket.

Vehicle movement is controlled with the steering wheel, accelerator, and brake pedals. The forks and boom are controlled with either levers or a joystick and switches. Study the operator's manual to become familiar with the machine you will be operating.

Safety considerations when operating a forklift include keeping the forklift in good working condition, obeying all safety rules, being aware of other people and equipment in the same area where you are operating, and not taking chances. Perform inspections and maintenance daily to keep the forklift in good working order. One of the primary safety considerations for forklift operations is the rated capacity of the forklift. Know the weight of the load and the capacity of the machine. You must be able to read and interpret a load chart. Exceeding the rated capacity of the machine can cause injury and death.

Always position the forklift so that it is square and level to the load. Pick a load up slowly and tilt the forks backward to cradle the load. Lower the load before traveling to the unloading area. Use extra caution when picking up elevated loads.

Notes

Typical Periodic Maintenance Requirements

Maintenance Interval Schedule

Note: All safety information, warnings, and instructions must be read and understood before you perform any operation or any maintenance procedure.

Before each consecutive interval is performed, all of the maintenance requirements from the previous interval must also be performed.

The normal oil change interval is every 500 service hours. If you operate the engine under severe conditions or if the oil is not Caterpillar oil, the oil must be changed at shorter intervals. Refer to the Operation and Maintenance Manual, "Engine Oil and Filter - Change" for further information. Severe conditions include the following factors: High temperatures, continuous high loads, and extremely dusty conditions .

Refer to the Operation and Maintenance Manual, "S·O·S Oil Analysis" in order to determine if the oil change interval should be decreased. Refer to your Caterpillar dealer for detailed information regarding the optimum oil change interval.

The normal interval for inspecting and adjusting the clearance between the wear pads and the boom is 500 service hours. If the machine is working with excessively abrasive material then the clearance may need to be adjusted at shorter intervals. Refer to the Operation and Maintenance Manual, "Boom Wear Pad Clearance - Inspect/Adjust" for further information.

When Required

Battery - Recycle
Battery or Battery Cable - Inspect/Replace
Boom Telescoping Cylinder Air - Purge
Boom and Frame - Inspect
Engine Air Filter Primary Element - Clean/Replace
Engine Air Filter Secondary Element - Replace
Engine Air Filter Service Indicator - Inspect
Engine Air Precleaner - Clean
Fuel System - Prime
Fuel System Primary Filter - Replace
Fuel System Secondary Filter - Replace
Fuel Tank Cap and Strainer - Clean
Fuses and Relays - Replace
Oil Filter - Inspect
Radiator Core - Clean
Radiator Screen - Clean
Transmission Neutralizer Pressure Switch - Adjust
Window Washer Reservoir - Fill
Window Wiper - Inspect/Replace

Every 10 Service Hours or Daily

Backup Alarm - Test
Boom Retracting and Boom Lowering with Electric Power - Check
Braking System - Test
Cooling System Coolant Level - Check
Cooling System Pressure Cap - Clean/Replace
Engine Oil Level - Check
Fuel System Water Separator - Drain
Fuel Tank Water and Sediment - Drain
Indicators and Gauges - Test
Seat Belt - Inspect
Tire Inflation - Check
Transmission Oil Level - Check
Wheel Nut Torque - Check
Windows - Clean

Every 50 Service Hours or 2 Weeks

Axle Support - Lubricate
Bearing (Pivot) for Axle Drive Shaft - Lubricate
Boom Cylinder Pin - Lubricate
Boom Pivot Shaft - Lubricate
Brake Control Linkage - Lubricate
Carriage Cylinder Bearing - Lubricate
Carriage Pivot Pin - Lubricate
Compensating Cylinder Bearing - Lubricate
Cylinder Pin (Grapple Bucket) - Lubricate
Cylinder Pin and Pivot Pin (Bale Handler) - Lubricate
Cylinder Pin and Pivot Pin (Multipurpose Bucket) - Lubricate
Cylinder Pin and Pivot Pin (Utility Fork) - Lubricate
Fork Leveling Cylinder Pin - Lubricate
Frame Leveling Cylinder Pin - Lubricate
Pulley for Boom Extension Chain - Lubricate
Pulley for Boom Retraction Chain - Lubricate
Quick Coupler - Lubricate
Stabilizer and Cylinder Bearings - Lubricate

Initial 250 Service Hours (or after rebuild)

Boom Wear Pad Clearance - Inspect/Adjust
Service Brake - Adjust

Initial 250 Service Hours (or at first oil change)

Hydraulic System Oil Filter - Replace

Every 250 Service Hours or 3 Months

Axle Breathers - Clean/Replace
Belts - Inspect/Adjust/Replace
Boom Chain Tension - Check/Adjust
Differential Oil Level - Check
Drive Shaft Spline - Lubricate
Drive Shaft Universal Joint - Lubricate

206A01.EPS

Engine Oil Sample - Obtain
Final Drive Oil Level - Check
Hydraulic System Oil Level - Check
Longitudinal Stability Indicator - Test
Transfer Gear Oil Level - Check
Transmission Breather - Clean

Every 500 Service Hours

Differential and Final Drive Oil Sample - Obtain

Every 500 Service Hours or 6 Months

Boom Wear Pad Clearance - Inspect/Adjust
Engine Oil and Filter - Change
Fuel System Primary Filter - Replace
Fuel System Secondary Filter - Replace
Fuel Tank Cap and Strainer - Clean
Hydraulic System Oil Filter - Replace
Hydraulic System Oil Sample - Obtain
Hydraulic Tank Breather - Clean
Service Brake - Adjust
Transmission Oil Sample - Obtain

Every 1000 Service Hours or 1 Year

Differential Oil - Change
Engine Valve Lash - Check
Final Drive Oil - Change
Rollover Protective Structure (ROPS) and Falling
  Object Protective Structure (FOPS) - Inspect
Transfer Gear Oil - Change
Transmission Oil - Change
Transmission Oil Filter - Replace

Every 2000 Service Hours or 3 Months

Axle Breathers - Clean/Replace
Hydraulic System Oil Level - Check

Every 2000 Service Hours or 2 Years

Fuel Injection Timing - Check
Hydraulic System Oil - Change

Every 3 Years After Date of Installation or Every 5 Years After Date of Manufacture

Seat Belt - Replace

Every 3000 Service Hours or 3 Years

Boom Chain - Inspect/Lubricate
Cooling System Coolant Extender (ELC) - Add
Cooling System Water Temperature Regulator -
  Replace
Engine Mounts - Inspect

Every 6000 Service Hours or 6 Years

Cooling System Coolant (ELC) - Change

206A02.EPS

OSHA Inspection Checklist

Operator's Daily Checklist - Internal Combustion Engine Industrial Truck - Gas/LPG/Diesel Truck

Record of Fuel Added

Date		Operator		Fuel	
Truck#		Model#		Engine Oil	
Department		Serial#		Radiator Coolant	
Shift		Hour Meter		Hydraulic Oil	

SAFETY AND OPERATIONAL CHECKS (PRIOR TO EACH SHIFT)

Have a **qualified** mechanic correct all problems.

Engine Off Checks	OK	Maintenance
Leaks – Fuel, Hydraulic Oil, Engine Oil or Radiator Coolant		
Tires – Condition and Pressure		
Forks, Top Clip Retaining Pin and Heel – Check Condition		
Load Backrest – Securely Attached		
Hydraulic Hoses, Mast Chains, Cables and Stops – Check Visually		
Overhead Guard – Attached		
Finger Guards – Attached		
Propane Tank (LP Gas Truck) – Rust Corrosion, Damage		
Safety Warnings – Attached (Refer to Parts Manual for Location)		
Battery – Check Water/Electrolyte Level and Charge		
All Engine Belts – Check Visually		
Hydraulic Fluid Level – Check Level		
Engine Oil Level – Dipstick		
Transmission Fluid Level – Dipstick		
Engine Air Cleaner – Squeeze Rubber Dirt Trap or Check the Restriction Alarm (if equipped)		
Fuel Sedimentor (Diesel)		
Radiator Coolant – Check Level		
Operator's Manual – In Container		
Nameplate – Attached and Information Matches Model, Serial Number and Attachments		
Seat Belt – Functioning Smoothly		
Hood Latch – Adjusted and Securely Fastened		
Brake Fluid – Check Level		
Engine On Checks – Unusual Noises Must Be Investigated Immediately	OK	Maintenance
Accelerator or Direction Control Pedal – Functioning Smoothly		
Service Brake – Functioning Smoothly		
Parking Brake – Functioning Smoothly		
Steering Operation – Functioning Smoothly		
Drive Control – Forward/Reverse – Functioning Smoothly		
Tilt Control – Forward and Back – Functioning Smoothly		
Hoist and Lowering Control – Functioning Smoothly		
Attachment Control – Operation		
Horn and Lights – Functioning		
Cab (if equipped) – Heater, Defroster, Wipers – Functioning		
Gauges: Ammeter, Engine Oil Pressure, Hour Meter, Fuel Level, Temperature, Instrument Monitors – Functioning		

206A03.EPS

Trade Terms Introduced in This Module

Articulated steering: A steering mode where the front and rear wheels may move in opposite directions, allowing for very tight turns, also known as four-wheel steering or circle steering.

Circle steering: A steering mode where the front and rear wheels may move in opposite directions, allowing for very tight turns. Also known as four-wheel steering or articulated steering.

Crab steering: A steering mode where all wheels may move in the same direction, allowing the machine to move sideways on a diagonal, also known as oblique steering.

Four-wheel steering: A steering mode where the front and rear wheels may move in opposite directions, allowing for very tight turns, also known as articulated steering or circle steering.

Oblique steering: A steering mode where all wheels may move in the same direction, allowing the machine to move sideways on a diagonal. Also known as crab steering.

Powered industrial trucks: An OSHA term for several types of light equipment that includes forklifts.

Telehandler: A type of powered industrial truck characterized by a boom with several extendable sections known as a telescoping boom. Another name for a shooting boom forklift.

Additional Resources

This module is intended to be a thorough resource for task training. The following reference work is suggested for further study. This is optional material for continued education rather than for task training.

Forklift Safety: A Practical Guide to Preventing Powered Industrial Truck Incidents and Injuries, 2nd Edition, 1999. George Swartz. Lanham, MD: Government Institutes.

Figure Credits

Sellick Equipment Ltd., 206F01, 206F27

Yale Materials Handling Corporation, 206F02

Reprinted courtesy of Caterpillar Inc., 206F03-206F26, 206F28-206F55, Appendix A, Appendix B

CONTREN® LEARNING SERIES — USER UPDATE

The NCCER makes every effort to keep these textbooks up-to-date and free of technical errors. We appreciate your help in this process. If you have an idea for improving this textbook, or if you find an error, a typographical mistake, or an inaccuracy in NCCER's Contren® textbooks, please write us, using this form or a photocopy. Be sure to include the exact module number, page number, a detailed description, and the correction, if applicable. Your input will be brought to the attention of the Technical Review Committee. Thank you for your assistance.

Instructors – If you found that additional materials were necessary in order to teach this module effectively, please let us know so that we may include them in the Equipment/Materials list in the Annotated Instructor's Guide.

Write: Product Development and Revision
National Center for Construction Education and Research
P.O. Box 141104, Gainesville, FL 32614-1104

Fax: 352-334-0932

E-mail: curriculum@nccer.org

Craft _____ Module Name _____

Copyright Date _____ Module Number _____ Page Number(s) _____

Description _____

_____

_____

_____

(Optional) Correction _____

_____

_____

(Optional) Your Name and Address _____

_____

_____

_____

Heavy Equipment Operations Level Two

22207-06

Excavation Math

22207-06
Excavation Math

Topics to be presented in this module include:

Overview

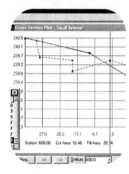

Which bucket should I use to load this material? How many truck loads of dirt must be removed from the site? Can I use this forklift to move this load? These may not seem like math problems; however, math is used to solve these and many other questions heavy equipment operators face everyday.

Can you look at a site plan and estimate the number of truck loads of soil that must be removed? Excavation math helps you answer these and other questions. First you must learn the basic formulas and then you can apply them in the right order to get the correct result. Just like operating a new piece of equipment, skill comes through a step-by-step approach and practice.

Objectives

When you have completed this module, you will be able to do the following:

1. Identify basic geometric shapes.
2. Calculate the surface area of squares, rectangles, triangles, trapezoids, and circles using formulas.
3. Calculate the volume of cubes, rectangular objects, prisms, and cylinders.
4. Calculate the excavation volume of a job using information supplied on the building plans.
5. Calculate the weight of an excavation from its volume.

Trade Terms

Average
Parallel

Parallelogram
Quadrilateral

Required Trainee Materials

1. Pencil and paper
2. Ruler
3. Calculator

Prerequisites

Before you begin this module, it is recommended that you successfully complete *Core Curriculum*, *Heavy Equipment Operations Level One*, *Heavy Equipment Operations Level Two*, Modules 22201-06 through 22206-06.

This course map shows all of the modules in the second level of the *Heavy Equipment Operations* curriculum. The suggested training order begins at the bottom and proceeds up. Skill levels increase as you advance on the course map. The local Training Program Sponsor may adjust the training order.

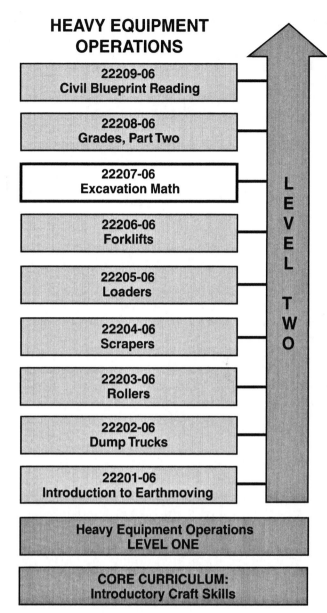

HEAVY EQUIPMENT OPERATIONS

22209-06
Civil Blueprint Reading

22208-06
Grades, Part Two

22207-06
Excavation Math

22206-06
Forklifts

22205-06
Loaders

22204-06
Scrapers

22203-06
Rollers

22202-06
Dump Trucks

22201-06
Introduction to Earthmoving

LEVEL TWO

Heavy Equipment Operations
LEVEL ONE

CORE CURRICULUM:
Introductory Craft Skills

207CMAP.EPS

1.0.0 ◆ INTRODUCTION

This module expands on the material you learned in *Introduction to Construction Math* in the *Core Curriculum*. In that module, you studied whole numbers, fractions, decimals, and the metric system. You also learned that volume is the measure of how much space an object or material, such as a box or soil, takes up. In this module, you will learn more about calculating volume. If necessary, you may want to review all or part of the material covered in *Introduction to Construction Math* before proceeding with the material covered here.

During excavations, heavy equipment operators need to understand how to calculate the volume of material to be removed from or brought into a site. Once you calculate the volume of material, you will need to calculate its weight. Knowing the volume and weight of excavation material is necessary for many reasons, but the most important is safety. Because heavy equipment is rated by weight and volume, you need to know what your machine can safely handle. Overloading a vehicle makes it harder to operate and places workers in danger of an accident. In addition, overloading a vehicle can damage it, requiring costly repairs.

After safety, the next most important reason you need to be able to calculate weight and volume is for efficient scheduling of workers and equipment. In the *Introduction to Earthmoving* module, you learned each machine has a cycle time that is used to estimate how long it will take to complete a job. For example, compare the volume and weight capacities of a Caterpillar® 5130 excavator shovel shown in *Table 1*. If you are working on a job where you are excavating dry or loose soil, a general-purpose bucket may fulfill the job requirements. But if you are working with wet soil or rock, another bucket may be required.

As you study this module, resist the desire to skip practice problems. Math is just like anything else—you become skillful with practice. Initially, calculations may seem difficult, but if you practice, you will soon improve.

In excavations, you will be concerned with determining the volume of a cut or fill area. To determine the volume of any excavation, you will first calculate the area of a shape and then you will use the area to calculate the volume of an object.

Table 1 Excavator Bucket Capacities

Bucket Type	Volume Capacity (heaped, cu ft)	Weight Capacity (pounds)
General purpose	13.7	30,319
Rock	13.7	34,288
High-density rock	11.0	23,153

2.0.0 ◆ WORKING WITH FORMULAS AND EQUATIONS

You will solve many excavation problems by using formulas. Formulas are equations made up of letters, which represent values, and symbols (mathematical signs such as + and × that tell you what to do). Study the following formula for calculating the area of a circle:

$$a = \pi r^2$$

In this case, the letter *a* means area, π means pi (pronounced "pie"), and the letter *r* means radius. To read this formula, you would say, "Area is equal to pi times radius **squared**." When you see an expression such as πr^2 in a formula, it means that you must multiply the values. In this case, you would multiply the radius squared by π.

Some values in formulas are variables. This means that the value is not a single number, so any number can replace a variable. In this sample formula, r is a variable. No matter what size a circle is, and thus no matter what its radius is, we can determine its area with this formula.

Some values, however, are constant. Pi is a constant. Its value is always 3.14159, often rounded off to 3.14. Whenever you see the word pi or its symbol (π) in a formula, you know to replace it with 3.14.

Equations are collections of numbers, symbols, and mathematical operators connected by equal signs (=). Consider the following equation for calculating the area of a rectangle:

$$\text{Area} = L \times W$$

In this case, the letter L means length and the letter W means width. The formula means multiply the length by the width. It can also be written as LW, without the multiplication sign. No multiplication sign is required when the intended relationship between symbols and letters is clear. For example, 2L means two times L.

2.1.0 Sequence of Operations

Complicated equations must be solved by performing the indicated operations in a prescribed order: parentheses, exponents, multiply and divide, and add and subtract (PEMDAS). Always move from left to right when performing multiplication and division, and addition and subtraction. For example, the following equation can result in a number of answers if the PEMDAS order is not followed:

$$(3 + 3) \times 2 - 6 \div 3 + 1 = ?$$

Step 1   Parentheses:
$$\underline{(3 + 3)} \times 2 - 6 \div 3 + 1 = ?$$

Step 2   Multiply and divide:
$$\underline{6 \times 2} - 6 \div 3 + 1 = ?$$
$$12 - \underline{6 \div 3} + 1 = ?$$

Step 3   Add and subtract:
$$12 - 2 + 1 = ?$$
$$10 + 1 = ?$$

Result   11

When none of the numbers are grouped within parentheses, the process is as follows:

$$3 + 3 \times 2 - 6 \div 3 + 1 = ?$$

Step 1   Multiply and divide:
$$3 + \underline{3 \times 2} - 6 \div 3 + 1 = ?$$
$$3 + 6 - \underline{6 \div 3} + 1 = ?$$

Step 2   Add and subtract:
$$\underline{3 + 6} - 2 + 1 = ?$$
$$\underline{9 - 2} + 1 = ?$$
$$\underline{7 + 1} = ?$$

Result   8

Many of the equations you will be using to calculate area and volume will require that you multiply a series of numbers. Since the operation is the same, the order in which you perform the operation is not important. For example, compare the following equations:

$$\underline{3 \times 4} \times 2 \times \tfrac{1}{2} = ?$$
$$\underline{12 \times 2} \times \tfrac{1}{2} = ?$$
$$\underline{24 \times \tfrac{1}{2}} = ?$$
Result 12

$$3 \times 4 \times \underline{2 \times \tfrac{1}{2}} = ?$$
$$3 \times \underline{4 \times 1} = ?$$
$$\underline{3 \times 4} = ?$$
Result 12

$$3 \times \underline{4 \times 2} \times \tfrac{1}{2} = ?$$
$$3 \times \underline{8 \times \tfrac{1}{2}} = ?$$
$$\underline{24 \times \tfrac{1}{2}} = ?$$
Result 12

2.2.0 Squares and Square Roots

The formula for the area of a circle is one of the formulas that involve squared numbers. You will also work with formulas in which you must find the square root of a given number.

A square is the product of a number or quantity multiplied by itself. For example, the square of 6 means 6 × 6. To denote a number as squared, simply place the exponent 2 above and to the right of the base number. An exponent is a small figure or symbol placed above and to the right of another figure or symbol to show how many times the latter is to be multiplied by itself. For example:

$$6^2 = 6 \times 6 = 36$$

$$6^3 = 6 \times 6 \times 6 = 216$$

The square root of a number is the divisor which, when multiplied by itself (squared), gives the number as a product. Extracting the square root refers to a process of finding the equal factors which, when multiplied together, return the original number. The process is identified by the radical symbol [√]. This symbol is a shorthand way of stating that the equal factors of the number under the radical sign are to be determined. Finding the square root is necessary in many calculations, including those involving right triangles.

For example, $\sqrt{16}$ is read as the square root of 16. The number consists of the two equal factors 4 and 4. Thus, when 4 is squared, it is equal to 16. Again, squaring a number simply means multiplying the number by itself.

The number 16 is a perfect square. Numbers that are perfect squares have whole numbers as the square roots. For example, the square roots of perfect squares 4, 25, 36, 121, and 324 are the whole numbers 2, 5, 6, 11, and 18, respectively.

Squares and square roots can be calculated by hand, but the process is very time consuming and subject to error. Most people find squares and square roots of numbers using a calculator. To find the square of a number, the calculator's square key [x^2] is used. When pressed, it takes the number shown in the display and multiplies it by itself. For example, to square the number 4.235, you would enter 4.235, press the [x^2] key, then read 17.935225 on the display.

Similarly, to find the square root of a number, the calculator's square root key [√] or [√$_x$] is used. When pressed, it calculates the square root of the number shown in the display. For example, to find the square root of the number 17.935225, enter 17.935225, press the [√] or [√$_x$] key, then read 4.235 on the display.

> **NOTE**
> On some calculators, the [√] or [√$_x$] key must be pressed before entering the number.

2.3.0 Using Formulas to Solve Problems

To solve problems using formulas, enter the measurements into the equation that makes up the formula. For example, let's say you need to know the area of a round access cover located in one area of a job site. See *Figure 1*. You measure across the cover and determine that its diameter is four feet. Its radius (r) is half that, or two feet. To determine the area of the round access cover, you can plug this number into the formula for the area of a circle (area = πr^2) in place of the letter r. Look at the formula with the numbers that represent the variable (r) and the constant (π) plugged into the equation:

$$a = \pi r^2$$

$$a = (3.14)(2^2)$$

Notice that 3.14 and 2^2 have been placed inside parentheses. When numbers are enclosed in parentheses, you must finish any calculations inside each set of parentheses before completing any other calculations. In this case, you will first find 2^2 by multiplying 2×2:

$$a = (3.14)(4)$$

$$a = 12.56 \text{ square feet}$$

The parentheses in this equation are a type of grouping symbol. Grouping symbols are just like punctuation in writing. Imagine how hard it would be to read this module if there were no periods or commas to tell you where to stop or pause. Grouping symbols help you make sense of an equation, just like punctuation helps you make sense of a sentence. Grouping symbols tell you which numbers belong together and which functions to perform first. It's important to pay attention to how terms are grouped in a formula and to do the calculations in the right order. In more complex problems, additional types of grouping symbols, such as square brackets [ ] and braces { }, are used.

2.4.0 Angles

Angles are measured in degrees. You can easily relate angles to a circle. As shown in *Figure 2*, a circle is 360 degrees, and it can be evenly divided into four parts, each containing 90 degrees. An angle is made when two straight lines meet (*Figure 3*). The point where they meet is called a vertex (point B in *Figure 3*). The two lines are the sides of the angle. These lines are called the rays of the angle. The angle is the amount of opening that exists between the rays. It is measured in degrees. Two ways are commonly used to identify angles. One is to assign a letter to the angle, such as angle D shown in *Figure 3*. This is written: $\angle D$. The other way is to name the two end points of the rays and put the vertex letter between them; for example, $\angle ABC$. When you show the angle measure in degrees, it should be written inside the angle, if possible. If the angle is too small to show the measurement, you may put it outside of the angle and draw an arrow to the inside.

There are several kinds of angles:

- *Right angle* – This angle has rays that are perpendicular to one another (*Figure 4*). The measure of this angle is always 90 degrees. The right angle is used often in construction, so you must remember that it is 90 degrees.

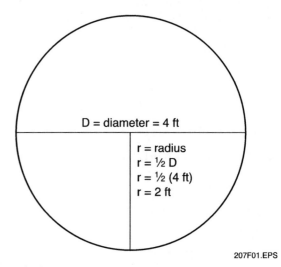

Figure 1 ◆ Access cover.

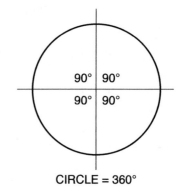

Figure 2 ◆ 360 degrees in a circle.

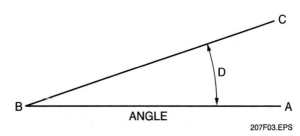

Figure 3 ◆ Angle.

- *Straight angle* – This angle does not look like an angle at all. The rays of a straight angle lie in a straight line, and the angle measures 180 degrees.
- *Acute angle* – An angle less than 90 degrees.
- *Obtuse angle* – An angle greater than 90 degrees, but less than 180 degrees.
- *Adjacent angles* – When three or more rays meet at the same vertex, the angles formed are adjacent (next to) one another. In *Figure 5*, the angles $\angle ABC$ and $\angle CBD$ are adjacent angles. The ray BC is common to both angles.

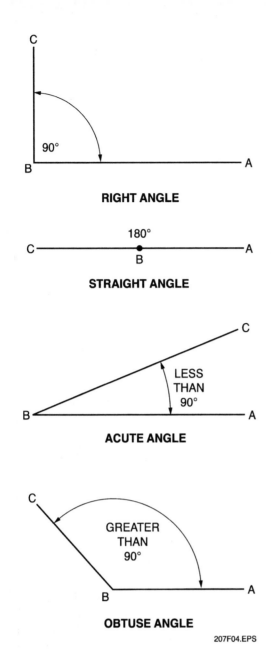

Figure 4 ◆ Right, straight, acute, and obtuse angles.

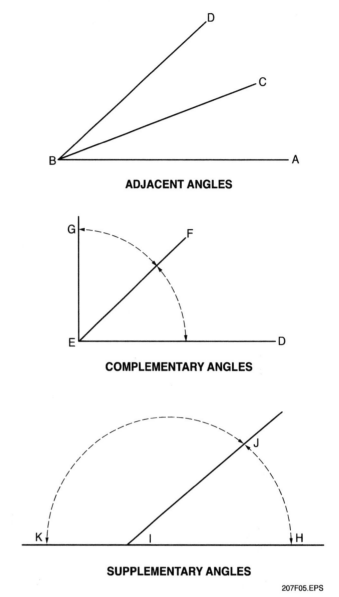

Figure 5 ◆ Adjacent, complementary, and supplementary angles.

- *Complementary angles* – Two adjacent angles that have a combined total measure of 90 degrees. In *Figure 5*, ∠DEF is complementary to ∠FEG.
- *Supplementary angles* – Two adjacent angles that have a combined total measure of 180 degrees. In *Figure 5*, ∠HIJ is supplementary to ∠JIK.

3.0.0 ◆ AREA

Area is the measurement of the amount of space on a flat surface. Houses and apartments are advertised using area—area tells you how much floor space is available. Area is measured in square units, such as square inches, feet, yards, and miles (in the metric system, square centimeters, meters, and kilometers). This unit of measure is often written with the abbreviation inches^2, feet^2, meters^2, etc.

A square inch is the area in a shape that is 1 inch long and 1 inch wide. Shapes that are measured in square units are called two-dimensional, because they have only two measurements. *Figure 6A* shows a block that has an area of 1 square inch. The block is 1 inch long and 1 inch wide so it is a 1-inch square and has an area of 1 square inch. *Figure 6B* shows a larger block. That block is 2 inches long and 1 inch wide and contains two 1-inch squares, so it has an area of 2 square inches.

You have probably already guessed that *Figure 6C* shows a block that has an area of 4 square inches, because it is 4 inches long and 1 inch wide and contains four 1-inch squares.

See *Figure 6D*. The block still contains four 1-inch squares so it has an area of 4 square inches just like

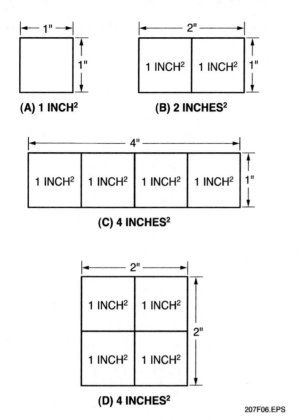

(A) 1 INCH² **(B) 2 INCHES²**

(C) 4 INCHES²

(D) 4 INCHES²

207F06.EPS

Figure 6 ◆ Square inches.

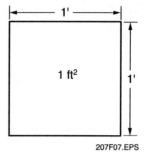

207F07.EPS

Figure 7 ◆ One square foot.

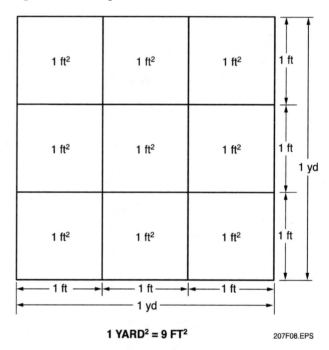

1 YARD² = 9 FT²   207F08.EPS

Figure 8 ◆ One square yard.

Figure 6C, but this block is 2 inches long and 2 inches wide. (You will learn about how to calculate the area of objects in later sections, but for now, concentrate on understanding the idea of square inches.) In *Figure 6D*, you can see that the shape of the surface does not matter—it is still the same area as the shape in *Figure 6C*.

So far, you have learned about area in terms of square inches. Now consider a shape that is one foot long and one foot wide. Just as a square inch is the area on the surface of a shape that is 1 inch long and 1 inch wide, a square foot is the area on the surface of a shape that is 1 foot long and 1 foot wide. See *Figure 7*.

When a shape is 1 yard long and 1 yard wide, it has the area of 1 square yard. Since 1 yard is equal to 3 feet, a square yard is 3 feet long and 3 feet wide. See *Figure 8*. When a square yard is divided into square foot blocks, there are 9 square foot blocks, so 1 square yard is equal to 9 square feet.

Because one foot equals 12 inches, a square foot is 12 inches long and 12 inches wide, and one square foot equals 144 square inches. See *Figure 9*. You may count the number of blocks if you wish.

3.1.0 Area Exercises

You can measure any flat shape in square units. To help you to understand this, perform the following exercise on a separate piece of paper.

EACH BLOCK IS 1 INCH HIGH AND 1 INCH WIDE

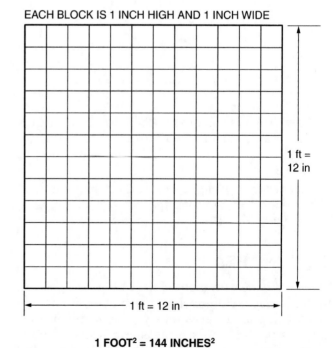

1 FOOT² = 144 INCHES²

207F09.EPS

Figure 9 ◆ 144 square inches.

1. Use a ruler to draw a block 4 inches long and 1 inch wide just like the one shown in *Figure 6C*. This block has an area of 4 square inches like the one shown in *Figure 6C*.

2. Use a ruler to draw a block 2 inches long and 2 inches wide just like the one shown in *Figure 6D*. This block has an area of 4 square inches like the one shown in *Figure 6D*.

3. Divide the figure drawn in exercise 1 in half so that there are two blocks 2 inches long and 1 inch wide. Since the whole block has an area of 4 square inches, half of the block must have an area of 2 square inches.

4. Divide the figure drawn in exercise 2 in half by drawing a line from its upper right corner to its lower left corner. Since the whole block has an area of 4 square inches, half of the block must have an area of 2 square inches.

5. Compare the lower half of the figure drawn in exercise 4 to the right half of the one drawn in exercise 3. Both have the same area of 2 square inches, but the shapes are quite different.

3.2.0 Squares and Rectangles

Squares and rectangles are quadrilaterals. Quad means four and lateral means side, so a **quadrilateral** is a four-sided closed shape. *Figure 10* shows some common quadrilaterals. All quadrilaterals have four corners with angles that add up to 360 degrees. Squares and rectangles are also **parallelograms** because they have two pairs of opposite **parallel** sides with angles that add up to 360 degrees. Of the shapes shown in *Figure 10*, all but the trapezoid are parallelograms.

A rectangle is a four-sided shape joined so that four 90-degree angles are formed. See *Figure 11*. The sum of the four angles in any rectangle is 360 degrees. A rectangle has two pairs of equal sides. The longer side is called the length and is designated with the letter *L*, while the shorter side is called the width and is designated with the letter *W*. The area of a rectangle is calculated by multiplying the length and width (L × W or LW).

A rectangle with a length of 3 inches and width of 6 inches has an area of 18 square inches. It is calculated as follows:

Area = L × W

= 3 inches × 6 inches

= 18 inches^2

When you are calculating the area of any object, the numbers multiplied together must be in the same units. For example, a rectangle with a length of 3 inches and a width of 1 foot has an area of 36

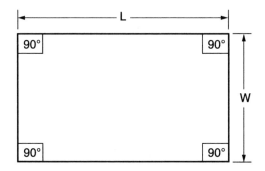

RECTANGLE AREA = LENGTH × WIDTH OR LW

207F11.EPS

Figure 11 ◆ Rectangle.

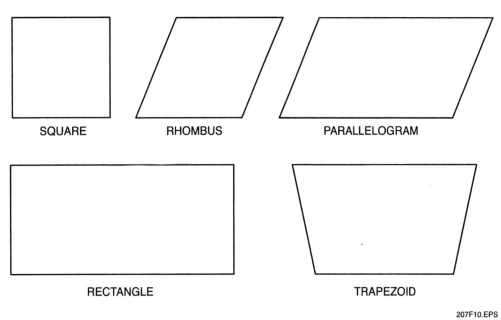

SQUARE RHOMBUS PARALLELOGRAM

RECTANGLE TRAPEZOID

207F10.EPS

Figure 10 ◆ Quadrilaterals.

square inches, because 1 foot equals 12 inches. It is calculated as follows:

$$\text{Area} = \text{L} \times \text{W}$$
$$= 3 \text{ inches} \times 1 \text{ foot}$$
$$= 3 \text{ inches} \times 12 \text{ inches}$$
$$\text{(convert 1 foot to 12 inches)}$$
$$= 36 \text{ inches}^2$$

A square is similar to a rectangle. See *Figure 12*. It has four sides that are joined to form four 90-degree angles. The sum of these angles is 360 degrees—just like a rectangle—but all sides of the square are the same length. The sides of the square are labeled with a single letter. In this module, the sides of a square will be labeled with the letter *e*. The area of a square is calculated by multiplying two sides together ($e \times e$ or e^2).

A square that is 3 inches long has an area of 9 square inches, which is calculated as follows:

$$\text{Area} = e^2$$
$$= 3 \text{ inches} \times 3 \text{ inches}$$
$$= 9 \text{ inches}^2$$

A square that is 12 inches long has an area of 144 square inches, which is calculated as follows:

$$\text{Area} = e^2$$
$$= 12 \text{ inches} \times 12 \text{ inches}$$
$$= 144 \text{ inches}^2$$

The numbers quickly become large when you work with inches, so you will probably want to work with the smallest numbers you can by converting inches into feet or even yards. In the example above, you used 12 inches to arrive at the answer of 144 square feet. By converting the 12 inches to one foot, you can calculate the area as follows:

$$\text{Area} = e^2$$
$$= 1 \text{ foot} \times 1 \text{ foot}$$
$$= 1 \text{ foot}^2$$

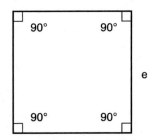

SQUARE AREA = $e \times e$ OR e^2

207F12.EPS

Figure 12 ◆ Square.

These two calculations tell you that 1 square foot is the same as 144 square inches. If you need to convert square feet to square inches, you must multiply the square feet by 144 as follows:

$$1 \text{ foot}^2 \times 144 = 144 \text{ inches}^2$$
$$2 \text{ feet}^2 \times 144 = 288 \text{ inches}^2$$
$$3 \text{ feet}^2 \times 144 = 432 \text{ inches}^2$$

If you need to convert square inches to square feet, you must divide the square inches by 144 as follows:

$$144 \text{ inches}^2 \div 144 = 1 \text{ foot}^2$$
$$432 \text{ inches}^2 \div 144 = 3 \text{ feet}^2$$
$$792 \text{ inches}^2 \div 144 = 5\frac{1}{2} \text{ feet}^2$$

Complete the following three exercises by drawing the shape described and then calculating the area of each shape. Hint: always convert the measurements into the same units before you calculate the area.

1. The building plans call for a building that is 40 feet long and 30 feet wide.
2. Outside the building, the parking pad is 15 feet square.
3. The driveway leading up to the building is 12 yards long and 12 feet wide.

3.3.0 Triangles

A triangle is a three-sided figure that has three angles. The angles in the triangle may vary, but the sum of the angles is always 180 degrees. In construction you will use the following types of triangles (see *Figure 13*):

- *Right triangle* – A right triangle has one 90-degree angle.
- *Equilateral triangle* – An equilateral triangle has three equal angles and three sides of equal length.
- *Isosceles triangle* – An isosceles triangle has two equal angles and two sides of equal length. An isosceles triangle can be divided into two equal right triangles.
- *Scalene triangle* – A scalene triangle has no equal angles or side lengths.

The right triangle is one of the most frequently used triangles in construction. A right triangle must have one 90-degree angle and the sum of the other two angles is 90 degrees. A right triangle is created when a square or rectangle is divided in half diagonally so that you have two identical triangles, each with half the area of the rectangle. See *Figure 14*.

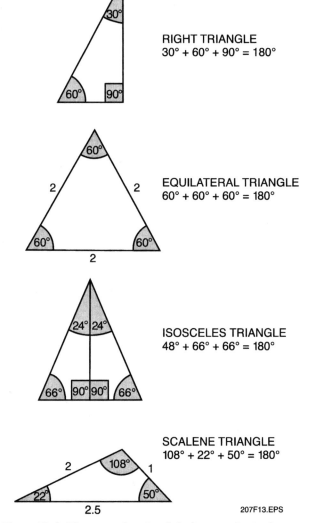

RIGHT TRIANGLE
30° + 60° + 90° = 180°

EQUILATERAL TRIANGLE
60° + 60° + 60° = 180°

ISOSCELES TRIANGLE
48° + 66° + 66° = 180°

SCALENE TRIANGLE
108° + 22° + 50° = 180°

207F13.EPS

Figure 13 ♦ The sum of a triangle's three angles is always 180 degrees.

A triangle's length is called the base, which is abbreviated with the letter B. Its width is called the height, which is abbreviated with the letter H, so the formula to calculate the area of a triangle is ½BH. See *Figure 15*. This formula is used to calculate the area of all triangles.

In *Figure 16*, the four triangles all have the same base (7 inches) and the same height (6 inches), so the area of each triangle is the same (21 square inches), even though their appearances are very different. The area is calculated as follows:

$$\text{Area} = \tfrac{1}{2}BH$$

$$= \tfrac{1}{2}(7 \times 6)$$

$$= \tfrac{1}{2}(42)$$

$$= 21 \text{ inches}^2$$

Some trainees have trouble understanding how the same formula can be used to calculate the area of all triangles, but the exercises at the end of this section will help you to understand.

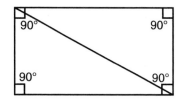

RECTANGLE
Surface Area = L × W so the Surface Area of these triangles must be ½ (L × W)

207F14.EPS

Figure 14 ♦ One rectangle, two triangles.

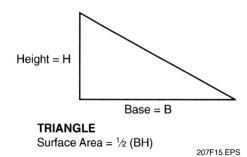

TRIANGLE
Surface Area = ½ (BH)

207F15.EPS

Figure 15 ♦ Triangle area.

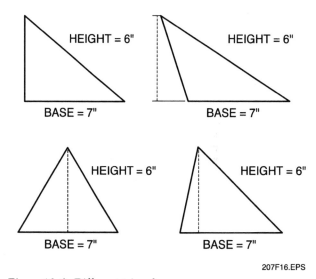

207F16.EPS

Figure 16 ♦ Different triangles, same area.

Triangles are related to excavation. See *Figure 17*. On this highway job, you need to remove part of the existing ground to make an inslope and backslope. The slope stake directs a cut of 3 feet and a grade of 3:1. You know from *Grades, Part One* that this means the ground must drop 1 foot every 3 feet from the stake until the cut depth is 3 feet. In this example, assume that you will cut the backslope the same way. Look carefully at *Figure 17*. The cut that you are directed to make forms an inverted isosceles triangle.

Think about the stake directions. See *Figure 18*. The maximum cut is 3 feet, so the height of the triangle must be 3 feet. A 3:1 slope means the ground must drop 1 foot every 3 feet from the stake. So to the point of maximum cut the horizontal distance is 9 feet (3 feet + 3 feet + 3 feet). Since you are cutting the backslope the same as the inslope, that side is 9 feet, too. So the base of the triangle is 18 feet. You now have all of the information you need to calculate the area of the proposed cut, which is as follows:

Base = 18 feet

Height = 3 feet

Area = ½BH

= ½(18 × 3)

= ½(54)

= 27 feet²

You can use this process to calculate the area of any cut that forms a triangle. See *Figure 19*. On this job, you need to cut the inslope just as you did in the above example, but the backslope has a 6:1 slope from the ditch. The inslope is cut 9 feet from the stake to the maximum depth of 3 feet, just as in the previous example. The backslope needs a 6:1 ratio, so the level of the ground must decrease 1 foot for every 6 feet from the stake to the maximum cut of 3 feet. This means that it is 18 feet (6 + 6 + 6) from the backslope stake to the maximum cut. Look carefully at *Figure 19*. The cut that you are directed to make forms an inverted scalene triangle.

The base of the triangle is 27 feet and the height is 3 feet. See *Figure 20*. You now have all of the

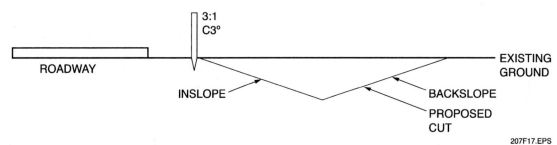

Figure 17 ◆ Proposed cut.

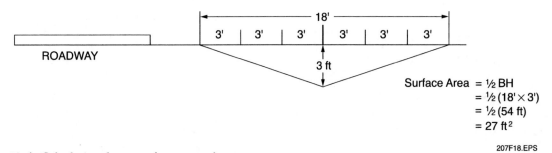

Figure 18 ◆ Calculating the area of a proposed cut.

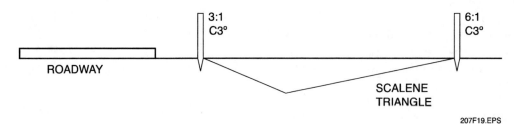

Figure 19 ◆ Another proposed cut.

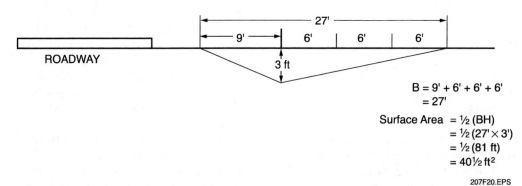

Figure 20 ◆ Calculating the area of another proposed cut.

information you need to calculate the area of the proposed cut, which is as follows:

Base   = 27 feet

Height = 3 feet

Area   = ½BH

     = ½(27 × 3)

     = ½(81)

     = 40½ feet^2

3.3.1 Triangle Area Exercises

The following exercises will help you to better understand area. Read each question carefully and draw figures as necessary.

1. Find the following for *Figure 21*:
   a. Length = _____
   b. Width   = _____
   c. Area   = _____

2. Find the following for *Figure 22*:
   Triangle 1:
   a. Base   = _____
   b. Height = _____
   c. Area   = _____

   Triangle 2:
   d. Base   = _____
   e. Height = _____
   f. Area   = _____

   Triangles 1 and 2:
   g. Total area = _____

3. Find the following for *Figure 23*:
   Triangle 1:
   a. Base   = _____
   b. Height = _____
   c. Area   = _____

   Triangle 2:
   d. Base   = _____
   e. Height = _____
   f. Area   = _____

   Triangle 3:
   g. Base   = _____
   h. Height = _____
   i. Area   = _____

   Triangles 1, 2, and 3:
   j. Total area = _____

4. Find the following for *Figure 24*:
   Triangle 1:
   a. Base   = _____
   b. Height = _____
   c. Area   = _____

   Triangle 2:
   d. Base   = _____
   e. Height = _____
   f. Area   = _____

   Triangle 3:
   g. Base   = _____
   h. Height = _____
   i. Area   = _____

   Triangles 1, 2, and 3:
   j. Total area = _____

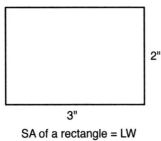

SA of a rectangle = LW

207F21.EPS

Figure 21 ◆ Triangle Exercise 1.

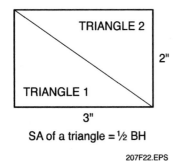

SA of a triangle = ½ BH

207F22.EPS

Figure 22 ◆ Triangle Exercise 2.

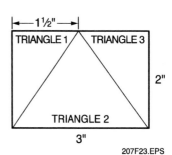

207F23.EPS

Figure 23 ◆ Triangle Exercise 3.

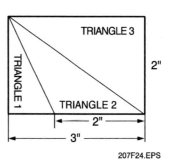

207F24.EPS

Figure 24 ◆ Triangle Exercise 4.

3.4.0 Trapezoid

A trapezoid is also a quadrilateral. It has four sides with two parallel sides. The angles formed by the sides add up to 360 degrees. *Figure 25* shows some examples of trapezoids. Note that the two parallel lines are called Base 1 and Base 2. You may be unfamiliar with the word trapezoid, but you are familiar with the shape. *Figure 26* shows the cross-section of a roadway with the shoulder slope, ditch, and backslope. The shape made by the two slopes and ditch is a trapezoid.

Any trapezoid can be divided into a rectangle (or square) and at least one triangle. See *Figure 27*. You could find the area of a trapezoid by calculating the areas of the rectangle and triangle and adding the results together (see *Figure 28*), but there is an easier way. You can **average** the lengths of the two bases and then multiply the average by the height. The formula to calculate the area of a trapezoid is as follows:

Area = ½(Base 1 + Base 2) × Height

Using the measurements of Base 1 = 8 inches, Base 2 = 6 inches, and Height = 5 inches given in

Figure 29, which are the same as the ones used in *Figure 27*, the area of the trapezoid is calculated as follows:

$$
\begin{aligned}
\text{Area} &= \tfrac{1}{2}(\text{Base 1} + \text{Base 2}) \times \text{Height} \\
&= \tfrac{1}{2}(8 + 3) \times 5 \\
&= \tfrac{1}{2}(11) \times 5 \\
&= 5\tfrac{1}{2} \times 5 \\
&= 27\tfrac{1}{2} \text{ inches}^2
\end{aligned}
$$

It may help you to remember the trapezoid formula by relating it to the formula for a rectangle. The rectangle in *Figure 30* has a length of 3 inches and a width of 2 inches, so it has an area of 6 square inches. Using the trapezoid formula for area, you need to use the length as Base 1 and Base 2, and calculate the average of the bases as follows:

½(3 + 3) = 3

Then use the width as the height and calculate the area as follows:

3 × 2 = 6 inches²

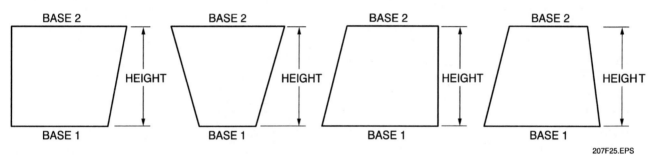

Figure 25 ◆ Trapezoids.

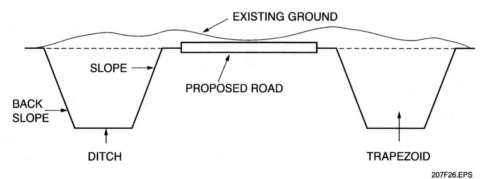

Figure 26 ◆ Trapezoids on a roadway.

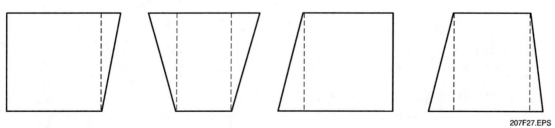

Figure 27 ◆ Trapezoids contain triangles and rectangles.

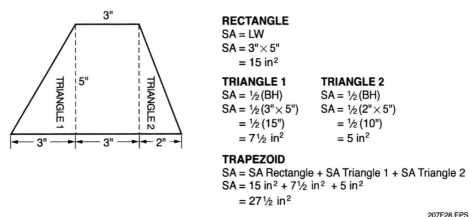

RECTANGLE
SA = LW
SA = 3" × 5"
= 15 in²

TRIANGLE 1
SA = ½ (BH)
SA = ½ (3" × 5")
= ½ (15")
= 7½ in²

TRIANGLE 2
SA = ½ (BH)
SA = ½ (2" × 5")
= ½ (10")
= 5 in²

TRAPEZOID
SA = SA Rectangle + SA Triangle 1 + SA Triangle 2
SA = 15 in² + 7½ in² + 5 in²
= 27½ in²

207F28.EPS

Figure 28 ◆ Trapezoid area using triangles and rectangles.

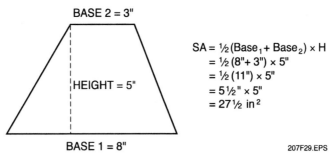

$$SA = ½ (Base_1 + Base_2) × H$$
$$= ½ (8" + 3") × 5"$$
$$= ½ (11") × 5"$$
$$= 5½" × 5"$$
$$= 27½ \text{ in}^2$$

207F29.EPS

Figure 29 ◆ Trapezoid area using formula.

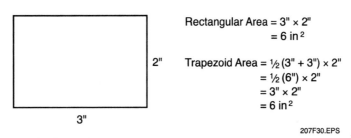

Rectangular Area = 3" × 2"
= 6 in²

Trapezoid Area = ½ (3" + 3") × 2"
= ½ (6") × 2"
= 3" × 2"
= 6 in²

207F30.EPS

Figure 30 ◆ Rectangle and trapezoid relationship.

3.4.1 Trapezoid Exercises

Refer to *Figure 31* to complete the following exercise. Remember to convert units of measure as required.

Scenario: You are working on a highway job and need to cut a shoulder slope, ditch, and backslope into existing ground. The cross-section plan is shown in *Figure 31*. The cut is 3 feet, the ditch is 2 feet wide, and the slope for the shoulder slope and backslope is 1:1 to the edges of the ditch. (Hint: the plan specifies that a 3-foot cut be made from the shoulder into the existing ground; 1:1 means that for every one foot of travel there is a one foot drop in elevation.)

1. Base 1 = _____
2. Base 2 = _____
3. Height = _____
4. Area = _____

3.5.0 Circles

A circle is a single curved line that connects with itself (*Figure 32*). A circle also has these other properties:

- All points on a circle are the same distance (equidistant) from the point at the center.
- The distance from the center to any point on the curved line, called the radius (r), is always the same.

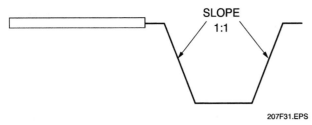

207F31.EPS

Figure 31 ◆ Trapezoid exercise cross-section plan.

- The shortest distance from any point on the curve through the center to a point directly opposite is called the diameter (d). The diameter is therefore equal to twice the radius (d = 2r).
- The distance around the outside of the circle is called the circumference. It can be determined by using the equation: circumference = πd, where π is a constant approximately equal to 3.14 and d is the diameter.
- A circle is divided into 360 degrees. Therefore, one degree = $\frac{1}{360}$ of a circle. The degree is the unit of measurement commonly used in construction for measuring the size of angles.
- The total measure of all the angles formed by all consecutive radii equals 360 degrees.

The formula to calculate the area of a circle is πr^2. The π symbol represents pi, which is a constant of 3.14, and the r means radius. The area of a circle with a radius of 6 inches is calculated as follows:

$$\begin{aligned} \text{Area} &= \pi r^2 \\ &= (3.14)(6^2) \\ &= (3.14)(36) \\ &= 113.04 \text{ inches}^2 \end{aligned}$$

The area of a circle with a radius of 10 feet is calculated as follows:

$$\begin{aligned} \text{Area} &= \pi r^2 \\ &= (3.14)(10^2) \\ &= (3.14)(100) \\ &= 314 \text{ feet}^2 \end{aligned}$$

The area of a circle with a diameter of 10 feet is calculated as follows:

$$\begin{aligned} \text{Area} &= \pi r^2, \text{ and radius} = d \div 2, \text{ so } r = 10 \div 2 = 5 \text{ feet} \\ &= (3.14)(5^2) \\ &= (3.14)(25) \\ &= 78\tfrac{1}{2} \text{ feet}^2 \end{aligned}$$

In earthwork, it will sometimes be easier to get the circumference of a circle than the diameter or radius—such as when you need to fill in a large depression. The formula for finding the diameter of a circle with the circumference is: c ÷ π, where c is circumference and π is 3.14. The area of a circle with a circumference of 31 feet is calculated as follows:

$$\begin{aligned} \text{Diameter} &= \text{circumference}/\pi \\ &= 31/3.14 \\ &= 9.87 \text{ feet} \\ \text{Radius} &= 9.87/2 \\ &= 4.94 \text{ feet} \end{aligned}$$

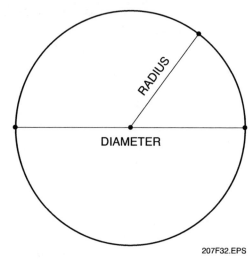

Figure 32 ◆ Circle.

$$\begin{aligned} \text{Area} &= \pi r^2, \text{ and} \\ &= (3.14)(4.94^2) \\ &= (3.14)(24.4) \\ &= 76.63 \text{ feet}^2 \end{aligned}$$

3.5.1 Circle Exercises

Complete the following exercises. Hint: when the circle diameter or circumference is given, be sure to calculate the radius.

1. What is the area of a circle with a diameter of 10 inches? _____
2. What is the area of a circle with a radius of 100 feet? _____
3. What is the area of a circle with a circumference of 628 feet? _____
4. What is the area of a circle with a radius of 7 feet? _____
5. What is the area of a circle with a diameter of 12 inches? _____

4.0.0 ◆ VOLUME

Volume is the amount of space inside a three-dimensional object. You are already familiar with three-dimensional objects—boxes, trash cans, coffee cups, and water pipes are three-dimensional objects. Any vessel that can hold a substance is a three-dimensional object, so it can be measured in terms of volume.

Three-dimensional objects have three measurements—length, width, and depth. The length is abbreviated with the letter *L*, width with the letter *W*, and depth with the letter *D*. Volume is measured in cubic units, so an object that is 1 inch in length, width, and depth has a volume of 1 cubic inch (see *Figure 33*), which can be abbreviated inch³ and is calculated as follows:

Volume = L × W × D

= 1 inch × 1 inch × 1 inch

= inch³

When two 1 cubic inch objects are placed together (see *Figure 34*), the volume is 2 cubic inches and is calculated as follows:

Volume = L × W × D

= 2 inches × 1 inch × 1 inch

= 2 inches³

Just as with area, the same volume measurement can take many shapes. See *Figure 35A*. A 2 cubic inch object can be divided into two equal parts horizontally (*Figure 35B*), vertically (*Figure 35C*), and diagonally (*Figure 35D*) to form three differently shaped objects each with a volume of 1 cubic inch.

When calculating the volume of an object, all the numbers must be in the same unit of measure. That is, all numbers must be in inches, feet, or yards. It is a good idea to use the smallest measure you can. For example, 1 yard equals 3 feet, and 3 feet equal 36 inches, so the volume of a box that has a length, width, and depth of 1 yard can be calculated as follows:

Volume in yards:

Volume = 1 yd × 1 yd × 1 yd

= 1 yd³

Volume in feet:

Volume = 3 × 3 × 3

= 27 ft³

Volume in inches:

Volume = 36 × 36 × 36

= 46,656 in³

All of the above represent the same volume, but as you can see, it is much easier to work with the yard measure because it has the smallest numbers.

4.1.0 Cubes and Rectangular Objects

When a square is the base of a three-dimensional object, all the dimensions are equal and the object is called a cube. See *Figure 36*. This object's volume is calculated by multiplying its length by its width by its depth. Since all of the cube's dimensions are equal, each side can be abbreviated with the same letter (*e*) and the formula can be written e^3. The cube shown in *Figure 32* has a length of 2 feet, width of 2 feet, and depth of 2 feet, so its volume is calculated as follows:

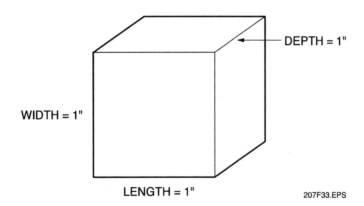

Figure 33 ◆ One cubic inch.

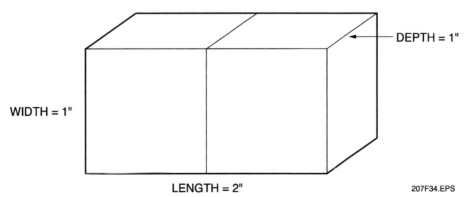

Figure 34 ◆ Two cubic inches.

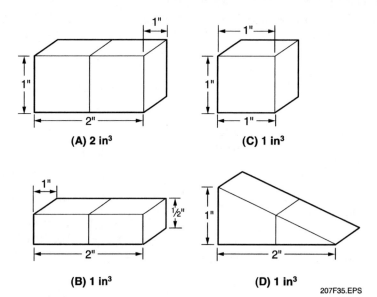

Figure 35 ◆ A cubic inch can be different shapes.

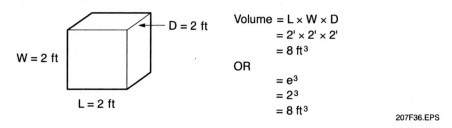

Volume = L × W × D
= 2' × 2' × 2'
= 8 ft³

OR

= e³
= 2³
= 8 ft³

207F36.EPS

Figure 36 ◆ Cube.

Volume = L × W × D or e³

= 2 × 2 × 2 or 2³

= 8 feet³

When a rectangle is the base of a three-dimensional object, it is called a rectangular object. See *Figure 37*. This object's volume is calculated by multiplying its length by its width by its depth, just like a cube. Since each dimension of a rectangle can be different, the length is abbreviated with the letter *L*, the width with the letter *W*, and the depth with the letter *D*. The object shown in *Figure 37* has a length of 4 feet, width of 2 feet, and depth of 2 feet, so its volume is calculated as follows:

Volume = L × W × D

= 4 × 2 × 2

= 16 feet³

You may have noticed that the formula for a cube and a rectangular object contains the formula for area (L × W), so to find the volume of a cube, you can multiply its area by its depth. This will be true for most three-dimensional objects that you will use to estimate excavations.

4.2.0 Prisms

A prism is a multi-sided three-dimensional object that must meet all of the following requirements:

• It has two bases.
• The bases are parallel.
• The bases are the same shape.
• The bases are the same size.
• The remaining sides must be parallelograms.

All of the objects shown in *Figure 38* are prisms. You will notice that rectangular objects and cubes are prisms. The other prisms you need to know about to estimate excavations have bases of triangles and trapezoids.

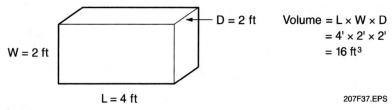

Figure 37 ◆ Rectangular object.

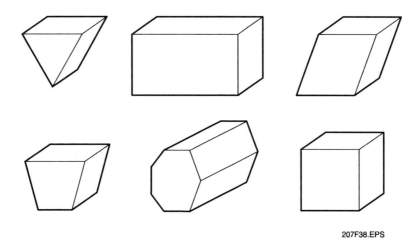

207F38.EPS

Figure 38 ◆ Prisms.

You have already learned that to find the volume of a rectangular object you need to multiply the area of the rectangle by the depth of the object. The same is true with prisms. To find the volume of a prism, first define the shape of the base, find the area of the base, and then multiply the area by the depth. Use the following procedure when you need to calculate the volume of a prism:

Step 1 Identify the shape of the base.

Step 2 Identify the dimensions of the shape.

Step 3 Calculate the area of the base.

Step 4 Calculate the volume of the prism.

Look carefully at the shape in *Figure 39A* and then perform steps 1 through 4.

Step 1 Identify the shape of the base. The base shape is a triangle.

Step 2 Identify the dimensions of the shape. The dimensions are Base = 3 inches, Height = 2 inches, and Depth = 4 inches.

Step 3 Calculate the area of the base.

Area = ½BH

= ½(3 × 2)

= ½(6)

= 3 inches^2

Step 4 Calculate the volume of the prism.

Volume = area × depth

= 3 inches^2 × 4 inches

= 12 inches^3

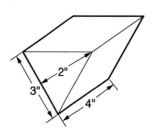

1. Base = Triangle
2. Triangle B = 3", and H = 2"
3. Surface Area = ½BH
    = ½ (3" × 2")
    = ½ (6")
    = 3 in^2

4. Prism D = 4"
5. Volume = 3 in^2 × 4"
    = 12 in^3

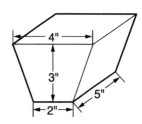

1. Base = Trapezoid
2. Trapezoid B_1 = 2", B_2 = 4", and H = 3"
3. Surface Area = ½ (2" + 4") × 3"
    = ½ (6") × 3"
    = 3" × 3"
    = 9 in^2

4. Prism D = 5"
5. Volume = 9 in^2 × 5"
    = 45 in^3

207F39.EPS

Figure 39 ◆ Finding the volume of a prism.

Repeat the procedure for the shape shown in *Figure 39B*.

Step 1 Identify the shape of the base. The base shape is a trapezoid.

Step 2 Identify the dimensions of the shape. The dimensions are Base 1 = 2 inches, Base 2 = 4 inches, Height = 3 inches, and Depth = 5 inches.

Step 3 Calculate the area of the base.

$$\text{Area} = \tfrac{1}{2}(\text{Base 1} + \text{Base 2}) \times \text{Height}$$

$$= \tfrac{1}{2}(2 + 4) \times 3$$

$$= \tfrac{1}{2}(6) \times 3$$

$$= 9 \text{ inches}^2$$

Step 4 Calculate the volume of the prism.

$$\text{Volume} = \text{Area} \times \text{Depth}$$

$$= 9 \text{ inches}^2 \times 5 \text{ inches}$$

$$= 45 \text{ inches}^3$$

4.3.0 Cylinders

A cylinder is a three-dimensional object with a circle as its base (*Figure 40*). The formula to calculate the area of a circle is πr^2, and the formula to calculate the volume of a cylinder is $\pi r^2 H$.

Cylinder A:

Diameter = 4 feet, so radius = 2 feet

Height = 4 feet

$$\text{Volume} = \pi r^2 H$$

$$= (3.14)(2^2)(4)$$

$$= (3.14)(4)(4)$$

$$= 50.24 \text{ feet}^3$$

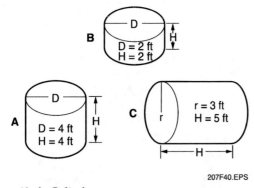

Figure 40 ◆ Cylinders.

Cylinder B:

Diameter = 2 feet, so radius = 1 foot

Height = 2 feet

$$\text{Volume} = \pi r^2 H$$

$$= (3.14)(1^2)(2)$$

$$= (3.14)(1)(2)$$

$$= 6.28 \text{ feet}^3$$

Cylinder C:

Radius = 3 feet

Height = 5 feet

$$\text{Volume} = \pi r^2 H$$

$$= (3.14)(3^2)(5)$$

$$= (3.14)(9)(5)$$

$$= 141.3 \text{ feet}^3$$

4.4.0 Volume Exercises

The following exercises will help you to better understand how to calculate the volume of an object.

1. Calculate the volume for an object that is 2 feet in length, width, and depth. _____

2. An object has a rectangular base with a area of 6 square feet and a depth of 11 feet. Calculate the volume of the object. _____

3. Calculate the volume of an object with a triangular base of 2 feet and a height of 6 feet. The object has a depth of 6 inches. _____

4. An object has a base shape of a triangle with a base of 6 inches and a height of 1½ inches. The object is 3 inches deep. Calculate the volume of the object. _____

5. An object has a trapezoidal base with the dimensions of Base 1 = 5 feet, Base 2 = 11 feet, and Height = 2 feet. The object's depth is 25 feet. Calculate the volume. _____

6. Calculate the volume of a cylinder with a diameter of 2 feet and a height of 6 feet. _____

7. Calculate the volume of the shape shown in *Figure 41*. _____

8. Calculate the volume of a cylinder that has a radius of 2 yards and a height of 3 yards. _____

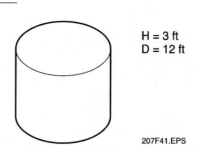

Figure 41 ◆ Cylinder Exercise 7.

5.0.0 ◆ ESTIMATING EXCAVATIONS

Excavations are measured in cubic yards of soil. You have already learned that 1 cubic yard contains 27 cubic feet and that those 27 cubic feet can take any three-dimensional shape and still have a volume of 1 cubic yard. See *Figure 42*. Each of the shapes shown in the figure has a volume of 1 cubic yard or 27 cubic feet. Since volume is the measurement of length, width, and depth, you must consider all three dimensions to calculate volume.

The capacity of heavy equipment is rated by both weight and volume. You have already learned in previous modules that operating an overloaded vehicle places you at risk for an accident, so it is important that you consider the weight of excavated material as well as the volume when you are loading your equipment. *Table 2* shows the weights of various materials. To calculate the weight of an excavation, multiply the volume in cubic yards by the material's weight per cubic yard.

Calculating volumes for cuts and fills can be time consuming, but by following a few easy steps, you can soon become skilled at it. First, to estimate excavation volume, assume that the ground is flat. (A way to calculate volume for uneven surfaces is described later in this module.) Second, study the area so that you can divide it into manageable shapes. Third, determine what information you need to make the volume calculations. Finally, gather the needed information. Once you know what shapes to use and have collected the dimensions, you can calculate the volume.

The construction plans shown in this section are simplified to make it easier for you to find the information that you need to calculate volumes. Once you gain some experience, you will be able to quickly find the information on actual building plans. Look at the highway plan in *Figure 43*. The cross-section is clearly a trapezoid. (It can also be divided into a rectangle and two triangles, but then three calculations need to be performed. There is only one calculation needed if a trapezoid is used.) The three-dimensional object of a trapezoid is a prism. See *Figure 44*.

5.1.0 Determining Required Information

Since you are assuming the existing grade is even, you can see that this job will involve only fill. To calculate the volume of a prism, you need the measurements for the following:

- Base 1
- Base 2
- Height
- Depth

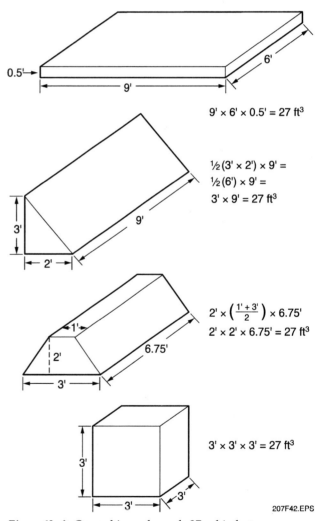

Table 2 Material Weights

Material	Weight in Pounds per Cubic Yard (27 Cubic Feet)
Clay, dry in lumps	1,701
Clay, compact	2,943
Earth, loamy, dry, loose	2,025
Earth, dry, packed	2,565
Earth, wet	2,970
Gravel, dry, loose	2,970
Gravel, dry, packed	3,051
Gravel, wet, packed	3,240
Limestone, fine	2,700
Limestone, 1½ to 2 inches	2,295
Limestone, above 2 inches	2,160
Sand, dry, loose	2,565
Sand, wet, packed	3,240

$9' \times 6' \times 0.5' = 27 \text{ ft}^3$

$\frac{1}{2}(3' \times 2') \times 9' =$
$\frac{1}{2}(6') \times 9' =$
$3' \times 9' = 27 \text{ ft}^3$

$2' \times \left(\frac{1' + 3'}{2}\right) \times 6.75'$
$2' \times 2' \times 6.75' = 27 \text{ ft}^3$

$3' \times 3' \times 3' = 27 \text{ ft}^3$

207F42.EPS

Figure 42 ◆ One cubic yard equals 27 cubic feet.

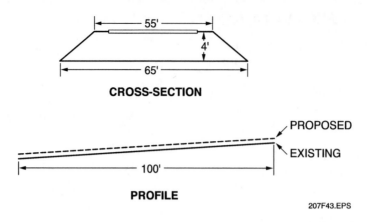

CROSS-SECTION

PROFILE

207F43.EPS

Figure 43 ◆ Highway plan.

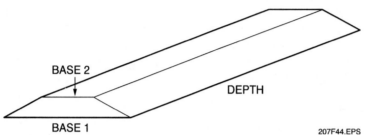

Figure 44 ◆ Highway plan as a prism.

207F44.EPS

5.2.0 Gathering Information

Carefully examine the diagrams shown in *Figures 43* and *44*. The cross-section is a trapezoid with the following measurements:

- Base 1 = 65 feet
- Base 2 = 55 feet
- Height = 4 feet

The profile shows that this section of roadway is 100 feet in length, so the trapezoid has a depth of 100 feet.

5.3.0 Calculating Excavations

To calculate the amount of fill needed to construct this roadway, you need to use the following dimensions:

Base 1 = 65 feet
Base 2 = 55 feet
Height = 4 feet
Depth   = 100 feet

The volume is calculated as follows:

Volume = ½ × (Base 1 + Base 2) × Height × Depth

= ½ × (65 + 55) × 4 × 100

= ½ × 120 × 4 × 100

= 24,000 feet³

= 24,000 ÷ 27 cubic feet/cubic yard

= 888.9 yards³

When the weight of the fill is 3,200 pounds per cubic yard, the fill weight is calculated as follows:

888.9 × 3,200 = 2,844,480 pounds

5.4.0 Estimating Excavations Examples

As stated earlier, the only way to become proficient at performing estimating calculations is to do them. The following examples will help explain how to perform calculations for other excavation jobs. The examples become increasingly complex, so be certain that you read each example and study the associated figures carefully to be sure you understand the result.

5.4.1 Excavating a Simple Foundation

Look at the foundation plan in *Figure 45*. The pad is a simple rectangular object. See *Figure 46*. The footer has the shape of a trapezoid, so its three-dimensional figure is a prism. (The footer can be divided into a rectangular solid and a prism, too).

Calculate the volume of the earth you will need to excavate so the building can be constructed. To calculate the total volume, you first need to calculate the volumes of all the objects in which the foundation has been divided and then add the volumes of the objects to arrive at the total excavation volume.

All the information you need to calculate soil excavations is usually not readily found on the

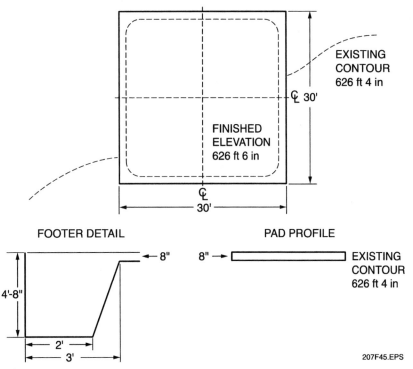

FOOTER DETAIL   PAD PROFILE

207F45.EPS

Figure 45 ◆ Foundation plan.

building plans. Sometimes you need to look at the information you have and calculate the information you need. Plans are drawn so that all construction tasks can be performed from a single set of plans. Do not let yourself become overwhelmed with all the information that you see on the plans. Look for only the information that you need.

See *Figure 45*. The measurements shown on these plans are in feet and inches rather than the typical decimal to make the example easier for you to understand. Since the pad is a rectangular object, you need the length, width, and depth of the excavation to calculate its volume. The length and width are the same as the pad dimensions, which are 30 feet by 30 feet. The pad is 8 inches thick. To calculate the beginning elevation, you need to know the pad's finished elevation, which

is 626 feet and 6 inches, and the existing elevation of the building site, which is 626 feet and 4 inches. See *Figure 47*. This means that the pad needs to begin at an elevation of 625 feet and 10 inches, which is 6 inches below the existing elevation, so the depth of the excavation is 6 inches.

You now have all of the information you need to calculate the excavation volume for the slab, which is as follows:

Volume = length × width × depth

Length = 30 feet

Width = 30 feet

Depth = 6 inches = 0.5 feet (convert inches to feet)

Volume = 30 × 30 × 0.5

Volume = 450 feet³ ÷ 27 feet³ (convert cubic feet to cubic yards)

Volume = 16.67 yards³

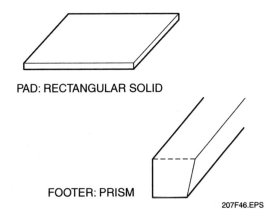

PAD: RECTANGULAR SOLID

FOOTER: PRISM

207F46.EPS

Figure 46 ◆ Foundation plan as a rectangular object and a prism.

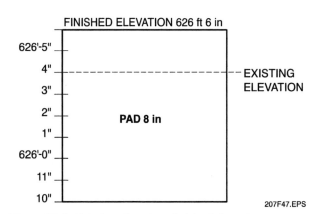

207F47.EPS

Figure 47 ◆ Existing elevation, finished elevation.

Looking at the footer plans (footer detail in *Figure 45*) and the prism that represents the footer (*Figure 46*), you see that the bottom base is 2 feet and the top base is 3 feet. The height is a little harder. The total height of the footer from its beginning to the finished surface of the pad is 4 feet 8 inches. The pad is 8 inches, so the height of the footer must be 4 feet. Therefore you know the following:

Base 1 = 2 feet

Base 2 = 3 feet

Height = 4 feet

You still need to know the lengths of the footers to calculate volume. See *Figure 48*. There is a footer along each edge of the pad, so there are four footers. Each footer is 3 feet wide, so each footer is 27 feet long (30 – 3 = 27). You now have all the information you need to calculate the volume of the footer prisms. It is as follows:

Base 1 = 2 feet

Base 2 = 3 feet

Height = 4 feet

Length = 27 feet

Volume = ½ × (Base 1 + Base 2) × Height × Depth

= ½ × (2 + 3) × 4 × 27

= ½ × 5 × 4 × 27

= 270 feet^3

= 270 ÷ 27 feet^3 (convert square feet to cubic yards)

= 10 yards^3 per footer

Total excavation volume is calculated by adding the four footer volumes and the pad volume as follows:

Total volume = (4 × 10 yards^3) + 16.67 yards^3

Total volume = 40 yards^3 + 16.67 yards^3

Total volume = 56.67 yards^3

Referring to *Table 2*, you can calculate the weight of the foundation excavation by multiplying the total excavation in cubic yards by the weight of the material per cubic yard. A comparison of the weight of loose earth, packed earth, and wet earth is as follows:

Earth (dry, loose) 2,025 lb/cu yd
56.67 × 2,025 = 114,756.75 lbs

Earth (dry, packed) 2,565 lb/cu yd
56.67 × 2,565 = 145,358.55 lbs

Earth (wet) 2,970 lb/cu yd
56.67 × 2,970 = 168,309.9 lbs

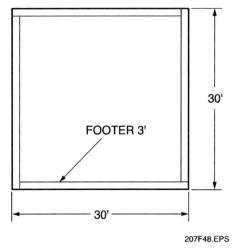

207F48.EPS

Figure 48 ◆ Footer length.

5.4.2 Excavating Slopes

The diagram in *Figure 49* shows a slope stake. It directs you to make a 2-foot cut at a 3:1 slope on level ground. You know from the *Grades Part One* module that this means for every 3 feet of horizontal travel, the elevation of the ground must drop 1 foot up to the desired 2-foot cut. This cut forms a triangular-shaped cut. The cut section is 100 feet long.

The prism's volume is calculated as follows (see *Figure 50*):

Height = 2 feet

Base = 6 feet

Depth = 100 feet

Volume = ½ × base × height × depth

= ½ × 6 × 2 × 100

= 600 feet^3

= 600 feet^3 ÷ 27 feet^3 (convert square feet to cubic yards)

= 22.22 yards^3

When the excavated material weights 1,200 pounds per cubic yard, the weight of the excavation material is calculated as follows:

22.22 × 1,200 = 26,664 pounds

5.4.3 Excavating a Complex Foundation

Study the foundation plan in *Figure 51*. This shape can be divided into three to five shapes. Practice looking at a complex shape until you can see many shapes. With a complex plan such as this one, you can save yourself time by calculating the area of the entire foundation before you calculate the volume.

Figure 52A shows three shapes: two trapezoids and one rectangle. *Figure 52B* shows five shapes:

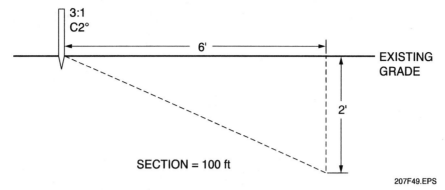

SECTION = 100 ft

207F49.EPS

Figure 49 ◆ Excavating a slope.

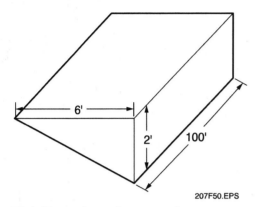

207F50.EPS

Figure 50 ◆ Excavating a slope as a prism.

two triangles and three rectangles. You will want to see as many shapes as possible, because building plans may not provide you with enough information to easily gather the measurements you need to calculate volume. In this case, the dimensions shown on *Figure 51* are sufficient for you to calculate the area of all of the shapes shown in *Figure 52A* and *Figure 52B*, so you will use *Figure 52A* because it has the fewest calculations.

When you are trying to gather information from the building plans, it is easy to become overwhelmed with all of the information on the diagram. Concentrate on one shape at a time. Look at

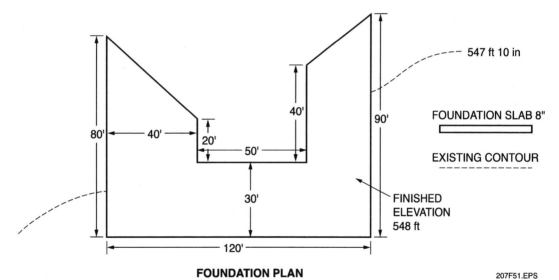

FOUNDATION PLAN

207F51.EPS

Figure 51 ◆ Complex foundation plan.

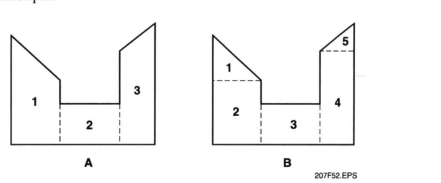

207F52.EPS

Figure 52 ◆ Complex foundation plan shapes.

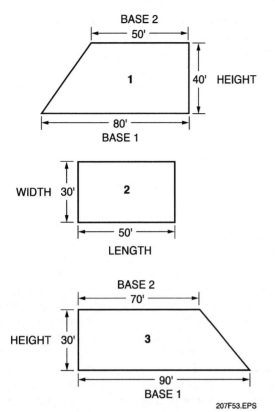

Figure 53 ◆ Three shapes.

Figure 51 and *Figure 53*. Shape 1 in *Figure 53* is a trapezoid; it is the left wing of the building shown in *Figure 51*. To calculate the area of shape 1, you need Base 1, Base 2, and Height. Looking at *Figure 51*, you can see that Base 1 is 80 feet. You can also see that the Height is 40 feet. Now you need Base 2, but its measurement is not obvious so you need to calculate it. The part of Base 2 represented by the solid line is 20 feet and the part of Base 2 represented by the dashed line is 30 feet, so the length is 50 feet.

The dimensions for shape 1 of *Figure 53* are as follows:

Base 1 = 80 feet

Base 2 = 50 feet

Height = 40 feet

The area for shape 1 is calculated as follows:

Area = ½(Base 1 + Base 2) × Height

= ½(80 + 50) × 40

= ½ × 130 × 40

= 2,600 feet²

If you are having trouble with this, ask your instructor for help. After he or she walks you through it once or twice, you will understand it.

Shape 2 in *Figure 53* is a little easier. Shape 2 is a rectangle and represents the main part of the building shown in *Figure 51*. You already used the width of the rectangle as part of Base 2 in shape 1,

so you know the width is 30 feet. The length of 50 feet is shown at the top of the main building in *Figure 51*, so now you have the information you need to calculate the area of shape 2, which is as follows:

Length = 50 feet

Width = 30 feet

Area = length × width

= 50 × 30

= 1,500 feet²

Shape 3 in *Figure 53* is another trapezoid and represents the right wing of the building shown in *Figure 51*. Base 1 is on the far right side of the foundation and is 90 feet. Base 2 is made of the width of shape 2 (30 feet) and the 40 feet measure on the left side of the wing (*Figure 51*), so Base 2 is 70 feet.

Now you need to calculate the height of shape 3. You know that the entire length of the face of the building is 120 feet (it is on *Figure 51*) and the face of the building is made up of the height of shape 1 (40 feet), the length of shape 2 (50 feet) and the height of shape 3. Since you know the dimensions of shapes 1 and 2 you can calculate the height of shape 3 as follows:

Face length = height shape 1 + length shape 2 + height shape 3

120 feet = 40 feet + 50 feet + height shape 3

120 feet = 90 feet + height shape 3

120 feet − 90 feet = 90 feet + height shape 3 − 90 feet

30 feet = height shape 3

So the height of shape 3 is 30 feet. You have all the information you need to calculate the area of shape 3, which is as follows:

Base 1 = 90 feet

Base 2 = 70 feet

Height = 30 feet

Area = ½(Base 1 + Base 2) × Height

So area = 2,400 square feet

You still need the depth of each object to calculate its volume. Look on the plans and find the thickness of the slab, the existing contour of the building site, and the finished elevation of the foundation. The slab is 8 inches thick and the finished elevation is 548 feet, so the slab must start at 547 feet 4 inches in elevation. Since the existing elevation is 547 feet 10 inches, 6 inches (or ½ foot) of earth needs to be excavated to achieve the finished elevation after the foundation is poured.

Now that you have the depth, you can calculate the volume of each object, which is shown below, and add the answers together to get the total excavation volume.

Shape 1:
   Volume = 2,600 feet² × ½ foot
   Volume = 1,300 feet³

Shape 2:
   Volume = 1,500 feet² × ½ foot
   Volume = 750 feet³

Shape 3:
   Volume = 2,400 feet² × ½ foot
   Volume = 1,200 feet³

Total volume:
   1,300 feet³ + 750 feet³ + 1,200 feet³ = 3,250 feet³
   3,250 feet³ ÷ 27 feet³ = 120.37 yards³

Another way to figure out total volume is to add together the area of all three shapes and then multiply the total area by the depth. Calculate this yourself. You will get a total area of 6,500 square feet.

6.0.0 ◆ COMPLEX CALCULATIONS

So far you have learned how to calculate excavation volumes for uniformly shaped objects, but most building sites are irregularly shaped.

Although it seems impossible to calculate actual volumes manually, it is possible, but it takes a great deal of time. There are a number of computer programs that can calculate these volumes quickly and accurately, saving time and money. *Figure 54* shows a screen from one such program.

If your employer does not have a computer program that you can use, you can still estimate complex excavations. You have already learned that you need to calculate the area of an object's base to calculate its volume. One of the fastest ways to find the area of an irregular shape is to draw a grid over it. *Figure 55* is a benched trench. Each block of the grid in *Figure 56* represents 1 square foot. To estimate the area of the benched trench, you need to count the number of blocks (estimating the size of the partial blocks).

Usually the ground at the building site is irregular (*Figure 57*), making it difficult to determine the excavation depth precisely. In this case you can make a more accurate estimation by averaging the elevations of the site. You do this by measuring the elevations at several points and then adding these numbers and dividing by the number of points measured. In *Figure 58*, 14 points are measured; their sum is 93, so the average depth of the excavation is 6.64 inches, which is computed as follows:

   93 ÷ 14 = 6.64 inches

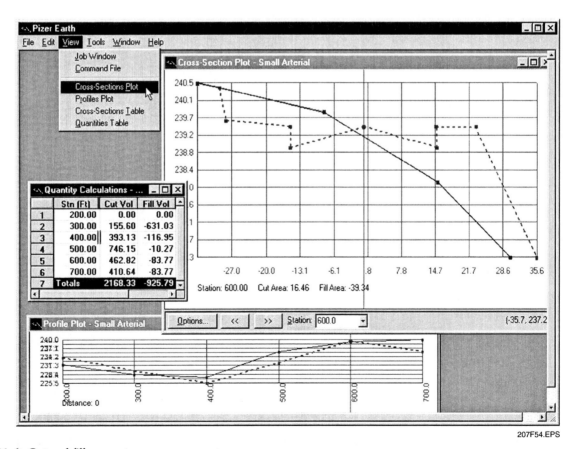

207F54.EPS

Figure 54 ◆ Cut and fill computer program.

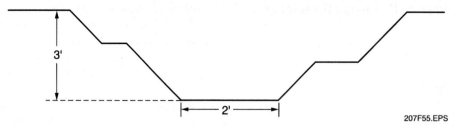

207F55.EPS

Figure 55 ◆ Benched trench.

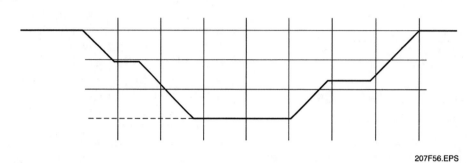

207F56.EPS

Figure 56 ◆ Benched trench with grid.

207F57.EPS

Figure 57 ◆ Irregular elevation.

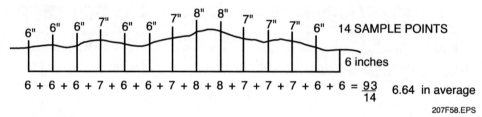

207F58.EPS

Figure 58 ◆ Irregular elevation averaged.

1. It is important to be able to calculate the volume and weight of cut and fill material because _____.
   a. the material cost is established by weight and volume
   b. all heavy equipment is rated by weight and volume
   c. it determines the number of workers needed on a job
   d. trucks use the least amount of fuel when overloaded

2. Volume is based on area.
   a. True
   b. False

3. When working with an equation that has numbers grouped in parentheses, you know that you need to perform that calculation _____.
   a. first
   b. second
   c. third
   d. last

4. The square root of the number 25 is _____.
   a. 5
   b. 25
   c. 50
   d. 625

5. The symbol π in an equation is considered a _____.
   a. variable
   b. root
   c. constant
   d. radius

6. A right angle is _____ degrees.
   a. 45
   b. 90
   c. 180
   d. 360

7. Two-dimensional objects can be measured in all of the following values *except* _____.
   a. square inches
   b. square feet
   c. square yards
   d. square radius

8. One square yard is equal to _____ square feet.
   a. 3
   b. 9
   c. 27
   d. 144

9. One side of a square is 6 inches, so its area is _____.
   a. 3 square feet
   b. 12 square inches
   c. 36 square inches
   d. not enough information given

10. A rectangle has a length of 6 inches, so its area is _____.
    a. 6 square inches
    b. 12 square inches
    c. 36 square inches
    d. not enough information given

11. Calculate the area for a triangle with a base of 7 inches and a height of 4 inches.
    a. 12 square inches
    b. 14 square inches
    c. 28 square inches
    d. Insufficient information

Figure 1

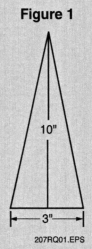

10"

3"

207RQ01.EPS

12. The area for the shape shown in *Figure 1* is _____ square inches.
    a. 3
    b. 15
    c. 30
    d. 45

Figure 2

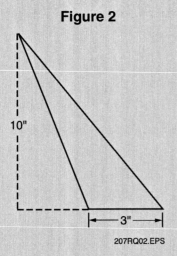

10"

3"

207RQ02.EPS

13. The area for the shape shown in *Figure 2* is _____ square inches.

a. 3
b. 15
c. 30
d. 45

Figure 3

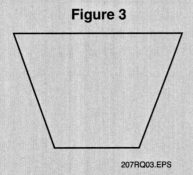

207RQ03.EPS

14. The shape shown in *Figure 3* is a _____.

a. rhombus
b. parallelogram
c. trapezoid
d. rectangle

15. A trapezoid's measurements are Base 1 = 2 inches, Base 2 = 4 inches, and Height = 4 inches. What is its area?

a. 3 square inches
b. 12 square inches
c. 30 square inches
d. 45 square inches

16. The distance from the center of a circle to the edge of its curved line is called the _____.

a. angle
b. circumference
c. diameter
d. radius

17. A circle has a diameter of 2 feet. What is its area?

a. 3.14 square feet
b. 6.28 square feet
c. 9.42 square feet
d. 45.5 square feet

18. A backhoe loader's bucket can hold 7.0 cubic feet or up to 134 pounds, so its volume is _____.

a. 7.0 cubic feet
b. 19 pounds per foot
c. 134 pounds
d. 938 cubics per pound

19. A cube that has a dimension of 1 yard has a volume of _____.

a. 1 cubic foot
b. 9 cubic feet
c. 27 cubic feet
d. not enough information given

20. A rectangular object has the dimensions of length = 2 feet, width = 4 feet, depth = 3 feet. It has a volume of _____.

a. 6 cubic feet
b. 8 cubic feet
c. 12 cubic feet
d. 24 cubic feet

21. A three-dimensional object with two identical triangular bases has the following dimensions: Base = 9 inches, Height = 10 inches, and Depth = 2 feet, so it has a volume of _____.

a. 90 cubic inches
b. 180 cubic inches
c. 1,080 cubic inches
d. 2,160 cubic inches

22. A three-dimensional object has a depth of 8 inches and a triangular base that has an area of 12 square inches, so it has a volume of _____.

    a. 8 cubic inches
    b. 20 cubic inches
    c. 48 cubic inches
    d. 96 cubic inches

23. A three-dimensional trapezoidal object has the following dimensions: Base 1 = 2 feet, Base 2 = 3 feet, Height = 2 feet, and Depth = 6 feet, so it has a volume of _____.

    a. 5 cubic feet
    b. 25 cubic feet
    c. 30 cubic feet
    d. 72 cubic feet

24. A cylinder has a diameter of 6 feet and a height of 2 feet, so its volume is _____.

    a. 12 cubic feet
    b. 28.3 cubic feet
    c. 56.5 cubic feet
    d. 226 cubic feet

25. A material has a weight of 1,700 pounds per cubic yard, and there are 54 cubic feet to move. What is the total weight of the material?

    a. 3,400 pounds
    b. 10,200 pounds
    c. 30,600 pounds
    d. 91,800 pounds

26. Wet excavation material is lighter than dry material.

    a. True
    b. False

27. An excavation has a volume of 203 cubic feet. The material weighs 1,200 pounds per cubic yard. What is the weight of the excavation?

    a. 2,700 pounds
    b. 9,000 pounds
    c. 81,120 pounds
    d. 243,600 pounds

28. A complex excavation has a depth of 6 inches and can be broken into three shapes with areas of 100, 226, and 300 square feet. The total volume of the excavation is ____ cubic feet.

    a. 313
    b. 626
    c. 3,756
    d. Insufficient information

29. To calculate the volume of excavation material for a simple foundation, you need all of the following information about the foundation *except* _____.

    a. thickness
    b. length
    c. width
    d. weight

30. A triangle-shaped slope excavation has a base of 4 feet, height of 6 feet, and length of 100 feet. Its volume is _____ cubic feet.

    a. 800
    b. 1,200
    c. 2,400
    d. Insufficient information

Summary

Calculating the volume and weight of excavations can be a time-consuming task, but it is necessary in order to determine the type of equipment you will need to use on the job. To determine the volume of any excavation, you will first calculate the area of a figure and then you will use the area to calculate the volume of an object.

Excavations are three-dimensional, but they are usually based on one of several common two-dimensional shapes, such as the square, rectangle, triangle, or circle. Once you have determined the area of the base shape, the volume is calculated by multiplying the base area by the object's depth. Whenever you need to calculate the volume of a complex object, you need to break the object into familiar shapes and then calculate the volume of each shape. Once you have the volume of each shape, it is only a matter of adding them together to calculate the total volume of the object. Cut and fill volumes are measured in cubic yards, so after you calculate the excavation volume, you may need to convert it to cubic yards.

Overloading a vehicle can place the operator at risk for an accident and damage the vehicle, so it is important to use equipment that is rated for the weight of the material. Each type of material has its own weight, so once you have calculated the volume of an excavation, you will need to calculate its weight. This is done by multiplying the total volume in cubic yards and the material weight per cubic yard.

Notes

Trade Terms Introduced in This Module

Average: The middle point between two numbers or the mean of two or more numbers. It is calculated by adding all numbers together, and then dividing the sum by the quantity of numbers added. For example, the average (or mean) of 3, 7, 11 is 7 (3 + 7 + 11 = 21; 21 ÷ 3 = 7).

Parallel: Two lines that are always the same distance apart even if they go on into infinity (forever is called infinity in mathematics).

Parallelogram: A two-dimensional shape that has two sets of parallel lines.

Quadrilateral: A four-sided closed shape with four angles whose sum is 360 degrees.

Additional Resources

This module is intended to be a thorough resource for task training. The following reference works are suggested for further study. These are optional materials for continued education rather than for task training.

Caterpillar Performance Handbook, Edition 27. A CAT® Publication. Peoria, IL: Caterpillar, Inc.

Construction Surveying and Layout. 1995. Wesley G. Crawford. West Lafayette, IN: Creative Construction Publishing.

Cut-and-Fill Software: The Real-World Experience, Grading and Excavation Magazine. September/October 2002. Penelope O'Malley. Forester Communications.

Excavating and Grading Handbook. 1987. Nicholas E. Capachi. Carlsbad, CA: Craftsman Book Company.

Excavators Handbook Advanced Techniques for Operators. 1999. Reinar Christian. Addison, IL: The Aberdeen Group, A division of Hanley-Wood, Inc.

Pipe & Excavation Contracting. 1987. Dave Roberts. Carlsbad, CA: Craftsman Book Company.

Figure Credits

Reprinted courtesy of Caterpillar Inc., 207T01

Earth Software by Pizer Inc., 207F54

The NCCER makes every effort to keep these textbooks up-to-date and free of technical errors. We appreciate your help in this process. If you have an idea for improving this textbook, or if you find an error, a typographical mistake, or an inaccuracy in NCCER's Contren® textbooks, please write us, using this form or a photocopy. Be sure to include the exact module number, page number, a detailed description, and the correction, if applicable. Your input will be brought to the attention of the Technical Review Committee. Thank you for your assistance.

Instructors – If you found that additional materials were necessary in order to teach this module effectively, please let us know so that we may include them in the Equipment/Materials list in the Annotated Instructor's Guide.

Write: Product Development and Revision
National Center for Construction Education and Research
P.O. Box 141104, Gainesville, FL 32614-1104

Fax: 352-334-0932

E-mail: curriculum@nccer.org

Craft   Module Name

Copyright Date   Module Number   Page Number(s)

Description

(Optional) Correction

(Optional) Your Name and Address

22208-06

Grades, Part Two

22208-06
Grades, Part Two

Topics to be presented in this module include:

Overview

Grades, Part One explained that a grade was the contour of the earth and explained how grades were measured and staked for excavation and fill. This module will expand on that base and include how to plan a job, how to use existing benchmarks to set grade stakes, how to set grades, and how to grade trenches for pipe. In addition, it will explain how positive drainage is maintained on most sites to control the flow of water.

The key to a successful and profitable operation is the effective use of resources and that requires planning. Becoming an effective scheduler and planner is an advancement opportunity for an equipment operator.

Objectives

When you have completed this module, you will be able to do the following:

1. Define selected terms associated with plan reading, grade setting, and drainage.
2. State how cycle time affects scheduling of earthwork.
3. Describe proper practices for setting grades from a bench mark.
4. Describe proper practices for setting grades using a laser level or string.
5. Describe various methods for keeping construction sites well drained.
6. Describe the work required for the basic grading operations.
7. Describe proper practices for setting the grade of a trench and drain pipe.
8. Interpret construction plans to determine grading requirements.

Trade Terms

Articulating frame
Balance point
Batter board
Bedding material
Boots
Laser
Stations
String line
Sump
Topographic survey

Required Trainee Materials

1. Pencil and paper
2. Appropriate personal protective equipment

Prerequisites

Before you begin this module, it is recommended that you successfully complete *Core Curriculum*; *Heavy Equipment Operations Level One*; and *Heavy Equipment Operations Level Two*, Modules 22201-06 through 22207-06.

This course map shows all of the modules in the second level of the *Heavy Equipment Operations* curriculum. The suggested training order begins at the bottom and proceeds up. Skill levels increase as you advance on the course map. The local Training Program Sponsor may adjust the training order.

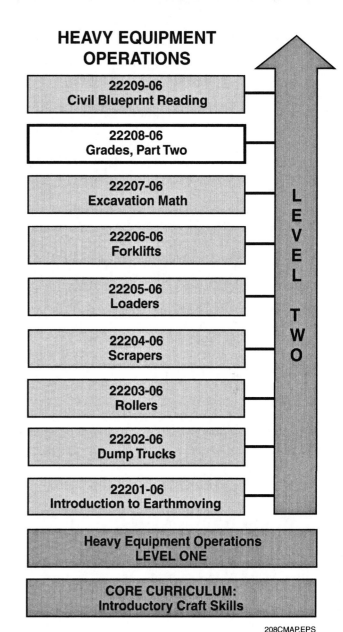

HEAVY EQUIPMENT OPERATIONS

22209-06
Civil Blueprint Reading

22208-06
Grades, Part Two

22207-06
Excavation Math

22206-06
Forklifts

22205-06
Loaders

22204-06
Scrapers

22203-06
Rollers

22202-06
Dump Trucks

22201-06
Introduction to Earthmoving

LEVEL TWO

Heavy Equipment Operations
LEVEL ONE

CORE CURRICULUM:
Introductory Craft Skills

208CMAP.EPS

1.0.0 ◆ INTRODUCTION

Earthmoving is a task anyone can do, but getting the job done quickly, profitably, and well takes a professional. The stability of any completed construction project depends on the stability of the ground on which it is built. It takes knowledge, skill, and experience to become a professional heavy equipment operator. In this module, you will not only be expanding your knowledge about earthwork, but you will also be applying knowledge from previous modules.

You may think that site preparation (*Figure 1*) is the first step to a construction project, but it really falls closer to the middle. Before ground is broken on a project, many hours of labor have already been invested. Engineers and architects develop plans for buildings or roadways. An appropriate building site is selected. Surveyors establish property boundaries and set grade markers so that the site can be prepared. It is at this point that you start your part in the construction project.

Just as with any other project, your first step is planning. After you have studied the construction plans, you need to determine what you need to do to complete your part of the job. Then you need to determine what resources you need to accomplish your part. It is only then that you can actually start working on the job.

2.0.0 ◆ PLANNING A JOB

Review the construction drawing set when you are first assigned to a job. The site plan will show the topographical features of a site, as well as project information such as the building outline, existing and proposed contours, and other information that is relevant to the project. Based on this information, survey teams will set grade stakes. Effective grading operations depend on proper staking of the project and the equipment operators' abilities to follow the grading instructions on the stakes. You will be performing your grading work after a survey crew places the initial grade stakes. Each phase of construction may require more grading work, so on these jobs the survey team will set new stakes. You may wish to review the information in the *Grades* module in *Heavy Equipment Level One* before you continue with this module.

To efficiently complete earthwork, you need the proper equipment, a plan to accomplish the work quickly, and skillful workers. The key to a successful and profitable operation is to effectively use resources. In this case, resources are manpower and equipment. These resources need to be scheduled so that work is always being accomplished and no workers or machines are sitting idle. On large jobs, scheduling will probably be done for you and you

Figure 1 ◆ Site preparation.

208F01.EPS

will follow the schedule. On small jobs, you may need to decide what has to be done and when it should be done to get the most out of your resources.

2.1.0 Scheduling Equipment

Even the most skilled heavy equipment operator cannot perform satisfactorily without the proper equipment. To determine what equipment is needed for a job, the grading supervisor needs to review the plans and examine the site. Earthwork operations on a job usually follow three-step pattern. The first step is clearing the area of unwanted vegetation and top soil. The second is to perform rough grading. The third step is to perform finish grading. Between each step, there may be other required excavation work such as trenching and excavation of bridge foundations or other structures. Each task may require a different type of equipment.

In order to determine the size, type, and amount of equipment needed to accomplish a job, you or your supervisor must be able to estimate the amount of soil that needs to be moved. As you learned in the module *Excavation Math*, soil volume is measured in cubic feet or yards, which can be used to determine equipment needs. Secondary to volume is weight—some material is extremely heavy for its volume, so sometimes a material's weight must be considered over volume when selecting equipment. *Table 1* shows the weights of various materials.

NOTE

Material weight varies with moisture content, compaction, and, where applicable, grain size. On-site testing must be performed when the exact weight of a material is required.

Table 1 Material Weights

Material	Weight in Pounds per Cubic Yard (27 Cubic Feet)
Clay, dry in lumps	1,701
Clay, compact	2,800
Earth, loamy, dry, loose	2,100
Earth, dry, packed	2,565
Earth, wet	2,970
Gravel, dry, loose	2,850
Gravel, dry, packed	3,050
Gravel, wet, packed	3,400
Limestone, fine	2,700
Limestone, 1½ to 2 inches	2,600
Limestone, above 2 inches	2,600
Sand, dry, loose	2,400
Sand, wet, packed	3,100

208T01.EPS

In the *Introduction to Earthmoving* module, you learned that each piece of equipment has its own duty cycle, which is the time it takes the equipment to complete one cycle of work and be ready to start again. The duty cycle for a dump truck is how long it takes to load it, travel to its destination, unload, and then travel back to get loaded again. The duty cycle for a backhoe is how long it takes to fill its bucket, travel to its destination, unload, and then get into position to load the bucket again.

Other things that must be considered to determine equipment needs are as follows:

- How far is the haul? Ideally, soil cut from a grade should be used close to the cut, since excessive travel wastes time and money.

- How much space is available at the site? Larger equipment is usually more efficient and cheaper to run than small equipment, but on sites with limited space, smaller equipment will be needed. If this is the case, expect higher costs.

- How steep are the slopes on the site? Steep slopes contribute to equipment tip-over. Tracked vehicles tend to be more stable on slopes than wheeled vehicles.

- Does the site have paved areas or underground structures that heavy equipment operators cannot avoid? The heavy weight of construction equipment can be damaging to existing site features. In addition, metal-tracked equipment cannot be operated on paved surfaces without laying some type of protective layer over the road surface.

Based on this type of information, you or your supervisor will determine the type of equipment that is needed to accomplish the job. The key to running an efficient operation is to have just enough equipment available so that no piece is sitting idle for any time. Of course, this is not truly possible, but by minimizing idle time, the earthmoving operation will be more profitable.

2.1.1 Dump Trucks

Dump trucks (*Figure 2*) are among the most widely used vehicles in construction work. They are built on a heavy-duty chassis and have a large bed that is used to move large amounts of loose material, such as sand, soil, gravel, and asphalt. Dump trucks can be quickly loaded using a chute or a piece of heavy equipment, such as an excavator. The beds can be raised and tilted back with a hoist that is mounted on the truck, so the driver can quickly unload the cargo without help. Dump trucks are used to haul away unsuitable or excess cut material and to haul in suitable fill at a construction site.

2.1.2 Scrapers

Scrapers (*Figure 3*) are self-loading machines that scrape material from the grade surface, haul it to the fill area, and then drop and spread the excavated material. Scrapers work best with light- and medium-weight materials that are nearly free of

208F02.EPS

Figure 2 ◆ Typical dump truck.

208F03.EPS

Figure 3 ◆ Typical scraper.

vegetation and rocks or boulders. They can be used in heavy or packed material that has been ripped or scarified before the scraper is used.

2.1.3 Bulldozers

A bulldozer is a tracked or wheeled vehicle that is mounted with a blade, which is used to excavate, cut, fill, backfill, and spread material (*Figures 4* and *5*). Bulldozers come in many sizes. Small dozers can be used for finish work or for working in tight spaces. Large dozers are generally used for clearing and rough grading of various large construction projects, although in some cases they are used for finishing work. Dozers that are tracked rather than wheeled provide good traction and stability and are well suited to work on rough irregular terrain and slopes.

2.1.4 Motor Graders

A motor grader (*Figure 6*) is a rubber-tired machine that is used to shape and finish materials in earthmoving excavation. Graders are used to cut ditches and level the surface of a grade. In addition, a grader can mix, place, and smooth materials on the ground or other surfaces, making it a very versatile piece of equipment.

208F04.EPS

Figure 4 ◆ Tracked dozer.

208F05.EPS

Figure 5 ◆ Typical wheeled bulldozer.

Graders are equipped with a moldboard, which pushes the graded material. The moldboard may be adjusted to the desired height and angle. Graders are often equipped with a scarifier or a ripper to loosen the top surface of material. Some graders have an **articulating frame** that allows them to easily go over rough terrain. The grader front axle is designed to allow the front wheels to tilt to the right or left under the control of the operator. This gives the grader good stability, helps the blade to bite into the surface being graded, and makes it easier for the operator to steer when cutting off to the side.

2.2.0 Scheduling Work

In most cases, travel time is the longest individual component in equipment cycle time. Effective grading supervisors know that to ensure a profitable operation, work must be scheduled to reduce equipment travel time. On some jobs, grading may be accomplished quickly with little regard to scheduling. But jobs that are extremely large or complex may require a significant amount of planning before work is started. Even on small jobs, the grading foreman or project engineer should go over the layout of the area with equipment operators so they understand the specific staking patterns on the job before beginning each earthmoving operation.

Unless the cut material is unsuitable for reuse, engineers usually design a job so that the fill requirements are met with the material cut from higher elevations. This is called a balanced job, because there is no fill and no spoil. To operate efficiently, the material from a cut needs to go into a nearby fill area. Traveling long distances to obtain or dump material wastes fuel and time, and cuts into profits.

2.3.0 Worker Skills

You cannot become proficient at leveling grades by reading a book. It takes experience. Good operators develop a feel for cutting grades by paying attention to ground conditions. The best you can gain from

208F06.EPS

Figure 6 ◆ Typical motor grader.

this module is knowledge of common industry practices and procedures used during earthmoving operations. It is important that you are aware of these practices and procedures so that you do not disrupt the workflow of a site while you are learning how to become a skilled operator.

Skilled operators make the job look easy, but they were beginners at one time too. Do not become discouraged while you are learning. Keep the following points in mind while you are learning to cut and fill grades:

- One of the most common problems new operators have is that they are reluctant to make deep cuts. Skilled operators use their equipment to the fullest, and they are able to cut to grade with a minimum of passes by keeping the blade or bucket full at most times.
- After you have made a pass, evaluate your work. Did you leave too many uncut areas that will require a second or third pass? If so, adjust your blade or bucket attachment.
- When you are cutting close to grade, it may be wise to make smaller cuts and more passes, rather than risk cutting too much. It is better to make a second pass with a blade than it is to cut too much material, which will require that fill and a compactor be brought in to fix the cut.
- Always follow your site plans. Many sites will have a stormwater control plan, which was discussed in *Introduction to Earthmoving*, to help prevent erosion and sedimentation. It is your responsibility to know and follow the plans. When you are not sure, ask your supervisor.
- Never cut in an area that does not need grading. This can increase sedimentation and erosion.
- Even if your site does not have a stormwater control plan, avoid driving your equipment across graded areas to avoid increased sedimentation.
- When performing rough grading or filling, be sure to leave enough material to trim during finish grading.
- When filling areas, thin layers of fill are easier to spread, have fewer air holes, and compress easier than thick layers of fill.

3.0.0 ◆ SETTING GRADE STAKES

Prior to construction, a **topographic survey** and property survey of the site are performed. This information is used to develop the project plans. The plans include information about elevation requirements at the construction site. Based on the plan, stakes are set by the project survey team to inform workers of the earthwork needed to bring the building site to the specified elevations. Methods for staking grades vary greatly with the type,

location, and size of project. Methods used by different engineering and construction organizations will also vary; some companies use electronic equipment, while others still use older manual instruments. Regardless of how the stakes are set, unless the project is very small, they will be set in reference to bench marks.

3.1.0 Bench Marks

Bench marks are installed at an exact location and are used to show that location's precise elevation in relation to sea level. On building projects, bench marks are used as reference points from which grade stakes are set. Bench marks may be permanent or temporary and can be set by the federal or local governments or by the project survey team. One important feature about bench marks is that they must not be disturbed. It is very important that you know the location of any bench marks in the area so that you can avoid them with your equipment.

3.1.1 State Plane Coordinates

Bench marks on the state plane coordinate system are permanent reference points set by the federal government. They are permanent markers placed in specially designed locations and have specific reference information engraved on a bronze cap (*Figure 7*). These markers are part of the National Geodetic Survey, which is part of the National Oceanic and Atmospheric Administration (NOAA). These markers are frequently accompanied by a witness post, such as the one shown in *Figure 8*, to make them easier to find.

The markers provide an accurate reference point for other surveying activities within that general area. Many such markers are located throughout the United States, usually on the ground set in concrete to help preserve them, and it is important that the markers are not moved or otherwise disturbed. If you notice that a permanent marker is damaged, such as the one shown in *Figure 9*, it may not be used as a bench mark. It is important that you report it to your supervisor or the project engineer. He or she will notify the proper agency so that it may be reset.

States may set similar markers in accordance with their own specifications. State markers are also intended as permanent markers, so if you find one in the work area that is not recorded on the project plan, do not disturb it. Put some type of stake or marker next to it so that you will be able to find it again, and notify the project engineer. He or she will notify the proper authorities to see if the monument needs to be relocated by a survey crew before any further work is done in that area.

208F07.EPS

Figure 7 ◆ Bench marks of the National Geodetic Survey.

208F08.EPS

Figure 8 ◆ Witness post.

3.1.2 Permanent Project Bench Marks

Sometimes a project is developed over a long period. If additional construction is scheduled in the future, there needs to be some control point established close to the construction site so that new construction work can be easily referenced to the previous work. If no other bench marks are close by, then a permanent project bench mark must be established. This bench mark may be a

208F09.EPS

Figure 9 ◆ Damaged bench mark.

special cap of some type set in concrete on the ground or on the side of a structure, such as a building or bridge abutment. *Figure 10* shows an example of a permanent project bench mark that was been set for a roadway project.

3.1.3 Temporary Project Bench Marks

Temporary project bench marks can be placed in the ground, on a structure, or on a large sturdy tree. They are used only for the specific project. The location and elevation of these bench marks is usually derived from permanent bench marks on the state plane coordinate system described earlier. *Figure 11* shows an example of how temporary project bench marks are placed.

3.2.0 Highway and Other Horizontal Construction

For rough grading on highway and other horizontal construction, there are several methods to set grade control depending on the type of construction and the surrounding terrain. The types of stakes used in highway construction will depend on the number of lanes and the existence of curbs and medians.

3.2.1 Setting Highway Grade Control

For road construction, right-of-way stakes or reference stakes that are located outside the construction limits are the first stakes set by the project survey team. They are used by the contractor as a reference to perform rough grading. These stakes are located outside the work area, so it is unlikely that they will be damaged when the area is being prepared for construction.

Figure 10 ◆ Permanent project bench marks.

Figure 11 ◆ Temporary project bench marks.

The center line stakes are set by the survey crew using bench marks and control points. Because the location and elevation of each bench mark is known, all center line stakes can be referenced directly or by referencing one stake to the stake behind it until the center line is tied into another control point further up the line. This tie-in to a second control point provides a double-check on the center line staking process, as well as any other stakes that must be precisely set. *Figure 12* shows an example of a highway plan and profile sheet that includes references to a bench mark used on this job. The bench mark is located at **Station** 109+16.00.

All members of the construction team use the drawing set, so sometimes the plan and profile drawings look confusing (refer to *Figure 12*). The profile portion of the road project is shown at the bottom of the sheet. On the right side of the sheet, the drawing shows the center line profile of the road. The drawing ends at station 115+00 with an elevation of about 2,842 feet (read from the elevation scale located on the right side of the drawing). The center drawing also shows the center line profile of the road. It starts at station 115+00 at an elevation of about 2,842 feet. The center line profile ends at station 110+20.89, which is written on the drawing, at an elevation of about 2,847 feet. The left side of the drawing does not show the roadway center line. The two profile curves shown on this portion of the drawing represent the sides of the road. The bottom drawing is the right side of the road and the top one is the left side of the road.

If right-of-way stakes are used as the reference for rough grading, they can be placed relative to the center line stakes or referenced directly from the bench marks. They are placed at the right-of-way limits at the distance from the center line shown in the plans. Because the stakes are placed at the right-of-way limit, there is little chance that they will be disturbed during earthmoving opera-

tions. These stakes are usually placed at each station (every 100 feet) but may be placed at closer intervals on corners and alignment changes.

The cut or fill amount is indicated on each stake. This is the vertical distance between the profile grade stake and a line on the stake. *Figure 13* shows a typical cross-section of a cut. The stake shown in the figure identifies the station, amount of cut, and the point from which the cut or fill is measured. Most slope stakes would be marked in a similar manner.

Slope stakes, which can be used as cut and fill stakes, are set by the survey team at points where the cut slopes and fill slopes intersect a hinge point. For rough grading, it is unlikely that the survey team will set cut and fill stakes, but will probably use some other type of marker to guide equipment operators in the initial cut and fill operation. Such temporary markers have no information written on them, but should be familiar to the equipment operator. If you are confused about the stakes, talk to our supervisor before you begin grading work. Final grading stakes will show information about the elevation and offset distance.

To visualize the grade for a center line, you should look at the profile sheet in the plans. *Figure 14* shows an example of the profile sheet and the cut and fill areas of the roadway. From Station 10 to Station 40+31.05, the proposed grade is higher than the existing grade, so the area needs to be filled. At Station 40+31.05 the existing grade crosses the proposed grade and both are maintained until Station 50+10.00. This area is called the **balance point**—it needs neither cut nor fill. At Station 50+10.00, the existing elevation is greater than the proposed elevation, which will require a cut to achieve the desired elevation.

The slopes of the proposed grade will also be shown, along with the reference axis. The vertical axis on the profile sheet will show the elevation.

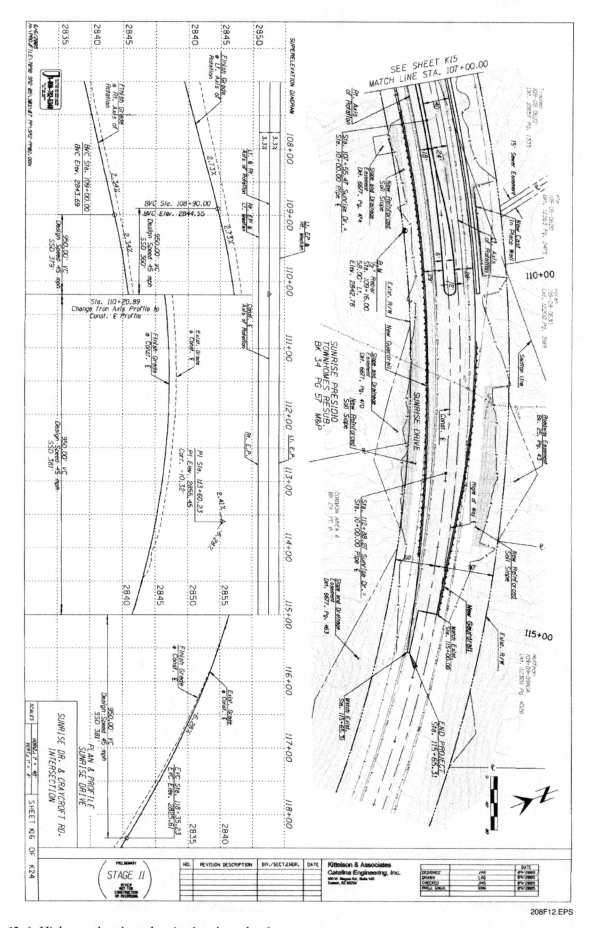

Figure 12 ◆ Highway plan sheet showing bench mark reference.

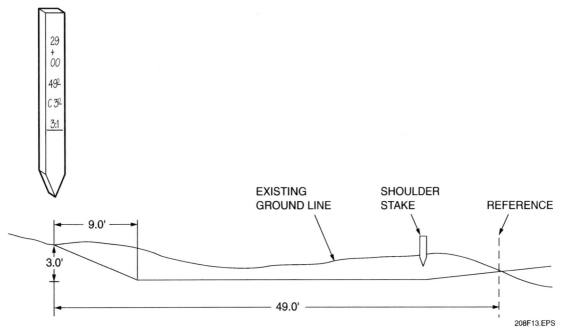

Figure 13 ◆ Cross-section of a cut.

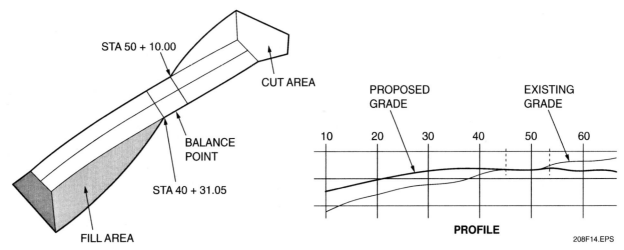

Figure 14 ◆ Example of cut and fill sections.

Usually, the elevation is shown at a much greater scale than the horizontal axis. The horizontal axis will show the stations along the center line. Therefore, you can identify the points on the natural ground where the elevation of the proposed grade is the same as the natural ground.

3.2.2 Finished Grade Reference Points

When the roadbed is close to the proposed grade, the survey crew places finish grade stakes to guide the final grading operation. If any type of stabilization of the subgrade is required, then the subgrade will have to be mixed, regraded, and compacted to the proper density. After compaction, the subgrade must again be shaped to conform to the finished grades and cross-sections shown in the plans. This finishing work is usually performed by motor graders. You will learn about finish grading in the next level of *Heavy Equipment Operations*.

By setting a **string line** at a convenient height above each grade stake, measurements can easily be made down to the finished subgrade surface. This process was described in *Grades, Part One*. To check the lines of the completed subgrade, little flags or other markers can be attached to the string line at the edge of the pavement, shoulder, or other appropriate points.

3.3.0 Building Foundations and Pads

Site plans vary for buildings and other structures. A large building or development will usually have a detailed plan that shows existing elevations and the finished elevation of the structure's foundation. All site plans should show the general outline of the structure and the property boundaries (*Figure 15*). On small jobs, a site plan may not have detailed elevations, so you may need to ask the site supervisor for verbal instructions before you begin earthwork operations.

Study the site plan before you begin work to be sure that you know the location of property boundaries. Do not cross a property line without permission. The site plan may also show the existing elevation of the property, so you or the site supervisor can determine the amount of excavation needed to ensure that the finished elevation of the building foundation is correct.

For structures with foundations, temporary stakes or markers are usually set at the corners as a rough guide for starting the excavation. These markers are set outside the construction limits or area of disturbance, but close enough to be convenient. Permanent control points are set to allow for measurements during the construction process.

Stakes or other markers are set on important lines to mark clearly the limits of the work. For small building foundations and trenches, grade information is usually provided using **batter boards** like those shown in *Figure 16*. A batter board is a wooden board nailed to two substantial posts with the top of the board horizontal and its top edge either at grade or at some whole number of feet above or below grade. The alignment is often fixed by a nail driven in the top edge of the board. Between two batter boards, a strong cord or wire is stretched to define the foundation alignment and grade.

To establish grades and elevation points on the ground, stakes and/or batter boards are set. Grade stakes may be different from stakes used in showing alignment. If wood stakes are used, the vertical measurement can be taken from the top of the stake, a crayon mark on the side, or from the ground surface at the stake. To avoid confusion, only one method should be used for a particular job. When batter boards are used, the vertical measurements are usually taken from the top edge of the board where the string line is placed.

3.3.1 Establishing the Proper Grade

To establish the proper grade for an excavation or fill, the grade setter must locate a bench mark or hub set by the surveyor. (Typically, a surveyor will set hubs for a project.) The grade setter will then begin setting **boots**. A boot is a mark on a stake or lath at a convenient height that can be used as a reference elevation relative to the finished subgrade, pavement, or structure foundation. *Figure 17* shows a typical boot placed by the grade setter. The boot will be 3, 4, or 5 feet above the finished grade. It provides a convenient height for the grade setter to view with a hand level.

For subdivision, office park, and commercial sites, the grade setter will set boots based on the finished grade, then add the curb or road subgrade information when checking grade from these boots. After placing each grade mark, the cut needed to reach subgrade will be written on all the cut stakes. This way the equipment operator can see what cut is needed directly without having to add the curb and road section height to the surveyor's cut marks.

4.0.0 ◆ GRADING

Some grading operations are automated. A computer-controlled grader is driven over the work site, and the computer adjusts the level and angle of the blade based on a topographical survey and Global Positioning System (GPS) coordinates. Most grading operations are not automated. On these jobs, the successful and efficient completion of the job relies on the skills of the equipment operators. The information presented in this section can provide you with knowledge that can help you develop grading skills, but the only way that you can actually develop grading skills is through experience.

The equipment you use during grading operations will depend on many factors, including the job size and type, the physical environment, and the availability of equipment. There are jobs that will require a particular type or size of equipment such as when you need to get into a tight spot (*Figure 18*). Much of the time, one or more types of equipment can be used and it is the skill of the operator that ensures a job well-done. Graders are most commonly used in grading operations.

4.1.0 Motor Grader

Motor graders are specially designed to be used in almost all grading operations. All graders have blades that are used to cut and level grading material. The angle and height of the blade can be adjusted from the cab and the frame is designed to keep the blade stable even when driving over rough terrain. The grader's front wheels can be tilted to the right or left, which helps to increase the steering ability of the grader while cutting.

4.1.1 Grader Blade

The grader blade (*Figure 19*) is made up of a moldboard, endbits, and cutting edge. The moldboard is shaped so that grading material will roll and mix as it is cut. The end bits and cutting edge protect the moldboard edges from the abrasive action of the grading material. You need to inspect the blade daily and replace the endbits and/or cutting edge as necessary.

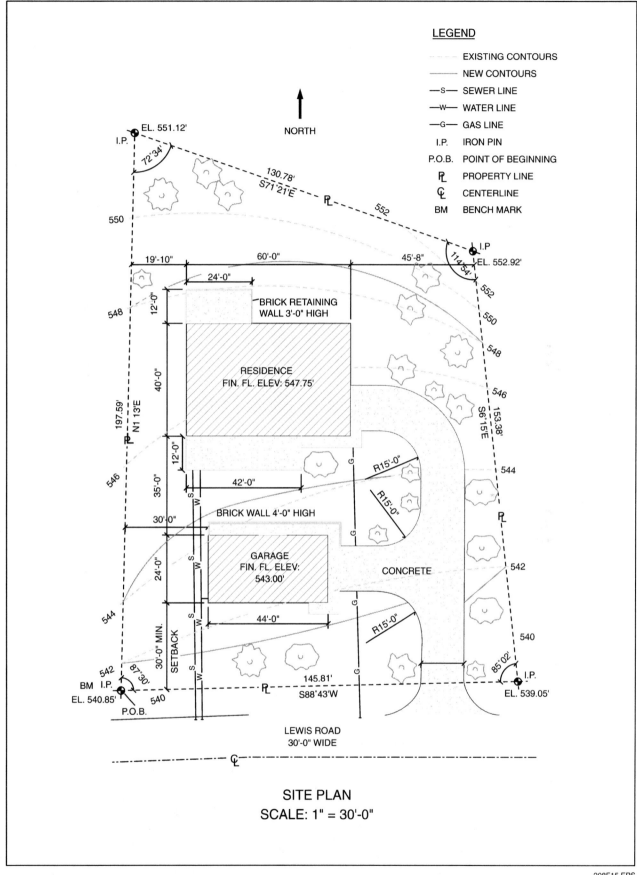

Figure 15 ◆ Site plan showing topographical features.

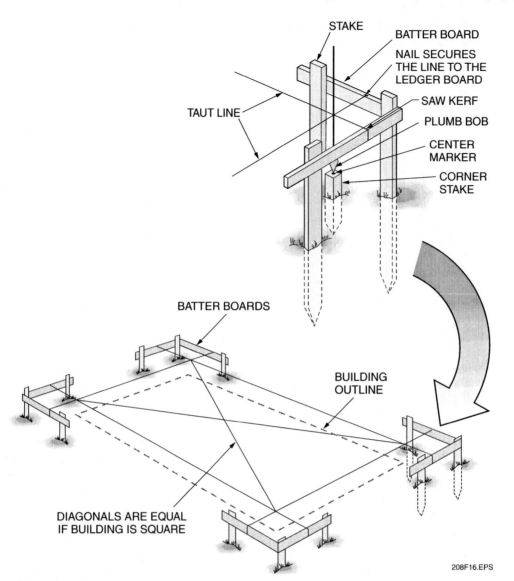

STAKE

BATTER BOARD

NAIL SECURES
THE LINE TO THE
LEDGER BOARD

SAW KERF

PLUMB BOB

CENTER
MARKER

CORNER
STAKE

TAUT LINE

BATTER BOARDS

BUILDING
OUTLINE

DIAGONALS ARE EQUAL
IF BUILDING IS SQUARE

208F16.EPS

Figure 16 ◆ Use of batter boards.

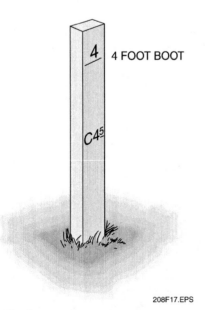

4 FOOT BOOT

4

C4⁵

208F17.EPS

Figure 17 ◆ Grade setter boot.

208F18.EPS

Figure 18 ◆ Working in a tight spot.

The grader blade may be positioned in the center of the grader frame or off to one side (*Figure 20*). The blade may be set perpendicular to the frame or it may be angled. When the blade is angled, the front edge is called the toe and the rear edge is the heel. Grading material spills off the heel of the blade.

> **CAUTION**
> When the blade is set to a sharp angle, it may hit a tire when the grader is turned, causing tire damage. Use caution when operating your grader.

The pitch of the blade can be adjusted for the desired results. For most grading operations, the blade will be upright (*Figure 21A*), but for more cutting, the blade may be pitched back slightly (*Figure 21B*). To increase the mixing of the graded material, the blade is pitched slightly forward (*Figure 21C*). Pitching the blade forward sharply (*Figure 21D*) will help to compact the graded material and ensure that low spots are filled.

208F19.EPS

Figure 19 ◆ Grader blade.

208F20.EPS

Figure 20 ◆ Grader with blade positioned slightly to the side.

4.1.2 Grader Wheels

The grader front wheels are positioned to help stabilize the grader while cutting, and to make it easier to steer while grading. The wheels are usually tilted in the direction of the heel of the blade. In *Figure 22*, the grader is forming the ditch slope. The blade is pushing dirt on the right side. The force of the dirt on the blade would normally tend to pivot the grader into the ditch. By tilting the wheels away from the ditch, the loader stays on a straight course.

4.1.3 Cutting a Ditch

Most grading operations require several passes to complete. As discussed previously, the most efficiency is achieved when the operator is skillful enough to keep the blade full and to complete the

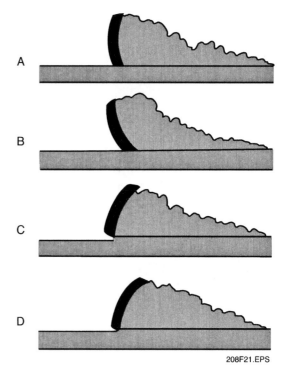

208F21.EPS

Figure 21 ◆ Blade pitch.

208F22.EPS

Figure 22 ◆ Tires tilted to cut a ditch.

operation with a minimum number of passes. Experienced operators can often cut a ditch with two or three passes. These operators are familiar with the equipment and have developed a sense for the soil conditions. Initially, it is best to remove the smallest amount of material on your first pass and then gradually increase the depth of your cuts until you are comfortable with your skills.

When cutting a ditch, the first pass is called the marking cut (*Figure 23A*). It is a 3- to 4-inch deep cut made with the toe of the blade and it is used as a guide to help you to cut a straight ditch. In the second cut, the blade is angled more and positioned over the marking cut (*Figure 23B*). When the second cut is made, more material is deposited onto the road from the heel of the blade. At some point, the cut graded material needs to be spread toward the center of the road—this is called shoulder pickup (*Figure 23C*). These steps are repeated until the ditch is cut to the desired grade.

5.0.0 ◆ KEEPING POSITIVE DRAINAGE

Proper drainage is a very important part of the construction process. It helps to maintain the stability of highway roadbeds and building foundations. Drainage is also necessary to reduce or eliminate erosion on embankments and slopes. If your job site has a stormwater control plan, it is your responsibility to ensure that you are familiar with its contents.

Drainage work required during construction falls into three broad categories: natural and constructed drainage; control ditches; and drains and collection systems. Natural drainage and constructed drains are placed to make sure water is drained away from the structure. Control ditches are constructed to keep water from entering the construction area and causing damage to work completed or under process. Drains and collection systems collect unwanted water within the construction site and provide a method of removal, allowing the soil or other material to dry out and not be saturated.

5.1.0 Natural Drainage and Drains

Roads are crowned and sloped so that water will drain from the road and minimize water ponding on the surface. Shoulders should be graded to slope as much or more than the road to keep water flowing to the ditches. For example, a paved roadway with an 11-foot lane and 4-foot shoulder should have a total crown (from center line to outside edge of shoulder) of not less than 3.5 inches. Roads with steeper grades may require higher crowns, because the water will tend to flow down the road rather than across the crown.

Ditches and channels must be constructed and maintained to avoid damage to the roadway. The primary function is to carry water away from the roadway for absorption, or to another area, such as a detention basin. Ditches must be properly shaped for safety, maintenance, and water-flow and erosion control. A ditch should be at least one foot below the bottom of the roadway base to properly drain the pavement.

It is very important that water flows through the ditches and does not pond. Ponding water may saturate the subsurface material beneath a roadway, preventing it from draining during a storm. Ditches with at least a one percent grade are required to ensure proper flow.

Figure 24 shows drainage problems that are very common to graded roads. Ponding water on this roadway surface is a result of poor grading and compaction and the lack of drainage ditches. Ponding water in ditches is the result of insufficient grading or poor definition of the grade line.

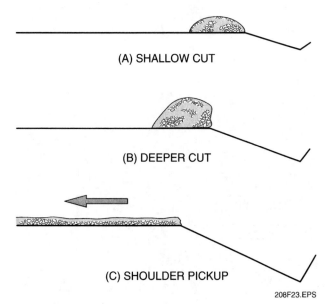

(A) SHALLOW CUT

(B) DEEPER CUT

(C) SHOULDER PICKUP

208F23.EPS

Figure 23 ◆ Cutting a ditch.

208F24.EPS

Figure 24 ◆ Drainage problems on graded road.

Springs or seepage areas under the road will require special treatment. For this problem use french drains (rock-filled trenches) or perforated pipes to drain subsurface water into ditches or streams. Pipe culverts should be opened as soon as they are finished to help control water flow.

Some soils used in embankments are subject to high capillary action, which means that water easily creeps up towards the surface; this creeping action is also called wicking. To reduce this process, a granular blanket of sand or gravel is placed over these soils. This blanket, between the embankment soils and the subgrade, will help protect the subgrade material from the water wicking upward. Requirements for a granular blanket will usually be specified in the plans.

5.2.0 Control Ditches

Control ditches (*Figure 25*), also called intercepting ditches, need to be placed wherever there is a possibility of water intrusion from a source outside the construction area. This is a precautionary

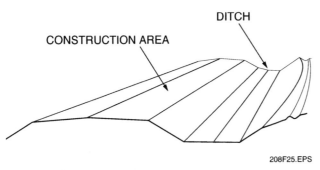

Figure 25 ◆ Control ditch.

action based on knowledge of the terrain as well as possible rain and flooding conditions. These ditches will drain water away from the construction site and deposit it in a nearby drainage ditch.

Intercepting ditches for highway construction should be constructed where the original ground outside of the finished cut sections slopes toward the center line. Watch for seeping water in cut sections, as this may indicate the need for a control ditch. The project engineer should be informed of these conditions, so intercepting ditches or other corrective action can be started.

5.3.0 Drains and Collection Systems

Excavation of pits and trenches can run into trouble because of unexpected pockets of water or a high groundwater table. When excavating below the water table, you normally will get some intrusion into your excavated area. Because this water is below the natural ground line, it cannot be channeled out by digging a control ditch or other drain.

Channels must be dug in the bottom of the excavation to collect the water and direct it into a **sump**, where it can be pumped out. The sump should be placed as close as possible to the greatest source of the flow. The channel depth below the excavation floor should be about 4 feet, but will depend on the pumping equipment being used. When a large amount of water flow exists, you should increase the number of sumps, not their size. *Figure 26* provides two examples of different types of drainage in an excavation.

The first example (*Figure 26A*) shows water coming from only one face of an excavation. A drain along this wall will channel the water into sumps at

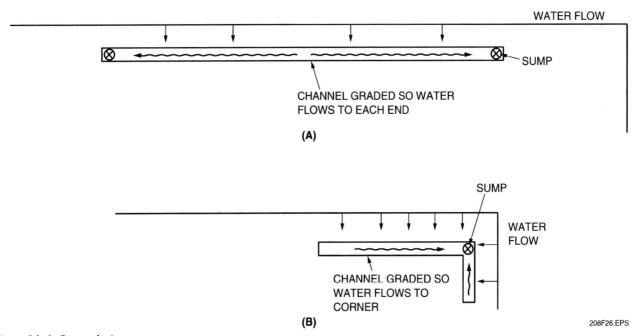

Figure 26 ◆ Sump drainage.

either end. Pump-out is done from the sumps. The second example (*Figure 26B*) is similar, except that the water comes off two faces of the excavation. In this case, the drains are constructed to channel water into a sump in the corner between the two faces. Pump-out is done from this one point.

When water is coming in from the bottom of the excavation, the bottom must be graded to drain to each side and then a layer of stone and damp-proof paper placed to provide a stable work platform. Channels are dug on the sides of the excavation to collect the water and the water is pumped out by the sumps. The sumps always need to be placed as close to the major source of water flow, using multiple sumps if necessary.

Soil conditions will dictate the method of constructing drains. Drainage layouts will be governed not only by soil conditions, but also by the location, direction, and quantity of flow. Drains should be limited in width and depth to minimum requirements. They should intercept the ground water as close to its source as possible.

6.0.0 ◆ TRENCHES

Getting the proper grade for pipe placement in trenches is very important because the drainage through the pipe is controlled by gravity. After the pipe is laid and the trench is backfilled, it is expensive to re-excavate when it will not drain properly. In addition, an uneven grade will put extra pressure on a pipe, causing it to break or leak at the joints.

Setting the proper grade for pipe or other conduits can be accomplished in different ways. The most popular method involves a **laser** level. Other methods include the use of a string line.

Regardless of the grading method, a survey crew will establish hubs that can be used as control points for setting conduit or pipe. Before these hubs are set, a decision is made about how far the offset should be and on which side of the trench it should be placed. The width of the trench and the direction the spoil will be thrown usually determines these two factors.

Often, some type of **bedding material** is used to help set the pipe to the proper grade. This material can be any material that does not readily compress, but it is important that the material uniformly support the pipe to avoid damaging it. Many workers find that it is helpful to use a shovel or long pole to position the pipe once it is in the trench.

Once the pipe is positioned, you can backfill the trench. Filling the trench with thin layers of fill rather than large loads will help to ensure that the pipe is not pushed off of grade. It also helps to reduce air pockets and decrease the need for compaction.

6.1.0 Setting Grade for Pipe Using a Laser

The center line for a pipe trench is located on the ground with stakes or other marks at 50-foot intervals where the grade is uniform. In rough terrain, stakes are set at a much closer interval. Hubs are set parallel to the center line but far enough away so they are not disturbed by the excavation process. These stakes are placed using manual transit or electronic survey device as discussed in the *Grades, Part One* module.

When the trench has been excavated, a laser level (*Figure 27*) can be used to provide a grade line at some elevation above the bottom of the trench. The pipe laser receiver is set up at one end of the segment, and the transmitter is set at the other end of the segment to check the grade. The laser level can be used first to check the rough grading of the trench, and then after the pipe has been placed in the trench, before the trench is backfilled.

> **CAUTION**
>
> A laser level is not a good choice if the work site is dusty or foggy. The laser beam is bent by dust and water particles and may read the incorrect grade. In addition, some laser levels will not operate well when part or all of the trench or pipe is very hot. Check the manufacturer's instructions for more information.

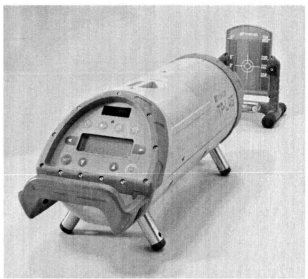

208F27.EPS

Figure 27 ◆ Using a pipe level.

6.1.2 Setting Grade Using a String Line

String lines can be used for setting grades if no laser level or other surveying instrument is available. String lines will be more effective in adverse environmental conditions such as heavy fog or dust that can hinder the operation of a laser level.

The initial setup for the excavation is similar to the procedure described earlier. Once the offset has been determined, the survey crew will place the required number of stakes. This information can now be transferred to the string line setup, as shown in *Figure 28*. You need to decide how high above the trench the string line should be placed over the top of the trench and how far apart the batter boards need to be spaced. A recommended maximum space for the batter boards is 25 feet. The string line will sag if support is spaced any further apart. The height of the string line will depend on the terrain and the amount of slope required at the bottom of the trench. The height of the string line is set by measuring a given elevation above the required grade of the trench plus the offset distance for the line.

To determine the exact elevation of the pipe, measure the distance from the string to the bottom of the trench, and subtract the height of the batter boards from the hub. Then subtract the result from the hub elevation. Repeat this process for each station.

When setting any string line, keep three stations set up along the line so you can sight down the line to correct errors or movement that might occur. As you sight down the string line, each station should blend with the next, with no sudden rise or dip from station to station. If you spot a sudden grade change, check all measurements. If you cannot find any error in your work, check the rate of slope shown on the plans. If the plans do not show a sudden grade change, it is possible the surveyors made an error that has been carried through to the grade line.

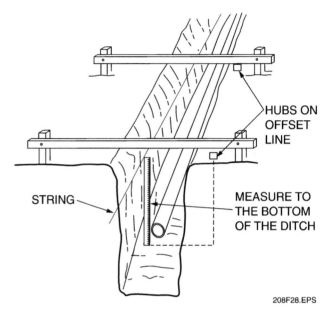

208F28.EPS

Figure 28 ◆ Using a string line to check grades.

1. The first step to any earthwork job is _____.
   a. staking the job site
   b. planning the task
   c. scheduling workers
   d. scheduling equipment

2. Which of the following is *not* part of the typical earthmoving pattern?
   a. Clearing and grubbing
   b. Rough grading
   c. Setting permanent bench marks
   d. Finish grading

3. The two most important factors in determining what equipment is needed for an earthmoving operation are the _____.
   a. excavation volume and weight
   b. size and location of the job site
   c. location of the job site and excavation volume
   d. size of the job and number of workers at the site

4. The cycle time for a dump truck is the time it takes to _____.
   a. get a load, travel to its destination, unload, and return for another load
   b. be loaded at an excavation site, and then unload at the destination site
   c. travel from the loading site to the dump site and back to the loading site
   d. to travel between the loading and unloading sites with a full load

5. Scrapers may be used to drop and spread excavated material.
   a. True
   b. False

6. When working on steeply sloped terrain, greater stability is provided by vehicles that are _____.
   a. heavy
   b. rubber-tired
   c. wide
   d. tracked

7. You can help to ensure that an earthmoving job is accomplished efficiently by increasing equipment travel time because this will reduce traffic at the work site.
   a. True
   b. False

8. The most common problem new operators have in earthwork operations is they _____.
   a. require too many passes to bring the site to the proper elevation
   b. cut too much material away so that the site plans need to be modified
   c. keep the blade or the bucket too full and need to make clean up passes
   d. ask their supervisors so many questions that they disrupt the work flow

9. Grade stakes are set by _____.
   a. the project survey team to inform equipment operators of grading needs
   b. government workers to inform the survey team of a site's exact elevation
   c. equipment operators to inform the project survey team of existing elevations
   d. government workers to inform equipment operators of grading needs

10. A permanent bench mark shows a point's _____.
    a. exact elevation
    b. precise location
    c. exact azimuth
    d. precise value

11. Bench marks may be set by all of the following *except* _____.
    a. the project survey team
    b. any local government
    c. the federal government
    d. Local Bench Mark Agency (LBA)

12. As a heavy equipment operator, you must check the site plans for bench marks so that you can _____.
    a. find them when you need them
    b. avoid them with your equipment
    c. help the survey team set grade stakes
    d. point them out to the project engineer

13. The posts the National Oceanic and Atmospheric Administration (NOAA) sets near its bench marks to make them easier to find are called _____.
    a. witness
    b. goal
    c. fence
    d. trial

14. Whenever you find a permanent bench mark that is damaged, you _____.
    a. need to stop work and repair it immediately
    b. must report it to the site supervisor or engineer
    c. can ignore it because there are plenty of other markers
    d. stop your work since the grade will be incorrect

15. A permanent project bench mark is set when additional construction is scheduled at a future date.
    a. True
    b. False

16. A temporary bench mark is often placed _____.
    a. on a structure or sturdy tree
    b. in a foundation of concrete
    c. on the site's office trailer
    d. in the general work area

17. When more than one bench mark is in the area, the second bench mark makes a good double-check for the center line stakes.
    a. True
    b. False

18. The site survey team set out stakes for rough grading that you do not recognize and you do not know how to perform your grading work. What do you do?
    a. Study the plans and try to figure out what to do for yourself.
    b. Study the site plans and ask your supervisor to help you.
    c. Ask a co-worker and then follow the instructions carefully.
    d. Since it is rough grading, you can clear and level the area.

19. When you are working on a grading job for a new building, it is important that you know where the property boundaries are so that you can _____.
    a. park your equipment over the line
    b. be sure to grade right up to the line
    c. stay 10 feet from the boundary line
    d. avoid crossing over the property line

20. To check the grade of an excavation, the grade setter can set grade boots that he or she can _____.
    a. sight with a hand level
    b. use with a pipe level
    c. use to check subgrades only
    d. look at through a laser level

21. The motor grader blade is made up of a moldboard, end pieces, and cutting edge. Every day the end pieces and cutting edge should be _____.
    a. replaced
    b. used
    c. inspected
    d. polished

22. The first pass when cutting a ditch is used _____.
    a. to make a marking cut
    b. for shoulder pick-up
    c. to level the ditch area
    d. to spread loose material

23. Drainage is important on a construction site because _____.
    a. flooding can cause work to stop and scheduling delays
    b. grading wet material takes longer than dry material
    c. it helps to maintain the stability of the building project
    d. a site will be fined if drainage is inadequate

24. Setting an incorrect grade for a drainage pipe can do all of the following *except* _____.
    a. assist drainage through the pipe
    b. apply extra pressure on the pipe
    c. cause it not to drain properly
    d. cause expensive re-excavation

25. A pipe laser may be used to set the grade for a pipe _____ it is laid in the trench.
    a. before and after
    b. before
    c. as
    d. after

Summary

Grading for a construction project requires many different operations. If the grading requirements are extensive or cover a large area, the operation will include clearing and grubbing, rough grading, and finish grading. Clearing and grubbing is the removal of all aboveground objects, roots, boulders, stumps, and other organic material, as well as the top few inches of soil. Rough grading includes moving excavated material to designated areas in order to bring the ground to the planned elevations. Tolerances for rough grading will vary, depending on the type of work and the client. Finish grading is the trimming and smoothing of placed material to the final required elevations within tolerances shown on the plans.

Equipment operators sometimes are required to read and use project plans in their work. At the beginning of a project, the grading supervisor or project engineer will usually discuss the grading and excavation operation with the equipment operators to familiarize them with the area. You should know the basic steps to go from natural ground to the finish grade, be able to follow the operation on the plans, and understand the work flow at the site. You will need to work with the plan and profile sheet, the cross-section plans, and the grading plans to get information about grading requirements on the job. Each of these sheets gives specific information about the grading requirements and the elevations of the grade, so you must be able to read and understand each part in order to do your job.

Most projects require the project survey team to set reference points, center lines, slope stakes, and grading stakes for equipment operators and other workers to follow. After these controls are set, the contractor may use his or her own personnel to set the detailed stakes that direct the grading operations. Equipment operators must be aware of the survey requirements and various survey activities going on throughout the duration of the project. They must understand the information on the stakes and be able to visualize the completed grading project based on the many stakes that are placed. In some cases, an equipment operator may be asked to assist a grade setter or surveyor in staking activities. Although you need to understand basic surveying functions, it is not your responsibility to set stakes independently.

Initial surveying requirements involve setting grades and control points from bench marks or known reference points. The bench marks are usually permanent monuments installed by federal or state governments. If one of these monuments is found on the construction site, it should not be disturbed until the proper authorities are notified and a decision is made about relocation.

Poor drainage on a construction site can cause major problems. If surface water is a problem, temporary drainage ditches or channels can be constructed to drain the water to a retaining ponds. If ground water problems are encountered in a cut area, control (intercept) ditches can be built to collect the outflow and reduce the possibility of damage to the construction site. In excavated pits or trenches, drains leading to sumps can be installed to channel water to sumps where it can be held until removed by pumping.

Notes

Trade Terms Introduced in This Module

Articulating frame: A frame that is hinged in the middle and is able to pivot for better traction and handling.

Balance point: The location on the ground that marks the change from a cut to a fill. On large excavation projects you may have several balance points.

Batter board: A wooden board erected horizontally over trenches and at corners of excavations to provide an attachment and support for a string line.

Bedding material: Select material that is used on the floor of a trench to support the weight of pipe. Bedding material serves as a base for the pipe.

Boots: A special name for laths that are placed by a grade setter to help control the grading operation. The boot can also be the mark on the lath, usually 3, 4, or 5 feet above the finish grade elevation, which can be easily sighted. This allows the grade setter to check the grade alone instead of having to use another person to hold a level rod on the top of the grade stake.

Laser: A beam of sharply focused light. Lasers are incorporated into many surveying instruments and replace many functions that levels used to perform. A laser level is commonly known as a laser.

Stations: Designated points along a line or a network of points used to survey and lay out construction work. The distance between two stations is normally 100 feet or 100 meters depending on the measurement system used.

String line: A tough cord or small diameter wire stretched between posts or pins to designate the line and elevation of a grade. String lines take the place of hubs and stakes for some operations.

Sump: A small excavation dug below grade for the purpose of draining or retaining subsurface water. The water is then usually pumped out of the sump by mechanical means.

Topographic survey: The process of surveying a geographic area to collect data indicating the shape of the terrain and the location of natural and man-made objects.

Resources & Acknowledgments

Additional Resources

This module is intended to be a thorough resource for task training. The following reference works are suggested for further study. These are optional materials for continued education rather than for task training.

Excavating and Grading Handbook, 1987. Nicholas E. Capachi. Carlsbad, CA: Craftsman Book Company.

Excavators Handbook Advanced Techniques for Operators, 1999. Reinar Christian. Addison, IL: The Aberdeen Group, A division of Hanley-Wood, Inc.

Basic Equipment Operator, 1994 Edition. John T. Morris (preparer), NAVEDTRA 14081, Naval Education and Training Professional Development and Technology Center.

Pipe & Excavation Contracting, 1987. Dave Roberts. Carlsbad, CA: Craftsman Book Company.

Moving the Earth, 1999. Herbert L. Nichols, Jr. and David A. Day. New York, NY: McGraw-Hill.

Figure Credits

Reprinted courtesy of Caterpillar Inc., 208F01, 208F03–208F06, 208F18, 208F20, 208F22

Topaz Publications, Inc.
208F02, 208F08, 208F10, 208F11

National Oceanic and Atmospheric Administration /Department of Commerce, 208F07, 208F09

Kittelson & Associates, 208F12

John Hoerlein, 208F15

Kennemetal EMCSG, 208F19

Curt Giovanine, 208F24

Topcon Positioning Systems, Inc., 208F27

The NCCER makes every effort to keep these textbooks up-to-date and free of technical errors. We appreciate your help in this process. If you have an idea for improving this textbook, or if you find an error, a typographical mistake, or an inaccuracy in NCCER's Contren® textbooks, please write us, using this form or a photocopy. Be sure to include the exact module number, page number, a detailed description, and the correction, if applicable. Your input will be brought to the attention of the Technical Review Committee. Thank you for your assistance.

Instructors – If you found that additional materials were necessary in order to teach this module effectively, please let us know so that we may include them in the Equipment/Materials list in the Annotated Instructor's Guide.

Write: Product Development and Revision
National Center for Construction Education and Research
P.O. Box 141104, Gainesville, FL 32614-1104

Fax: 352-334-0932

E-mail: curriculum@nccer.org

Craft _____ Module Name _____

Copyright Date _____ Module Number _____ Page Number(s) _____

Description _____

_____

_____

_____

(Optional) Correction _____

_____

_____

(Optional) Your Name and Address _____

_____

_____

22209-06

Civil Blueprint Reading

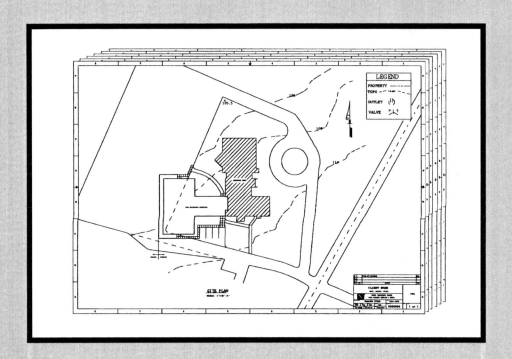

22209-06
Civil Blueprint Reading

Topics to be presented in this module include:

Overview

Questions arise on every job. How deep is this cut? How wide is this trench? A successful project must have a master plan to resolve any questions. In construction these master plans are the blueprints.

Equipment operators must be able to read civil blueprints to be able to understand how to plan and perform their job. Certain plans tell the operator where and how to perform earthmoving and other operations. Landscape plans provide the information for final grading.

Operators need to know how to read all of the plans in a drawing set. Staging the soils pile in an area slated for excavation would create additional and costly work. Being able to read and understand the drawings and plans will help an operator make wise choices in performing their tasks.

Objectives

When you have completed this module, you will be able to do the following:

1. Describe the types of drawings usually included in a set of plans, and list the information found on each type.
2. Identify the different types of lines used on drawings.
3. Define common abbreviations and symbols used on plans.
4. Read and interpret drawings to determine the type of excavations needed to prepare the site.
5. Describe the operator's duties to ensure that the job is completed safely and according to the plans.

Trade Terms

Change order
Contour lines
Easement
Elevation view
Front setback

Monuments
Plan view
Property lines
Request for information (RFI)

Required Trainee Materials

1. Pencil and paper
2. Appropriate personal protective equipment

Prerequisites

Before you begin this module, it is recommended that you successfully complete *Core Curriculum*; *Heavy Equipment Operations Level One*; and *Heavy Equipment Operations Level Two*, Modules 22201-06 through 22208-06.

This course map shows all of the modules in the second level of the *Heavy Equipment Operations* curriculum. The suggested training order begins at the bottom and proceeds up. Skill levels increase as you advance on the course map. The local Training Program Sponsor may adjust the training order.

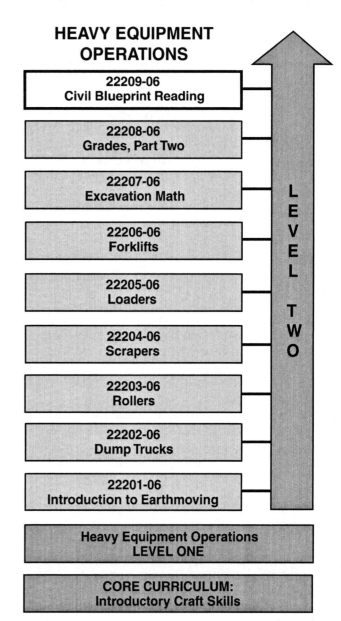

HEAVY EQUIPMENT OPERATIONS

22209-06
Civil Blueprint Reading

22208-06
Grades, Part Two

22207-06
Excavation Math

22206-06
Forklifts

22205-06
Loaders

22204-06
Scrapers

22203-06
Rollers

22202-06
Dump Trucks

22201-06
Introduction to Earthmoving

LEVEL TWO

Heavy Equipment Operations
LEVEL ONE

CORE CURRICULUM:
Introductory Craft Skills

209CMAP.EPS

1.0.0 ◆ INTRODUCTION

This module reviews and builds on the construction drawing (blueprint) material introduced in the *Core Curriculum*. It introduces techniques for reading construction drawings and specifications. Construction drawings tell how to build a specific roadway, building, or structure. Specifications may be used along with the construction drawings. They contain detailed written instructions that supplement the set of drawings. Drawings and specifications are part of the contract between the builder and the client, so they are legal documents.

During the planning phase of a new construction project, architects and/or engineers will develop detailed plans for the project using data gathered by the survey team. The team will verify the site boundaries, which are established by a licensed surveyor, and then set up the construction project's precise position in relation to the property boundaries. In addition, the survey crew will establish the exact location of road and utility right-of-ways (ROW), **easements**, and other important features of the project.

It is extremely important that you interpret construction drawings and specifications correctly. Failure to do so may result in costly rework and unhappy customers. Depending on the severity of the mistake, it can also expose you and your employer to legal liability.

The project plan will give you a good idea about your role in the project. Your duties will vary with the type and size of the construction project. Small highway construction jobs may require the constant use of heavy equipment for hauling fill, cutting grades, leveling the roadbed, and spreading paving materials, while residential or commercial building projects may require the intense use of heavy equipment at the beginning of the project to prepare the site for construction and at end of the project to complete final grades and landscaping tasks.

2.0.0 ◆ DRAWING SET

The set of detailed drawings or plans drawn by an architect and/or engineer shows all the information and dimensions necessary to build or remodel a structure. Copies of the architect's original drawings are made for contractors and others to use. These copies are called blueprints. The term blueprint is derived from a method of reproduction that was used in the past. A true blueprint shows the details of the structure as white lines on a blue background. Today, computers are used to design construction projects, and plans are printed with black ink on white paper. However, the term blueprint is still widely used when referring to any copies made from original drawings.

Most drawings are drawn to scale, which is prominently displayed on the drawing. The scale used depends on the size of the project. A large project often has a small scale, such as 1" = 100', while a small project might have a large scale, such as 1" = 10'. The dimensions shown on drawings can be in engineering scale, which is in feet and tenths of a foot, or architectural scale, which is in feet, inches, and fractions of an inch.

Construction drawings or architect's plans consist of several different kinds of drawings assembled into a set (*Figure 1*). For complete information about a structure to be conveyed to the reader, various drawings in the set illustrate the structure using a variety of different views.

A set of drawings also includes sheets that contain relevant written information, such as notes and equipment or material schedules. Notes often contain essential information, so it is very important to study all parts of the plans.

As you may have guessed, the plans for a roadway project differ significantly from those for a building or structure. Because roadways can stretch across many miles, these plans need to consider the topography and soil conditions over long distances, while building plans need to consider the terrain over a comparatively small area. You must understand how to read and interpret both types of plans to perform your job as a heavy equipment operator.

2.1.0 Title Sheets, Title Blocks, and Revision Blocks

A title sheet is normally placed at the beginning of a set of drawings or at the beginning of a major section of drawings. It provides an index to the other drawings; a list of abbreviations used on the drawings and their meanings; a list of symbols used on the drawings and their meanings; and various other project data, such as the project location, the size of the land parcel, and the building size. It is important that you use the title sheet(s) that come with the drawing set so that you understand the specific symbols and abbreviations used throughout the drawings. These symbols and abbreviations may vary from job to job.

A title block or box (*Figure 2*) is normally placed on each sheet in a set of drawings. It is usually located in the bottom right-hand corner of the sheet, but this location can vary. The drawing set should be folded so that the title block faces up.

The title block serves several purposes in terms of communicating information. Generally, it contains the name of the firm that prepared the drawings, the owner's name, and the address and name of the project. It also gives locator information, such as the title of the sheet, the drawing or sheet number, the date

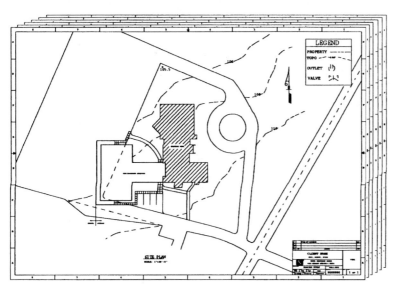

TITLE SHEET(S)

ARCHITECTURAL DRAWINGS

- SITE (PLOT) PLAN
- FOUNDATION PLAN
- FLOOR PLAN
- INTERIOR/EXTERIOR ELEVATIONS
- SECTIONS
- DETAILS
- SCHEDULES

STRUCTURAL DRAWINGS

PLUMBING PLANS

MECHANICAL PLANS

ELECTRICAL PLANS

209F01.EPS

Figure 1 ◆ Format of a working drawing set.

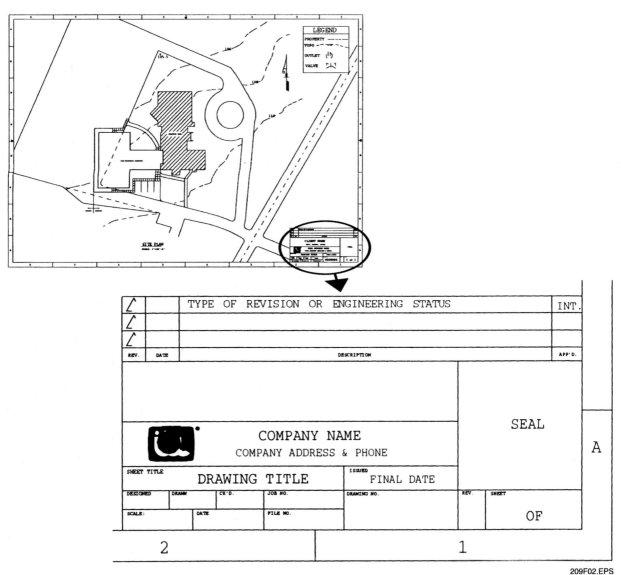

209F02.EPS

Figure 2 ◆ Title and revision blocks.

the sheet was prepared, the scale, and the initials or names of the people who prepared and checked the drawing.

A revision block is normally shown on each sheet in a set of drawings. Typically, it is located in the upper right-hand corner or bottom right-hand corner of the drawing, near or within the title block. It is used to record any changes (revisions) to the drawing. An entry in the revision block usually contains the revision number or letter, a brief description of the change, the date, and the initials of the person making the revision(s). When using drawings, it is essential to note the revision designation on each drawing and use only the latest issue; otherwise, costly mistakes will result. If in doubt about the revision status of a drawing, check with your supervisor to make sure that you are using the most recent version of the drawing. Also, check to see if a **request for information (RFI)** or sketches have been issued in the latest plan revision.

2.2.0 Highway Plans

Highway construction projects are called horizontal projects because they cover long distances and have almost no height. On highway jobs, the emphasis is on grading the roadbed and nearby areas according to engineering plans, which are called plan and profile sheets and cross-section sheets. Equipment operators and other construction workers use these plans to obtain grading information at various times during the construction project.

The type and amount of information on highway plans varies according to the complexity of the project, the existing terrain, and the amount of grading required. Often grade information will be transferred from the plans to the grade stakes, and you will read the grade stakes to accomplish your job. Other times you will need to read the plans to obtain grade information.

2.2.1 Plan and Profile Sheets

Figure 3 shows a typical highway construction plan sheet and *Figure 4* shows a typical profile sheet. These two drawings are often placed on the same sheet with the **plan view** on the top and profile on the bottom.

The plan view shown in *Figure 3* is the view you would see if you were on top of the project and looking down. The profile view shown in *Figure 4* is similar to a side **elevation view**. It shows the key elevations and slopes along the route center line. For some projects, a whole sheet may be required just to show the plan view and a separate sheet needed for the profile view. The main information shown on these sheets includes the following:

- *Direction* – The directional arrows always point north.
- *Station numbers* – These numbers are listed along the bottom axis of the profile sheet and on the plan sheet. Find the numbers 110 and 115 located at the top of *Figure 3*. These numbers correspond to the numbers shown at the bottom of *Figure 4*.
- *Elevations* – These are listed along the right side of the profile sheet according to the designated scale.
- *Natural ground* – The elevation of the natural ground is drawn as a continuous line on the profile sheet, while **contour lines** appear on the plan sheet.
- *Planned grade* – The planned elevation of the grade is drawn as a continuous line on the profile sheet.
- *Center line* – The center line of the roadway is plotted on the plan sheet.
- *Right-of-way* – The right-of-way limits are shown on the plan sheet.
- *Bench marks* – Bench marks, if any, are noted on the plan sheet.

Figure 4 shows a typical profile at the center line of a proposed highway. Elevations are shown on the left and right side of the sheet and station numbers are on the bottom of the sheet. The profile view provides information about the existing natural ground (or grade) and the planned final grade. It graphically shows each grade in relation to the other and gives a good picture of what types of excavations need to be done. In this case, all the work would be fill because the natural ground is below the required finished grade. When plotting the profile view, it is common to use a vertical scale much larger than the horizontal one to make the elevation differences very clear.

2.2.2 Cross-Sections

Typical cross-section sheets are views of the construction you would have if you cut through the area in a crosswise manner. For a highway, this would be like taking a knife and slicing across the road from one right-of-way line to the other and looking at the slice taken. The cross-section shows the layers of the road construction and the shapes of the side slopes and ditches.

Figure 5 shows a typical cross-section or template of a multi-lane divided highway. It also shows cross-sections for several additional features, such as the ditches and ramps. A very simple road that is straight and level would need only one typical cross-section sheet because it would be the same everywhere. However, most roads require many typical cross-section sheets because the terrain will vary from point to point. You will need to check this sheet frequently to get grade information

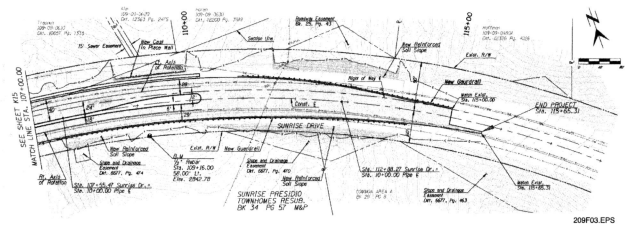

Figure 3 ◆ Typical plan sheet for highway construction.

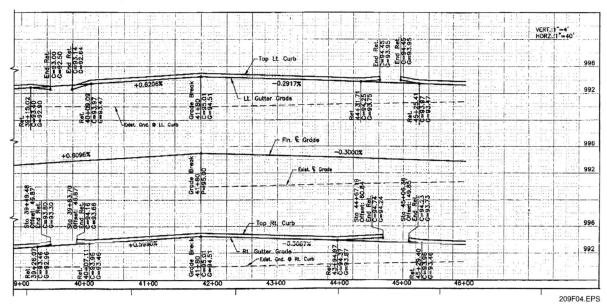

Figure 4 ◆ Profile sheet detail.

about the section of road on which you are working. In addition to these typical cross-sections, there may be an additional set of special cross-section sheets showing the natural ground and the shape of the final grade every 100 feet.

2.3.0 Building Plan Drawings

The types of written information and views normally contained in a drawing set for a building include the following:

- Title sheets, title blocks, and revision blocks
- Architectural drawings consisting of the following:
  - Plan views
  - Elevations
  - Sections
  - Details
  - Schedules
- Structural drawings
- Plumbing plans

- Mechanical plans
- Electrical plans

Plan view drawings are drawings that show the site looking down from above. The object is projected from a horizontal plane. Typically, plan view drawings are made to show the overall construction site (plot or site plan), the structure's foundation (foundation plan), and the structure's floor plans.

2.3.1 Site Plans

Man-made and topographic (natural) features and other relevant project information, including the information needed to correctly locate the structure on the site, are shown on a site plan (commonly called a plot plan). Man-made features include roads, sidewalks, utilities, and buildings. Topographical features include trees, streams, springs, and existing contours. Project information includes the building outline, general utility information, proposed sidewalks, parking areas, roads, landscape information,

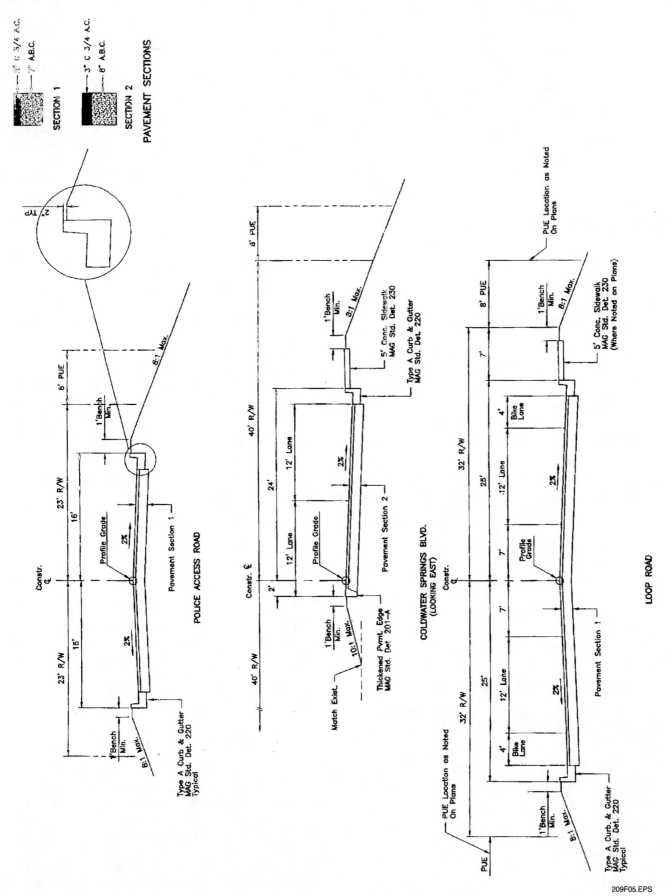

Figure 5 ◆ Cross-section.

209F05.EPS

proposed contours, and any other information that will convey what is to be constructed or changed on the site. A prominently displayed north direction arrow is included for orientation purposes on site plans. Sometimes a site plan contains a large-scale map of the overall area that indicates where the project is located on the site. *Figures 6* and *7* show examples of basic site plans.

Typically, site plans show the following types of detailed information:

- Coordinates of control points or property corners
- Direction and length of **property lines** or control lines
- Description, or reference to a description, for all control and property **monuments**
- Location, dimensions, and elevation of the structure on site
- Finished and existing grade contour lines
- Finished elevations of building floors
- Location of utilities
- Location of existing elements such as trees and other structures
- Location and dimensions of roadways, driveways, and sidewalks
- Names of all roads shown on the plan
- Locations and dimensions of any easements

Like other drawings, site plans are usually drawn to scale, which is prominently displayed on the drawing. The scale used depends on the size of the project. A project covering a large area typically will have a small scale, such as 1" = 100', while a project on a small site might have a large scale, such as 1" = 10'.

Normally, the dimensions shown on site plans are stated in feet and tenths of a foot (engineer's scale). However, some site plans state the dimensions in feet, inches, and fractions of an inch (architect's scale). Dimensions to the property lines are shown to establish code requirements. Frequently, building codes require that nothing be built on certain portions of the land. For example, local building codes may have a **front setback** requirement that dictates the minimum distance that must be maintained between the street and the front of a structure. Normally, side yards have a minimum width specified from the property line to allow for access to rear yards and to reduce the possibility of fire spreading to adjacent buildings.

The plans will show areas of easement on the property. Easements are legal rights of persons other than the owner to use the property. A property owner cannot build on an area where an easement has been identified. Examples of typical easements are the right of a neighbor to build a road; a public

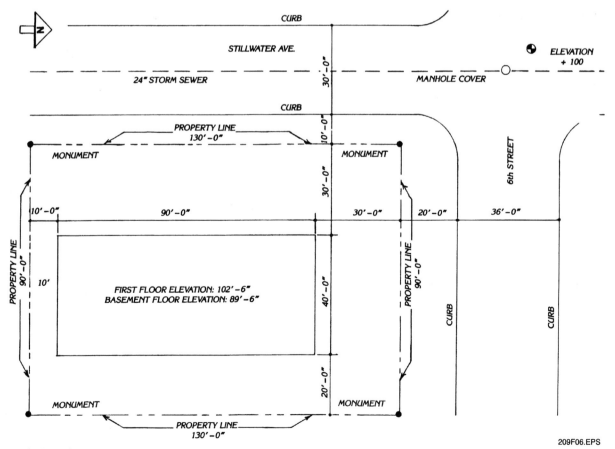

Figure 6 ◆ Site plan.

209F06.EPS

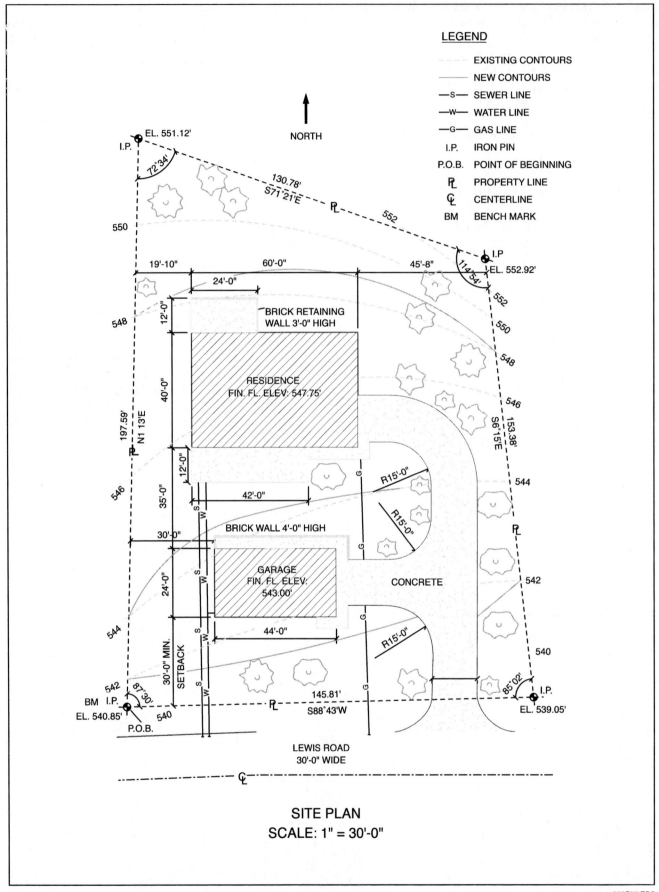

Figure 7 ◆ Site plan showing topographical features.

utility to install water, gas, or electric lines on the property; or an area set aside for drainage of groundwater.

Site plans show finished grades (also called elevations) for the site based on data provided by a surveyor or engineer. It is necessary to know these elevations for grading the lot and for construction of the structure. Finished grades are typically shown for all four corners of the lot, as well as other points within the lot. Finished grades or elevations are also shown for the corners of the structure and relevant points within the building.

Heavy equipment operators need to pay particular attention to existing and proposed contour lines on these plans. These lines tell you how deep your cuts will be or how much fill is required to bring the site to the correct elevation. You will be referring to the site plan often to perform your job.

In addition to using the site plan to get information about what you need to do, you will also be using the plan to get information about what *not* to do. Follow these rules while you are working:

- Do not cross property boundaries unless you know that the owner has given the managers of the project an easement. Heavy equipment can damage terrain and underground structures such as drainage pipes, culverts, and septic tanks.

- Do not operate heavy equipment in the location of a bench mark, monument, or control points until you are sure of its location. Damaging these references can cause costly delays.

- Do not operate heavy equipment near any utilities unless you are certain of their locations. Not only can hitting underground gas lines and power cables cause delay, it can be fatal.

Take a few moments to study the site plan shown in *Figure 7*. Find the North marker and review the legend to become familiar with the symbols used on the drawing. Look for property boundaries and marking; utilities; and existing and proposed grades (contours). If you are having trouble reading the plan, ask your instructor for help.

All the finished grade references shown are keyed to a reference point, called a bench mark or job datum. This is a reference point established by the surveyor on or close to the property, usually at one corner of the lot. At the site, this point may be marked by a plugged pipe driven into the ground, a brass marker, or a wood stake. The location of the bench mark is shown on the plot plan with a grade figure next to it. This grade figure may indicate the actual elevation relative to sea level, or it may be an arbitrary elevation, such as 100.00' or 500.00'. All other grade points shown on the site plan, therefore, will be relative to the bench mark. In *Figure 7*, this point is labeled P.O.B. for point of beginning and is located at the southwest corner of the property.

A site plan usually shows the finished floor elevation of the building. This is the level of the first floor of the building relative to the job-site bench mark. For example, if the bench mark is labeled 100.00' and the finished floor elevation indicated on the plan is marked 105.00', the finished floor elevation is 5' above the bench mark. During construction, many important measurements are taken from the finished floor elevation point.

On *Figure 7*, the bench mark is located at the southwest property corner and is 540.85' (P.O.B.). Since the residence finished floor elevation is 547.75', the finished floor elevation is 6.9' above the bench mark.

2.3.2 Foundation Plans

As applicable, foundation plans (*Figure 8*) give information about the location and dimensions of footings, grade beams, foundation walls, stem walls, piers, equipment footings, and windows and doors. The specific information shown on the plan is determined by the type of construction involved: full-basement foundation, crawl space, or a concrete slab-on-grade level (*Figure 9*).

The following are types of information normally shown on foundation plans for full-basement and crawl space foundations:

- Location of the inside and outside of the foundation walls
- Location of the footings for foundation walls, columns, posts, chimneys, fireplaces, etc.
- Walls for entrance platforms (stoops)
- Notations for the strength of concrete used for various parts of the foundation and floor
- Notations for the composition, thickness, and underlaying material of the basement floor or crawl space surface

The types of information normally shown on foundation plans for slab-on-grade foundations include the following:

- Size and shape of the slab
- Exterior and interior footing locations
- Loadbearing surface (fireplace, for example)
- Notations for slab thickness
- Notations for wire mesh reinforcing, fill, and vapor barrier materials

2.3.3 Floor Plans

The floor plan is the main drawing of the entire set. For a floor plan view, an imaginary line is cut horizontally across the structure at varying heights so all the important features such as windows, doors, and plumbing fixtures can be shown. For multistory

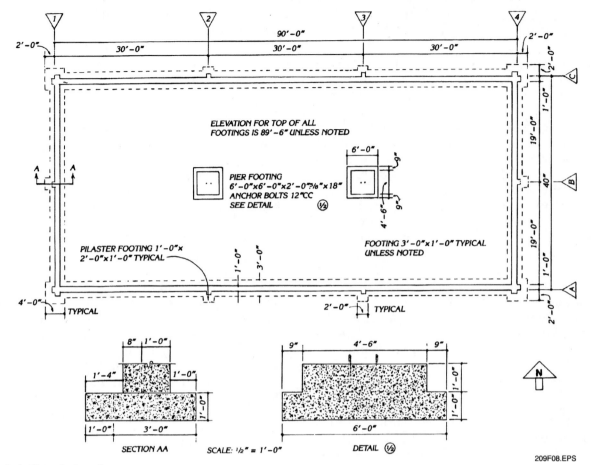

Figure 8 ◆ Foundation plan.

209F08.EPS

buildings, separate floor plans are normally drawn for each floor. *Figure 10* shows an example of a basic floor plan. Most of this information will not be used by heavy equipment operators, but will help you visualize the final stage of construction.

2.3.4 Roof Plans

When supplied, roof plans provide information about the roof slope, roof drain placement, and other pertinent information regarding ornamental sheet metal work, gutters, and downspouts. It is unlikely that you will need to refer to these plans to complete your work as a heavy equipment operator.

2.4.0 Elevation Drawings

Elevation drawings are views that look straight ahead at a structure. The object is projected from a vertical plane. Typically, elevation views are used to show the exterior features of a structure so that the general size and shape of the structure can be determined. Elevation drawings clarify much of the information on the floor plan. For example, a floor plan shows where the doors and windows are located in the outside walls; an elevation view of the same wall shows actual representations of these

doors and windows. *Figure 11* shows an example of a basic elevation drawing. Look for the existing and proposed grade elevations, identified with the arrows on *Figure 11*.

The following types of information are normally shown on elevation drawings:

- Grade lines
- Floor height
- Window and door types
- Roof lines and slope, roofing material, vents, gravel stops, and projection of eaves
- Exterior finish materials and trim
- Exterior dimensions

Unless one or more views are identical, four elevation views are generally used to show each exposure. With very complex buildings, more than four views may be required. Because elevation drawings often contain grade information, you may need to refer to them to complete your job.

2.5.0 Section Drawings

A section view or drawing shows how a particular feature looks inside or how something is put together internally. The feature is drawn as if a cut

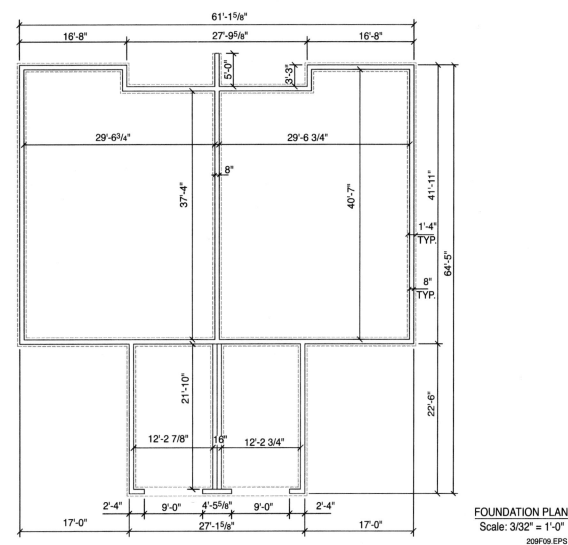

Figure 9 ◆ Slab-on-grade plan.

FOUNDATION PLAN
Scale: 3/32" = 1'-0"
209F09.EPS

has been made through the middle of it at a certain point. The location of the cut and the direction to be viewed are shown on the related plan view.

Section drawings that show a view made by cutting through the length of a structure are referred to as longitudinal sections. Those showing the view of a cut through the width or narrow portion of the structure are referred to as traverse sections. To show greater detail, section views are normally drawn to a larger scale than that used in plan views. The following types of information are normally shown by a section view:

- Details of construction and information about stairs, walls, chimneys, or other parts of construction that may not show clearly on a plan view
- Floor levels in relation to grade
- Wall thickness at various locations
- Anchors and reinforcing steel

Because section drawings often contain grade information, you may need to refer to them. See the line pointing to the existing grade on *Figure 12*.

2.6.0 Detail Drawings

Details are enlargements of special features of a building or of equipment installed in a building. They are drawn to a larger scale in order to make the details clearer. The detail drawings are often placed on the same sheet where the feature appears in the plan. *Figure 13* shows a typical detail view.

Typically, details may be drawn for the following objects or situations:

- Footings and foundations, including anchor bolts, reinforcing, and control joints
- Beams, floor joists, bridging, and other support members
- Sills, floor framing, exterior walls, and vapor barriers
- Floor heights, thickness, expansion, and reinforcing

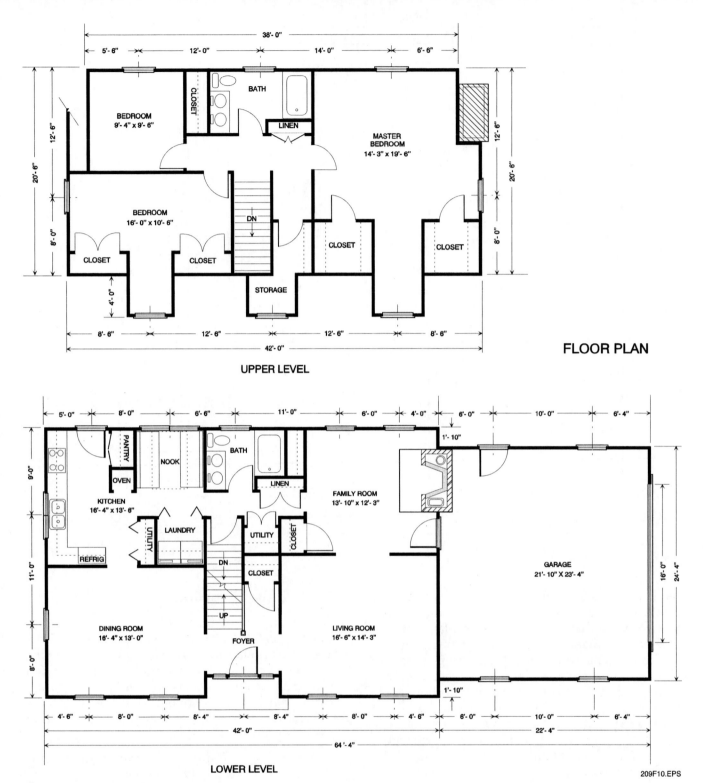

FLOOR PLAN

UPPER LEVEL

LOWER LEVEL

209F10.EPS

Figure 10 ◆ Floor plans for a building.

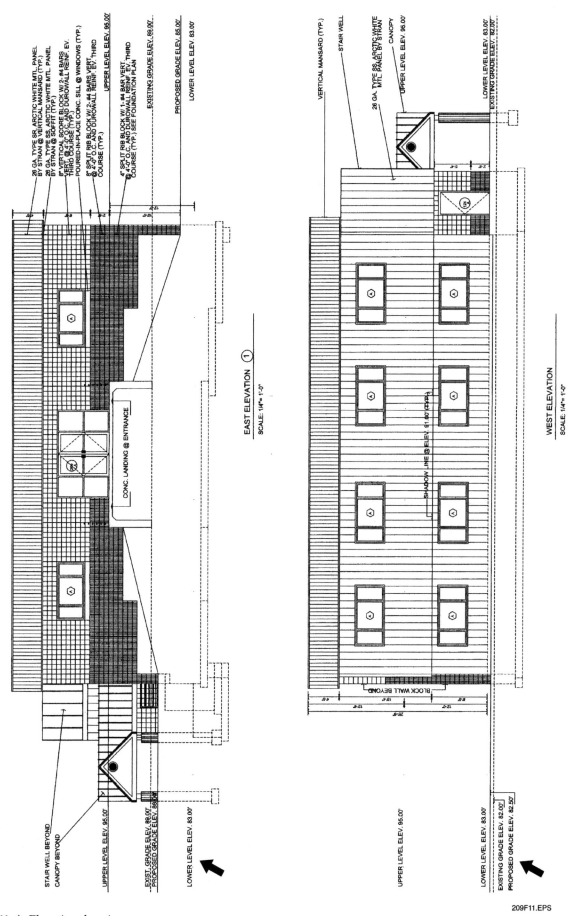

Figure 11 ◆ Elevation drawing.

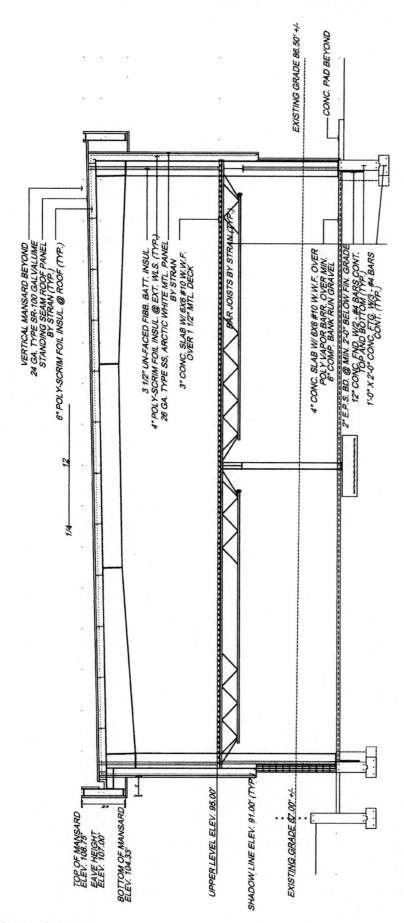

VERTICAL MANSARD BEYOND
24 GA. TYPE SR-100 GAL VALUME
STANDING SEAM ROOF PANEL
BY STRAN (TYP.)

6" POLY-SCRIM FOIL INSUL. @ ROOF (TYP.)

3 1/2" UN-FACED FIBB. BATT. INSUL.
4" POLY-SCRIM FOIL INSUL. @ EXT. WLS. (TYP.)
26 GA. TYPE SS, ARCTIC WHITE MTL. PANEL
BY STRAN

3" CONC. SLAB W/ 6X6 #10 W.W.F.
OVER 1 1/2" MTL. DECK

BAR JOISTS BY STRAN (TYP.)

EXISTING GRADE 86.50' +/-

CONC. PAD BEYOND

4" CONC. SLAB W/ 6X6 #10 W.W.F. OVER
POLY VAPOR BARR. OVER MIN.
6" COMP. BANK RUN GRAVEL

2" E.P.S. BD. @ MIN. 2'-0" BELOW FIN. GRADE
12" CONC. FND. W/2 - #4 BARS CONT.
TOP AND BOTTOM (TYP.)
1'-0" X 2'-0" CONC. FTG. W/3 - #4 BARS
CONT. (TYP.)

12

1/4

TOP OF MANSARD
ELEV. 108.75'

EAVE HEIGHT
ELEV. 107.00'

BOTTOM OF MANSARD
ELEV. 104.33'

UPPER LEVEL ELEV. 95.00'

SHADOW LINE ELEV. 91.00' (TYP.)

EXISTING GRADE 87.00' +/-

BUILDING SECTION
SCALE: 1/4"= 1'-0"

209F12.EPS

Figure 12 ◆ Section drawing.

2.7.0 Schedules

Schedules in a drawing set are tables that describe and specify the various types and sizes of materials used in the construction of a building. Commonly, there are finish schedules (*Figure 14*), as well as door and window schedules.

In the finish schedule for a structure, each room is identified by name or number. The material and finish for each part of the room (walls, floor, ceiling, base, and trim) are designated, along with any clarifying remarks. It is unlikely that you will need this information.

2.8.0 Structural Drawings

Structural drawings are created by a structural engineer and accompany the architect's plans. They are usually drawn for large structures such as office buildings or factories. They show requirements for structural elements of the building, including columns, floor and roof systems, stairs, canopies, and bearing walls.

Structural drawings contain details such as:

- Heights of finished floors and walls
- Height and bearing of bar joist or steel joist
- Locations of bearing steel materials
- Height of steel beams, concrete plank, concrete Ts, and poured-in-place concrete
- Bearing plate locations
- Location, size, and spacing of anchor bolts
- Stairways

2.9.0 Plumbing Plans

Plumbing plans show the layout of fixtures, water supply lines, and lines to sewage disposal systems. The plans may be included in the floor plan of a regular construction job or on a separate plan for a large commercial structure. When drawn as a separate plan, the plumbing plan details are usually overlaid on tracings of the various building floor plans from which unnecessary details have been omitted to allow the location and layout of the plumbing systems to show clearly. Because you may be digging the trench for the water service to enter and the sewage to leave the building, you will need to refer to this diagram. *Figure 15* is a plumbing plan showing the water and sewage plan for a large office building.

2.10.0 Mechanical Plans

Mechanical plans show the heating, ventilation, and air conditioning systems, as well as other mechanical systems for a building. For some residential jobs, the mechanical plan may be combined with the plumbing plan and show very little detailed information other than the locations of the main HVAC system components. It is unlikely that you will need to refer to these plans.

2.11.0 Electrical Plans

For smaller construction jobs, the electrical plans are usually shown on the architectural floor plans.

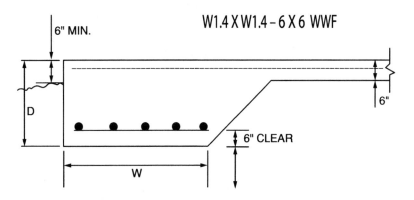

FOOTING DETAILS

FOOTING	W	D	REINFORCING	NOTES
F1	4'-0 x 4'-0 x 24"		6 - #5 e.w.	
F2	2'-0 x 2'-0 x 16"		2 - #5 e.w.	

209F13.EPS

Figure 13 ◆ Detail drawing for a footer.

For large commercial jobs, the electrical plans are typically stand-alone drawings that show only information about the electrical system installation. It is unlikely that you will need any of this information.

2.12.0 Shop Drawings

Shop drawings are specialized drawings that show how to fabricate and install components of a construction project. It is unlikely that you will need any of this type of information.

2.13.0 As-Built Drawings

As-built drawings are drawings that are formally incorporated into the drawing set to record changes. These drawings are marked up on the job by the various trades to show any differences between what was originally shown on a plan by the architect or engineer and what was actually built. Such changes result from the need to relocate equipment to avoid obstructions; to alter the location of a door, window, or wall, for some reason; or because the architect has changed a certain detail in the building design in response to customer preferences. On many jobs, any such changes to the design can only be made after a **change order** has been generated and approved by the architect or other designated person. Depending on the complexity of the change, as-built drawings are typically outlined with a unique design or marked in red ink to make sure they stand out. Changes must be dated and initialed by the responsible party.

Your supervisor or the project engineer will tell you if you need to deviate from the design plans for some reason, but it is part of your job to ensure that the changes get marked on the as-built drawings. One of the most important entries on these plans is the deviation of laying underground utilities. Whenever you work on a job that has an existing structure on it, an as-built drawing set should be used.

2.14.0 Soil Reports

Soil conditions are among the factors that determine the type of foundation best suited for a structure. This information can be vital in determining how

ROOM FINISH SCHEDULE

ROOMS	FLOOR: CARPET	FLOOR: CERAMIC TILE	FLOOR: RUBBER TILE	FLOOR: CONCRETE	CEILING: ACOUSTIC TILE	CEILING: DRYWALL	CEILING: PAINT	WALL: CERAMIC TILE	WALL: DRYWALL	WALL: PAINT	WALL: WALLPAPER	WALL: CERAMIC TILE	BASE: WOOD	BASE: RUBBER	BASE: CERAMIC TILE	TRIM: STAIN	TRIM: WOOD	TRIM: STAIN	TRIM: PAINT	REMARKS
ENTRY		✓			✓				✓	✓	✓				✓		✓	✓	✓	See owner for all painting
HALL	✓				✓				✓	✓					✓		✓	✓	✓	
BEDROOM 1	✓				✓				✓	✓					✓		✓	✓	✓	See owner for grade of carpet
BEDROOM 2	✓				✓				✓	✓					✓		✓	✓	✓	See owner for grade of carpet
BEDROOM 3	✓				✓				✓	✓					✓		✓	✓	✓	See owner for grade of carpet
BATH 1	✓	✓			✓			✓	✓	✓	✓	✓			✓			✓	✓	Wallpaper 3 walls around vanity
BATH 2		✓			✓			✓	✓	✓	✓	✓		✓				✓	✓	Water-seal tile / Wallpaper w/wall
UTIL + CLOSETS	✓		✓			✓	✓		✓	✓					✓	✓	✓	✓	✓	Use off-white flat latex
KITCHEN			✓		✓				✓	✓			✓				✓	✓		
DINING	✓				✓				✓	✓	✓				✓		✓	✓		
LIVING	✓				✓				✓	✓					✓		✓	✓		See owner for grade of carpet
GARAGE				✓		✓	✓			✓			✓				✓	✓		

Figure 14 ◆ Example of a finish schedule.

209F14.EPS

you do your job. Building a structure on soil where the soil conditions can cause a large amount of uneven settlement to occur can result in cracks in the foundation and structural damage to the rest of the building. Therefore, in designing the foundation for a structure, an architect must consider the soil conditions of the building site. Typically, the architect consults a soil engineer who makes test bores of the soil on the building site and analyzes the samples. The results of the soil analysis are summarized in a soil report issued by the engineer. This report is often included as part of the drawing set. When you use this information, consider all aspects of the soil report, including elevation of the water table. The type of soil on the job site will determine the types of equipment that you need to do your job. For example, a backhoe can easily excavate sandy soil, but hard-packed clay may require some other equipment to break it up before it can be excavated with a backhoe.

3.0.0 ◆ READING AND INTERPRETING DRAWINGS

To read and interpret the information on drawings, you need to learn the special language used in construction drawings. This section of the module describes the different types of lines, dimensioning, symbols, and abbreviations used on drawings.

3.1.0 Site/Plot Plans

Site/plot plans show the positions and sizes of all relevant structures on the site, as well as the features of the terrain. It would be difficult to show the amount of information on these drawings without using symbols and abbreviations. *Figure 16* shows some common symbols and *Table 1* shows some common abbreviations. It is important that you take time to become familiar with these symbols so that you can quickly identify them on plans.

Note that the contour line symbols shown on *Figure 16* are used to show changes in the elevation and contour of the land. The lines may be dashed or solid. Generally, dashed lines are used to show the natural or existing grade, and solid lines show the finished grade to be achieved during construction.

Each contour line across the plot of land represents a line of constant elevation relative to some point such as sea level or a local feature. *Figure 17* shows an example of a contour map for a hill. As shown, contour lines are drawn in uniform elevation intervals called contour intervals. Commonly used intervals are 1', 2', 5', and 10'.

On some plans and surveys, every fifth contour line is drawn using a heavier-weight line and is labeled with its elevation to help the user more easily determine the contour. This method of drawing contour lines is called indexing contours. The elevation is marked above the contour line, or the line is interrupted for it.

Table 1 Common Construction Abbreviations

Term	Abbreviation	Term	Abbreviation
Backsight	BS	Iron pipe	IP
Bench mark	BM	Irrigation pipe	irr.P
Boundary	Bndry.	Manhole	MH
Control Point	CP	Marker	mkr.
Curb	Cb.	Monument	mon.
Curb and gutter	G&G	Nail	N
Cut	C	North	N
Ditch	Dit.	Original ground	OG
Edge of gutter	EG	Pavement	pvmt.
Edge of pavement	EP	Pipe	P
Edge of shoulder	ES	Point	pt.
Elevation	el.	Power pole	PP
Fence	fe.	Right-of-way	R/W
Fence post	FP	Shoulder	shldr.
Fill	f	Slope stake	SS
Finish grade	FG	South	S
Footer	FDR	Station	sta.
Foresight	FS	Storm drain	SDr.
Found	fd.	Subgrade	SG
Foundation	fdn	Tack	tk.
Galvanized steel pipe	GSP	Telephone cable	tel. C.
Gas line	GL	Temporary bench mark	TBC
Hub & tack	H&T	Water line	WL

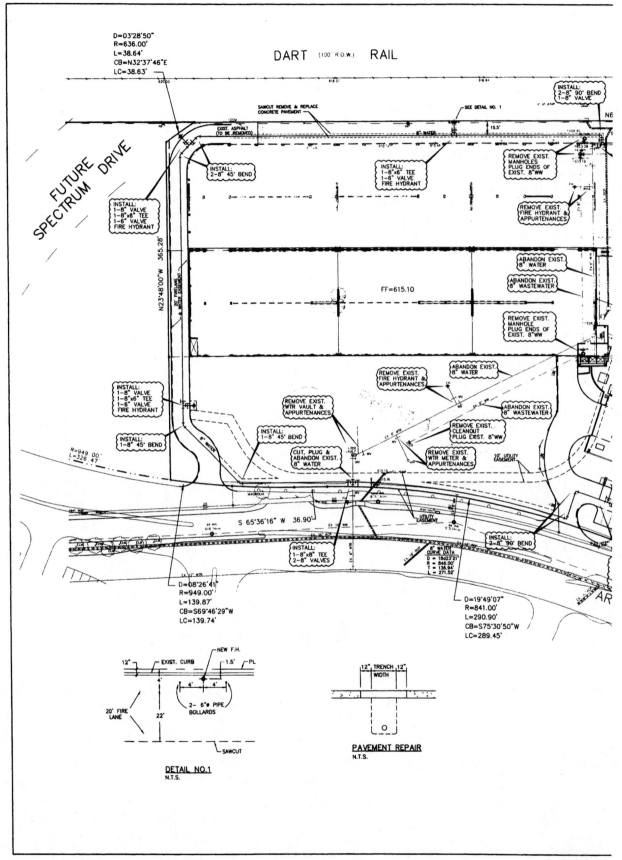

Figure 15 ◆ Plumbing plan. (1 of 2)

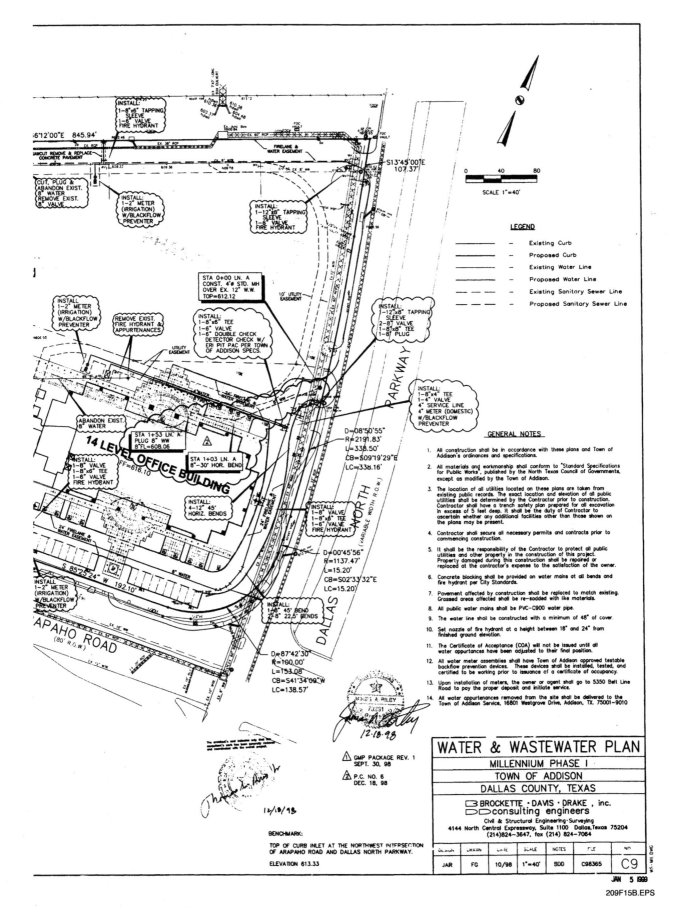

Figure 15 ◆ Plumbing plan. (2 of 2)

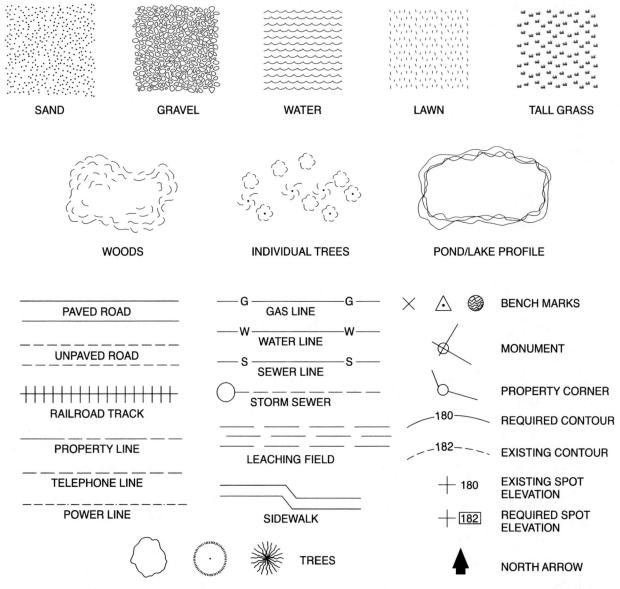

SAND

GRAVEL

WATER

LAWN

TALL GRASS

WOODS

INDIVIDUAL TREES

POND/LAKE PROFILE

PAVED ROAD

UNPAVED ROAD

RAILROAD TRACK

PROPERTY LINE

TELEPHONE LINE

POWER LINE

G —— GAS LINE —— G

W —— WATER LINE —— W

S —— SEWER LINE —— S

STORM SEWER

LEACHING FIELD

SIDEWALK

TREES

BENCH MARKS

MONUMENT

PROPERTY CORNER

—180— REQUIRED CONTOUR

—182--- EXISTING CONTOUR

180 EXISTING SPOT ELEVATION

182 REQUIRED SPOT ELEVATION

NORTH ARROW

209F16.EPS

Figure 16 ◆ Common site/plot plan symbols.

CONTOUR MAP OF HILL WITH 5' CONTOUR INTERVAL

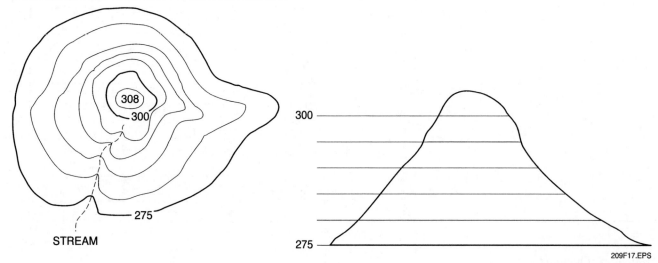

308

300

275

STREAM

300

275

209F17.EPS

Figure 17 ◆ Contour map of a hill.

As shown in *Figure 17*, contour lines can form a closed loop within the map. These lines represent an elevation (hill) or depression. If you start at any point on these lines and follow their path, you will eventually return to the starting point. A contour may close on a site plan or map, or it may be discontinued at any two points at the borders of the plan or map. Examples of this are shown on the site plan shown earlier in this module. Such points mark the ends of the contour on the map, but the contour does not end at these points. The contour is continued on a plan or map of the adjacent land. Some rules for interpreting contours include the following:

- Contour lines do not cross.
- Contour lines crossing a stream point upstream.
- The horizontal distance between contour lines represents the degree of slope. Closely spaced contour lines represent steep ground and widely spaced contour lines represent nearly level ground with a gradual slope. Uniform spacing indicates a uniform slope.
- Contour lines are at right angles to the slope. Therefore, water flow is perpendicular to contour lines.
- Straight contour lines parallel to each other represent man-made features such as terracing.

You need to learn how to read the information on site plans because you will perform most of your work based on this information.

3.2.0 Lines Used on Drawings

Many different types of lines are used to draw and describe a structure. Lines are drawn wide, narrow, dark, light, broken, and unbroken, with each type of line conveying a specific meaning. *Figure 18* shows the most common lines used on construction drawings. The description of each type of line is as follows:

- *Object lines* – Heavier-weight lines used to show the main outline of the structure, including exterior walls, interior partitions, porches, patios, sidewalks, parking lots, and driveways.
- *Dimension and extension lines* – Provide the dimensions of an object. An extension line is drawn out from an object at both ends of the part to be measured to indicate the part being measured. Extension lines are not supposed to touch the object lines. This is so they cannot be confused with the object lines. A dimension line is drawn at right angles between the extension lines and a number placed above, below, or to the side of it to indicate the length of the dimension line. Sometimes a gap is made in the dimension line and the number is written in the gap.

- *Leader line* – Connects a note or dimension to a related part of the drawing. Leader lines are usually curved or at an angle from the feature being distinguished to avoid confusion with dimension and other lines.
- *Center line* – Designates the center of an area or object and provides a reference point for dimensioning. Center lines are typically used to indicate the centers of roadways and the center of objects such as columns, posts, footings, and door openings. On roadways, the center line is a common reference point.
- *Cutting plane (section line)* – Indicates an area that has been cut away and shown in a section view so that the interior features can be seen. The arrows at the ends of the cutting plane indicate the direction in which the section is viewed. Letters identify the cross-sectional view of that specific part of the structure. More elaborate methods of labeling section reference lines are used in larger, more complicated sets of plans (*Figure 19*). The sectional drawing may be on the same page as the reference line or on another page.
- *Break line* – Shows that an object or area has not been drawn in its entirety.
- *Hidden line* – Indicates an outline that is invisible to an observer because it is covered by another surface or object that is closer to the observer.
- *Phantom line* – Indicates alternative positions of moving parts, such as a damper's swing, or adjacent positions of related parts. It may also be used to represent repeated details.
- *Stair indicator line* – A short line with an arrowhead that shows the ascent or descent of stairs on a floor plan.

3.3.0 Symbols Used on Drawings

Symbols are used in architectural plans and drawings to pictorially show different kinds of materials, fixtures, and structural members. Although most of these symbols are not relevant to heavy equipment operators, you must be able to recognize them. The meanings of symbols and the types used are not standardized and can vary from location to location. A set of drawings generally includes a sheet that identifies the specific symbols used and their meanings. When using any drawing set, you should always refer to this sheet of symbols to avoid making mistakes when reading the drawings.

Some symbols will give you a good idea of what the object it represents looks like, but others are used to show the position of the object. Examples of this type of symbol include door and window designators that refer to door and window schedules where the different types are described. Still other symbols are used to show the orientation of the

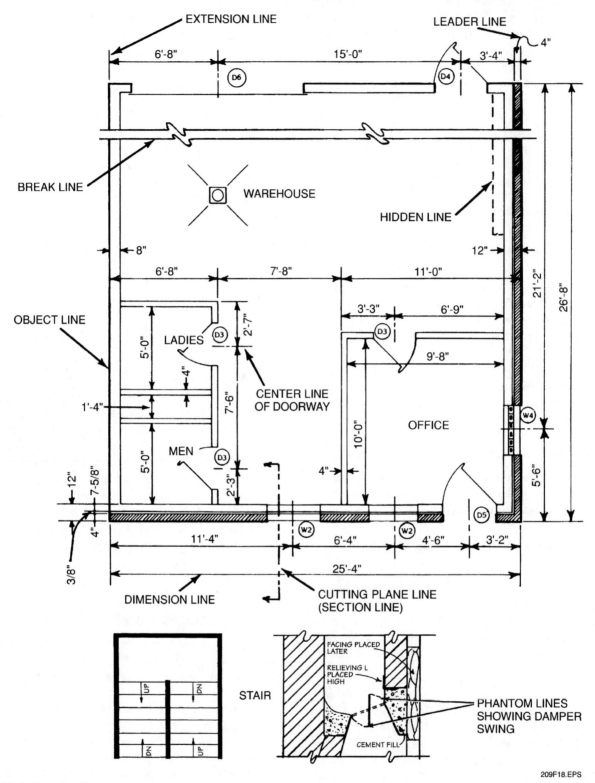

Figure 18 ◆ Drawing lines.

209F18.EPS

object, showing the direction or side (north, south, front, back, and so on).

For electrical, plumbing, and HVAC, there are symbols that show the types of equipment to be installed, such as switches, lavatories, and warm air supply ducts. Each trade has its own symbols and abbreviations.

Symbols used on drawings generally fall into the following categories:

- Material (general plan) symbols (*Figure 20*)
- Window and door symbols (*Figure 21*)
- Electrical symbols (*Figures* 22 and 23)
- Plumbing symbols (*Figure 24*)

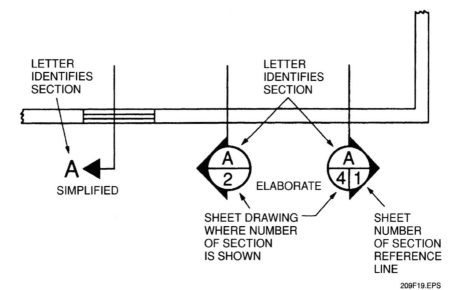

Figure 19 ◆ Methods of labeling section reference lines.

- HVAC symbols (*Figure 25*)
- Structural member symbols (*Figure 26*)
- Welding symbols (*Figure 27*)

3.4.0 Dimensioning

Dimensions given on drawings show actual sizes, distances, and heights of the objects and spaces being represented. Dimensions may be from outside to center, center to center, wall to wall, or outside to outside (*Figure 28*). In all cases, dimensions shown on drawings are given in full scale regardless of the fact that the plan shows an object or distance on a smaller scale.

Note that sectional and detail views may use a nominal size in labeling. Nominal sizes or dimensions are approximate or rough sizes by which lumber or block is commonly known and sold. For example, a nominal size 2 × 4 board is actually 1½" × 3½".

Measurements on diagrams are usually made on architectural drawings using an architect's scale rather than a standard ruler. Architect's scales are divided into feet and inches and usually consist of several scales on one rule. Plan drawings are created using a specified scale. Inches or fractions of an inch on the drawing are used to represent feet in the actual measurement of a building. For example, in a plan drawn to ¼" scale, ¼" on the drawing represents 1' of the building. The scale of a drawing is usually shown directly below the drawing. The same scale may not be used for all the drawings that make up a complete set of plans. Methods for using the architect's rule to scale a drawing were covered earlier in the *Core Curriculum* module *Introduction to Blueprints*.

NOTE

Always use the dimensions written on a drawing. Never attempt to calculate dimensions from a drawing using the scale. Reproduction methods used to make copies of drawings can introduce errors in the reproduced image, thus introducing errors in your calculation. If dimensions are not called out, talk to your supervisor.

Some common practices used for dimensioning drawings are listed here. Keep in mind that these are not rules. The practices in your area may be different.

- Architectural dimension lines are unbroken lines, with the dimensions placed above and near the center of the line.
- Dimensions over one foot are shown in feet and inches (not decimals). Dimensions less than one foot are shown in inches only. The common exception to this rule is the center-to-center distances for standard construction, such as for framing 16" on center (OC) or 24" OC.
- Dimensions are placed to read from the right or from the bottom of the drawing.
- Overall building dimensions go to the outside of all other dimensions.
- Room sizes can be shown by stating width and length.
- Rooms are sometimes dimensioned from the center lines of partition walls, but wall-to-wall dimensions are more common.
- Window and door sizes are usually shown in window and door schedules.

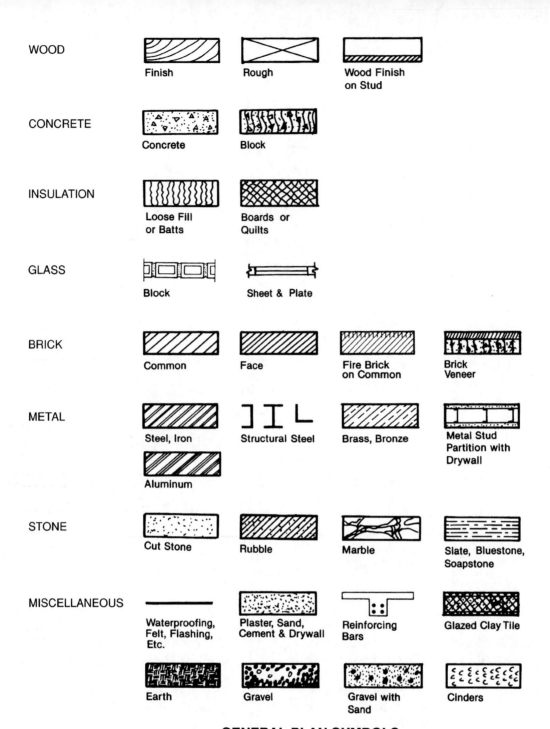

GENERAL PLAN SYMBOLS

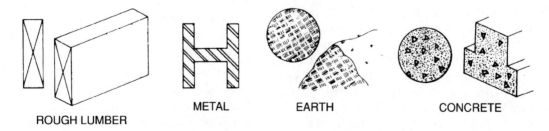

SECTION VIEW SYMBOLS

209F20.EPS

Figure 20 ◆ Material symbols.

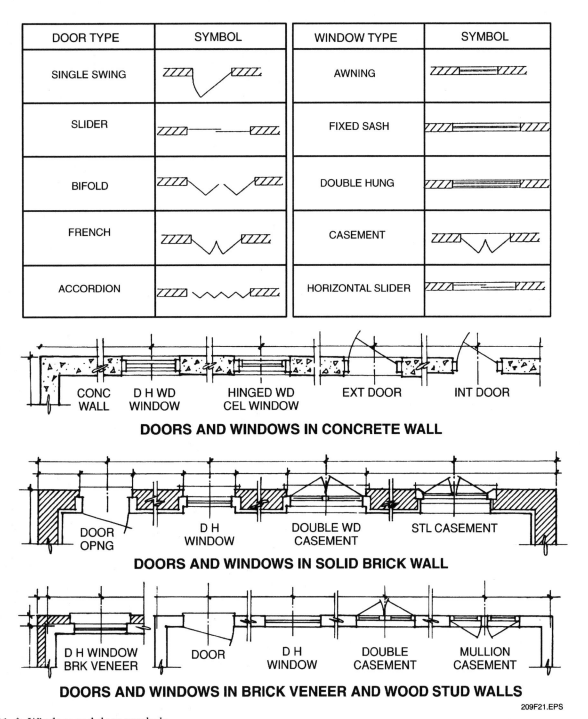

DOOR TYPE	SYMBOL	WINDOW TYPE	SYMBOL
SINGLE SWING		AWNING	
SLIDER		FIXED SASH	
BIFOLD		DOUBLE HUNG	
FRENCH		CASEMENT	
ACCORDION		HORIZONTAL SLIDER	

CONC WALL D H WD WINDOW HINGED WD CEL WINDOW EXT DOOR INT DOOR

DOORS AND WINDOWS IN CONCRETE WALL

DOOR OPNG D H WINDOW DOUBLE WD CASEMENT STL CASEMENT

DOORS AND WINDOWS IN SOLID BRICK WALL

D H WINDOW BRK VENEER DOOR D H WINDOW DOUBLE CASEMENT MULLION CASEMENT

DOORS AND WINDOWS IN BRICK VENEER AND WOOD STUD WALLS

209F21.EPS

Figure 21 ◆ Window and door symbols.

- Dimensions that cannot be shown on the floor plan because of their size are placed at the end of leader lines.
- When stairs are dimensioned, the number of risers is placed on a line with an arrow pointing either up or down.
- Architectural dimensions always refer to the actual size of the building, regardless of the scale of the drawings.

3.5.0 Abbreviations

Many written instructions are needed to complete a set of construction drawings. It is impossible to print out all such references, so a system of abbreviations is used. By using standard abbreviations, such as BRK for brick or CONC for concrete, the architect ensures that the drawings will be accurately interpreted. A brief list of some commonly used abbreviations is contained in the *Appendix*.

GENERAL OUTLETS

Junction Box, Ceiling

Fan, Ceiling

Recessed Incandescent, Wall

Surface Incandescent, Ceiling

Surface or Pendant Single
Fluorescent Fixture

SWITCH OUTLETS

Single-Pole Switch

Double-Pole Switch

Three-Way Switch

Four-Way Switch

Key-Operated Switch

Switch w/Pilot

Low-Voltage Switch

Door Switch

Momentary Contact Switch

Weatherproof Switch

Fused Switch

Circuit Breaker Switch

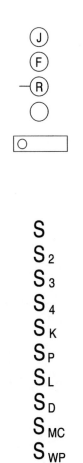

RECEPTACLE OUTLETS

Single Receptacle

Duplex Receptacle

Triplex Receptacle

Split-Wired Duplex Recep.

Single Special Purpose Recep.

Duplex Special Purpose Recep.

Range Receptacle

Switch & Single Receptacle

Grounded Receptacle

Duplex Weatherproof Receptacle

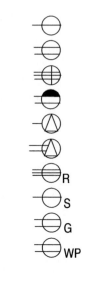

AUXILIARY SYSTEMS

Telephone Jack

Meter

Vacuum Outlet

Electric Door Opener

Chime

Pushbutton (Doorbell)

Bell and Buzzer Combination

Kitchen Ventilating Fan

Lighting Panel

Power Panel

Television Outlet

209F22.EPS

Figure 22 ◆ Electrical symbols.

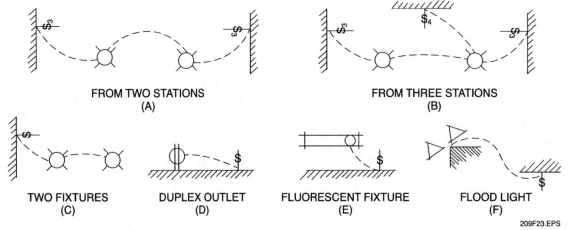

FROM TWO STATIONS
(A)

FROM THREE STATIONS
(B)

TWO FIXTURES
(C)

DUPLEX OUTLET
(D)

FLUORESCENT FIXTURE
(E)

FLOOD LIGHT
(F)

209F23.EPS

Figure 23 ◆ Electrical symbols showing control of an outlet.

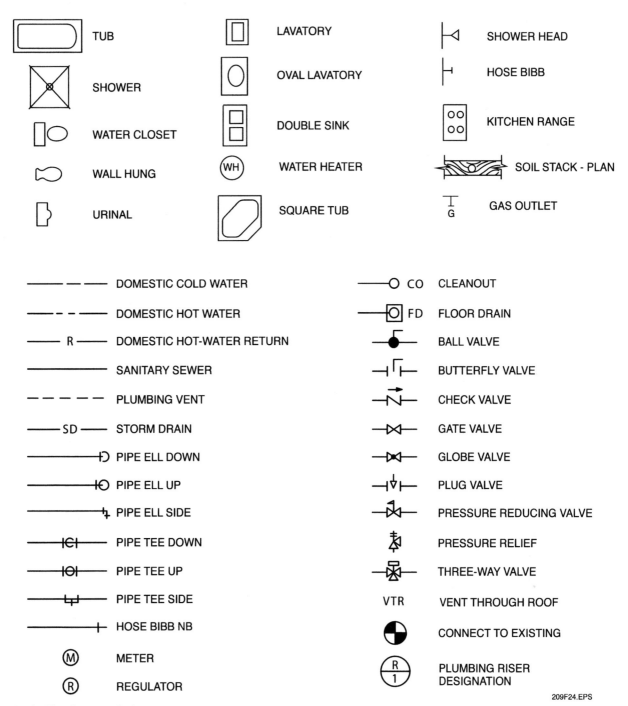

Figure 24 ◆ Plumbing symbols.

209F24.EPS

Note that some architects and engineers may use different abbreviations for the same terms. Normally, the title sheet in a drawing set contains a list of abbreviations used in the drawings. For this reason, it is important to get a complete set of drawings and specifications, including the title sheet(s), so that you can better understand the exact abbreviations used. Some practices for using abbreviations on drawings are as follows:

• Most abbreviations are capitalized.
• Periods are used when abbreviations might look like a whole word.

• Abbreviations are the same whether they are singular or plural.
• Several terms have the same abbreviations and can only be identified from the context in which they are found.
• Many abbreviations are similar.

4.0.0 ◆ SPECIFICATIONS

Specifications, commonly called specs, are written instructions provided by architectural and engineering firms to the general contractor and,

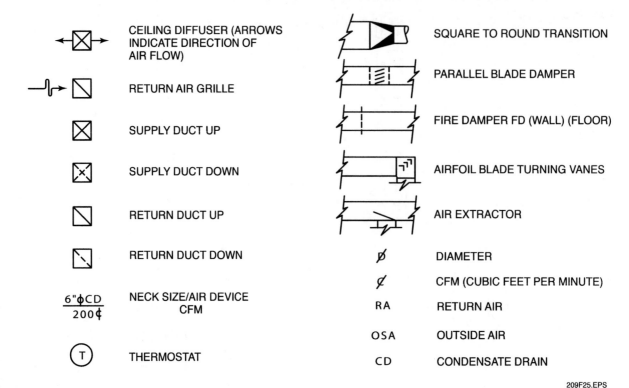

Figure 25 ◆ HVAC legend.

209F25.EPS

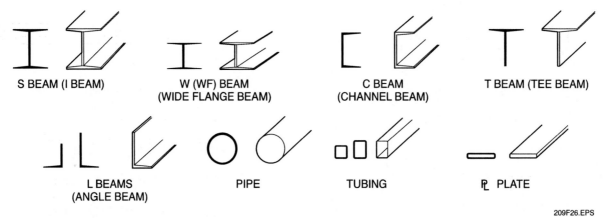

Figure 26 ◆ Structural member symbols.

209F26.EPS

consequently, to the subcontractors. Specifications are just as important as the drawings in a set of plans. They furnish what the drawings cannot in that they define the quality of work to be done and the materials to be used. Specifications serve several important purposes:

- Clarify information that cannot be shown on the drawings
- Identify work standards, types of materials to be used, and the responsibility of various parties to the contract
- Provide information on details of construction
- Serve as a guide for contractors bidding on the construction job
- Serve as a standard of quality for materials and workmanship

- Serve as a guide for compliance with building codes and zoning ordinances
- Provide the basis of agreement between the owner, architect, and contractors in settling any disputes

You will use specifications to determine the quality of fill, the depth of top soil, special finish grading requirements, and landscaping requirements. Specifications are legal documents. When there is a difference between the drawings and the specifications, the specifications normally take legal precedence over the working drawings. However, the plans are often more specific to the job than the specifications. Therefore, notes on the plans may be considered by the architect/owner to be the true intent. You must be very careful to

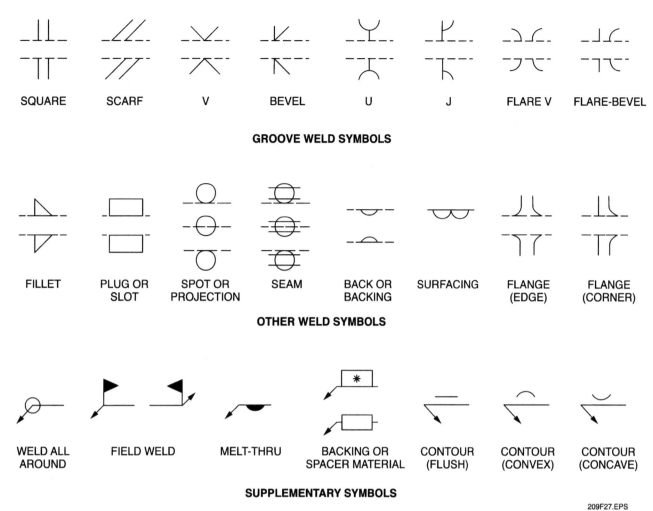

GROOVE WELD SYMBOLS

SQUARE SCARF V BEVEL U J FLARE V FLARE-BEVEL

OTHER WELD SYMBOLS

FILLET PLUG OR SLOT SPOT OR PROJECTION SEAM BACK OR BACKING SURFACING FLANGE (EDGE) FLANGE (CORNER)

SUPPLEMENTARY SYMBOLS

WELD ALL AROUND FIELD WELD MELT-THRU BACKING OR SPACER MATERIAL CONTOUR (FLUSH) CONTOUR (CONVEX) CONTOUR (CONCAVE)

209F27.EPS

Figure 27 ◆ Basic weld symbols.

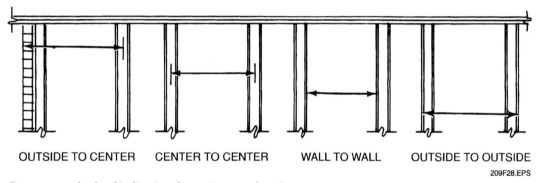

OUTSIDE TO CENTER   CENTER TO CENTER   WALL TO WALL   OUTSIDE TO OUTSIDE

209F28.EPS

Figure 28 ◆ Common methods of indicating dimensions on drawings.

watch for discrepancies between the plans and specifications and report them to your supervisor immediately.

4.1.0 Organization and Types of Specifications

Specifications consist of various elements that may differ somewhat for particular construction jobs. For small projects, they may be simple; for large projects, they may be complex. Basically,

two types of information are contained in a set of specifications: special and general conditions, and technical aspects of construction.

4.1.1 Special and General Conditions

Special and general conditions cover the non-technical aspects of the contractual agreements. Special conditions cover topics such as safety and temporary construction. General conditions cover the following points of information:

- Contract terms
- Responsibilities for examining the construction site
- Types and limits of insurance
- Permits and payments of fees
- Use and installation of utilities
- Supervision of construction
- Other pertinent items

The general conditions section is the area of the construction contract where misunderstandings often occur. Therefore, these conditions are usually much more explicit on large, complicated construction projects. Part of a typical residential material specification is shown in *Figure 29*.

The earthwork sections of this specification are shown as Item 2 (Excavation) and under the headings Other Onsite Improvements and Landscaping, Planting, and Finishing Grading.

NOTE

Residential specifications often do not spell out general conditions and are basically material specifications only.

4.1.2 Technical Aspects

The technical aspects section includes information on materials that are specified by standard numbers and by standard testing organizations such as ASTM International. The technical data section of specifications can be any of three types:

- *Outline specifications* – These specifications list the materials to be used in order of the basic parts of the job, such as foundation, floors, and walls.
- *Fill-in specifications* – This is a standard form filled in with pertinent information. It is typically used on smaller jobs.
- *Complete specifications* – For ease of use, most specifications written for large construction jobs are organized in the Construction Specification Institute format called the Uniform Construction Index. This is known as the CSI format and is explained in the next section.

4.2.0 Format of Specifications

The most commonly used specification-writing format used in North America is the *MasterFormat™*. This standard was developed jointly by the Construction Specifications Institute (CSI) and Construction Specifications Canada (CSC). In this format, the specifications are divided into a series of sections dealing with the construction requirements, products, and activities. Using this format

makes it easy to write and use the specification, and it is easily understandable by the different trades.

For many years prior to 2004, the organization of construction specifications and suppliers catalogs was based on a standard with 16 sections, otherwise known as divisions, where the divisions and their subsections were individually identified by a five-digit numbering system. The first two digits represented the division number and the next three individual numbers represented successively lower levels of breakdown. For example, the number 13213 represents division 13, subsection 2, sub-subsection 1 and sub-sub-subsection 3. In this older version of the standard, electrical systems, including any electronic or special electrical systems, were lumped together under Division 16 – *Electrical*. Today, specifications conforming to the 16 division format may still be in use.

In 2004, the *MasterFormat™* standard underwent a major change. What had been 16 divisions was expanded to four major groupings and 49 divisions with some divisions reserved for future expansion. The first 14 divisions are essentially the same as the old format. Subjects under the old Division 15 – *Mechanical* have been relocated to new divisions 22 and 23. The basic subjects under old Division 16 – *Electrical* have been relocated to new divisions 26 and 27. In addition, the numbering system was changed to 6 digits to allow for more subsections in each division, which allows for finer definition. In the new numbering system, the first two digits represent the division number. The next two digits represent subsections of the division and the two remaining digits represent the third level sub-subsection numbers. The fourth level, if required, is a decimal and number added to the end of the last two digits. This allows tasks to be divided into finer definitions. For example, Division 31 is entitled *Earthwork*, so much of your work will fall under this division. *Figure 30* shows part of Division 31. Use *Figure 30* to look up code 312219.13 and you will find it relates specifically to spreading and grading topsoil.

5.0.0 ◆ GUIDELINES FOR READING A DRAWING SET

The following general procedure is suggested as a method of reading a set of drawings for maximum understanding:

Step 1 Acquire a complete set of drawings and specifications, including the title sheet(s), so that you can better understand the abbreviations and symbols used throughout the drawings.

Form RD 1924-2
(Rev. 7-99)

UNITED STATES DEPARTMENT OF AGRICULTURE
U.S. DEPARTMENT OF HOUSING AND URBAN DEVELOPMENT-FEDERAL
HOUSING ADMINISTRATION
U.S. DEPARTMENT OF VETERANS AFFAIRS

FORM APPROVED
OMB NO. 0575-0042

☐ **Proposed Construction**

DESCRIPTION OF MATERIALS

No. _____
(To be inserted by Agency)

☐ **Under Construction**

Property address _____ City _____ State _Oklahoma_

Mortgagor or Sponsor _____ _____
(Name)   (Address)

Contractor or Builder _____ _____
(Name)   (Address)

INSTRUCTIONS

1. For additional information on how this form is to be submitted, number of copies, etc., see the instructions applicable to the FHA Application for Mortgage Insurance, VA Request for Determination of Reasonable Value or other, as the case may be.

2. Describe all materials and equipment to be used, whether or not shown on the drawings, by marking an X in each appropriate check-box and entering the information called for in each space. If space is inadequate, enter「See misc.」and describe under item 27 or on an attached sheet: THE USE OF PAINT CONTAINING MORE THAN THE PERCENT OF LEAD BY WEIGHT PERMITTED BY LAW IS PROHIBITED.

3. Work not specifically described or shown will not be considered unless

required, then the minimum acceptable will be assumed. Work exceeding minimum requirements cannot be considered unless specifically described.

4. Include no alternates,「or equal」phrases, or contradictory items. (Consideration of a request for acceptance of substitute materials or equipment is not thereby precluded.)

5. Include signatures required at the end of this form.

6. The construction shall be completed in compliance with the related drawings and specifications, as amended during processing. The specifications include this Description of Materials and the applicable building code.

1. **EXCAVATION:**
   Bearing soil, type _Firm clay; Note: Where fill is in excess of 18", concrete piers to be installed_
2. **FOUNDATIONS:** at 8' O.C. and the cost will be added to the contract.
   Footings: concrete mix _transite 14"x18" ftg_ ; strength psi _2500 PSI_ Reinforcing _(4) 5/8" steel rebar_
   Foundation wall: material _2500 PSI concrete_ _concrete_ Reinforcing _____
   Interior foundation wall: material _2500 PSI concrete_ Party foundation wall _2500 concrete_
   Columns: material and sizes _____ Piers: material and reinforcing _____
   Girders: material and sizes _____ Sills: material _W.Coast Utility Douglas Fir w/sill sealer_
   Basement entrance areaway _____ Window areaways _____
   Waterproofing _waterproof mix in concrete_ Footing drains _open mortar joints_
   Termite protection _Pretreat soil_ ___ _Chlordane and issue 5 year warranty._
   Basementless space _____ ; foundation vents _____
   Special _____ _turbed soil._

   ___tion construction: 6"x8" poured monolithic with the ___
   ___loor: 4" concrete slab with 6x6-W1.4xW1.4 WWF smooth trowel finish.

TERRACES:
   stoops: 4" concrete slab- smooth trowel finish.
   patio: 4" concrete slab- smooth trowel finish. - see plans for size.

GARAGES: automatic garage door opener
   foundation: 14"x 18" concrete footing with (4) 5/8" rebars; 6" concrete stem wall
   floor: 4" concrete with 6x6-W1.4xW1.4 WWF; smooth trowel finish floors.
   interior: 3/8" prefinished Sheetrock on walls; texture and paint 1/2" Sheetrock

WALKS AND DRIVEWAYS:
                                           on ceiling
   see
Driveway: width _plans_ ; base material _tamped earth_ ; thickness _4_ ì; surfacing material _concrete_ ; thickness _4_
Front walk: width _36"_ ; material _concrete_ ; thickness _4_ ì. Service walk: width ____ ; material ____ ; thickness ____
Steps: material _____ ; treads ____ ì; risers ____ ì. Check walls _____

OTHER ONSITE IMPROVEMENTS:
(Specify all exterior onsite improvements not described elsewhere, including items such as unusual grading, drainage structures, retaining walls, fence, railings, and accessor structures.)

   NOTE: All dimensions to be rechecked on site prior to beginning construction by
   builder and builder shall be responsible for the same.

LANDSCAPING, PLANTING, AND FINISH GRADING:
Topsoil ____ ì thick: ☐ front yard: ☐ side yards; ☐ rear yard to _____ feet behind main building.
Lawns *(seeded, sodded, or sprigged):*. ☐ front yard ____ ; ☐ side yards ____ ; ☐ rear yard ____
Planting: ☐ as specified and shown on drawings; ☐ as follows:
_____ Shade trees, deciduous. ____ î caliper.   _____ Evergreen trees ____ í to ____ í, B & B.
_____ Low flowering trees, deciduous. ____ í to ____ í   _____ Evergreen shrubs ____ í to ____ í, B & B.
_____ High-growing shrubs, deciduous. ____ í to ____ í   _____ Vines, 2-years _____
_____ Medium-growing shrubs, deciduous, ____ í to ____ í   _____
_____ Low-growing shrubs, deciduous. ____ í to ____ í   _____

IDENTIFICATION. This exhibit shall be identified by the signature of the builder, or sponsor, and/or the propsed mortgagor if the latter is known at the time of application.

Date _____ Signature _____

Signature _____

HUD-FHA 2005
VA Form 26-1852

4

209F29.EPS

Figure 29 ◆ Parts of a typical materials specification.

Step 2 Read the title block. The title block tells you what the drawing is about. Take note of the critical information such as the scale, date of last revision, drawing number, and architect or engineer. After you use a sheet from a set of drawings, be sure to refold the sheet with the title block facing up.

Step 3 Find the north arrow. Always orient yourself to the structure. Knowing where north is enables you to more accurately describe the location of the building and other structures.

Step 4 Always be aware that the drawings work together as a group. The reason the architect or engineer draws plans, elevations, and sections is that drawings require more than one type of view to communicate the whole project. Learn how to use more than one drawing when necessary to find the information you need.

Step 5 Check the list of drawings in your set. Note the sequence of the various plans. Some drawings have an index on the front cover. Notice that the prints in the set are of several categories:
- Architectural
- Structural
- Mechanical
- Electrical
- Plumbing

Step 6 Study the plot/site plan to determine property boundaries and carefully note the location of any bench marks. Further, determine the location of the building to be constructed, as well as the various utilities, roadways, and any easements. Note the various elevations and the existing and proposed contours.

Step 7 Check the floor plan for the orientation of the building. Observe the locations and features of entries, corridors, offsets, and any special features to get an idea of the finished construction.

Step 8 Check the foundation plan for the sizes and types of footings, reinforcing steel, and loadbearing substructures.

Step 9 Check the floor construction and other details relating to excavations.

Step 10 Study the plumbing, mechanical, electrical, and structural plans for features that affect earthwork.

Step 11 Check the notes on various pages, and compare the specifications against the construction details.

Step 12 Thumb through the sheets of drawings until you are familiar with all the plans and details.

Step 13 Recognize applicable symbols and their relative locations in the plans. Note any special excavation details.

Step 14 After you are acquainted with the plans, walk the site so that you can relate the plans to the site.

DIVISION 31 – EARTHWORK

31 00 00 EARTHWORK
31 01 00 Maintenance of Earthwork
31 01 10 Maintenance of Clearing
31 01 20 Maintenance of Earth Moving
31 01 40 Maintenance of Shoring and Underpinning
31 01 50 Maintenance of Excavation Support and Protection
31 01 60 Maintenance of Special Foundations and Load Bearing Elements
31 01 62 Maintenance of Driven Piles
31 01 62.61 Driven Pile Repairs
31 01 63 Maintenance of Bored and Augered Piles
31 01 63.61 Bored and Augered Pile Repairs
31 01 70 Maintenance of Tunneling and Mining
31 01 70.61 Tunnel Leak Repairs
31 05 00 Common Work Results for Earthwork
31 05 13 Soils for Earthwork
31 05 16 Aggregates for Earthwork
31 05 19 Geosynthetics for Earthwork
31 05 19.13 Geotextiles for Earthwork
31 05 19.16 Geomembranes for Earthwork
31 05 19.19 Geogrids for Earthwork
31 05 23 Cement and Concrete for Earthwork
31 06 00 Schedules for Earthwork
31 06 10 Schedules for Clearing
31 06 20 Schedules for Earth Moving
31 06 20.13 Trench Dimension Schedule
31 06 20.16 Backfill Material Schedule
31 06 40 Schedules for Shoring and Underpinning
31 06 50 Schedules for Excavation Support and Protection
31 06 60 Schedules for Special Foundations and Load Bearing Elements
31 06 60.13 Driven Pile Schedule
31 06 60.16 Caisson Schedule
31 06 70 Schedules for Tunneling and Mining
31 08 00 Commissioning of Earthwork
31 09 00 Geotechnical Instrumentation and Monitoring of Earthwork
31 09 13 Geotechnical Instrumentation and Monitoring
31 09 13.13 Groundwater Monitoring During Construction
31 09 16 Special Foundation and Load Bearing Elements Instrumentation and Monitoring
31 09 16.13 Foundation Performance Instrumentation
31 09 16.23 Driven Pile Load Tests
31 09 16.26 Bored and Augered Pile Load Tests

31 10 00 SITE CLEARING
31 11 00 Clearing and Grubbing
31 12 00 Selective Clearing
31 13 00 Selective Tree and Shrub Removal and Trimming
31 13 13 Selective Tree and Shrub Removal
31 13 16 Selective Tree and Shrub Trimming

31 - 1

209F30A.EPS

Figure 30 ◆ CSI Division 31 – *Earthwork* (1 of 2).

31 14 00   Earth Stripping and Stockpiling
    31 14 13   Soil Stripping and Stockpiling
        31 14 13.13   Soil Stripping
        31 14 13.16   Soil Stockpiling
        31 14 13.23   Topsoil Stripping and Stockpiling
    31 14 16   Sod Stripping and Stockpiling
        31 14 16.13   Sod Stripping
        31 14 16.16   Sod Stockpiling

31 20 00   EARTH MOVING
    31 21 00   Off-Gassing Mitigation
        31 21 13   Radon Mitigation
            31 21 13.13   Radon Venting
        31 21 16   Methane Mitigation
            31 21 16.13   Methane Venting
    31 22 00   Grading
        31 22 13   Rough Grading
        31 22 16   Fine Grading
            31 22 16.13   Roadway Subgrade Reshaping
        31 22 19   Finish Grading
            31 22 19.13   Spreading and Grading Topsoil
    31 23 00   Excavation and Fill
        31 23 13   Subgrade Preparation
        31 23 16   Excavation
            31 23 16.13   Trenching
            31 23 16.16   Structural Excavation for Minor Structures
            31 23 16.26   Rock Removal
        31 23 19   Dewatering
        31 23 23   Fill
            31 23 23.13   Backfill
            31 23 23.23   Compaction
            31 23 23.33   Flowable Fill
            31 23 23.43   Geofoam
        31 23 33   Trenching and Backfilling
    31 24 00   Embankments
        31 24 13   Roadway Embankments
        31 24 16   Railway Embankments
    31 25 00   Erosion and Sedimentation Controls
        31 25 13   Erosion Controls
        31 25 23   Rock Barriers
        31 25 53   Sedimentation Controls
        31 25 63   Rock Basins

31 30 00   EARTHWORK METHODS
    31 31 00   Soil Treatment
        31 31 13   Rodent Control
            31 31 13.16   Rodent Control Bait Systems
            31 31 13.19   Rodent Control Traps
            31 31 13.23   Rodent Control Electronic Syste ms
            31 31 13.26   Rodent Control Repellants
        31 31 16   Termite Control

209F30B.EPS

Figure 30 ◆ CSI Division 31 – *Earthwork* (2 of 2).

1. Failure to follow the construction plans and specifications can result in all of the following *except* _____.
   a. legal liability
   b. costly rework
   c. unhappy clients
   d. saving money

2. Elevations, planned grades, station numbers, and bench marks all appear on _____ construction plans.
   a. highway
   b. building
   c. certified
   d. modified

3. On *Figure 1*, the elevation of the finished roadway at station 45 is approximately _____ feet.
   a. 00.00
   b. 992
   c. 994
   d. 996

4. A typical cross-section sheet of highway plans shows the shapes of the side slopes and the ditches.
   a. True
   b. False

5. On a typical roadway project, you will need to check the cross-section sheet _____.
   a. to memorize the grade data since it is the same throughout the project
   b. often because grade details frequently change when the terrain changes
   c. to confirm the accuracy of grade figures with the project engineer
   d. for the most up-to-date cut and fill information for the current station

6. While working on a building project, you are having trouble maneuvering your bulldozer in a tight area near the property boundary, so you _____.
   a. cross the boundary and quickly finish your work, being careful not to damage the terrain
   b. ask the project engineer or your supervisor if you have permission to cross the boundary
   c. stop working in the area. The project engineer knows that the work cannot be completed
   d. cross the boundary and finish your work. You do not need permission to cross boundaries

Figure 1

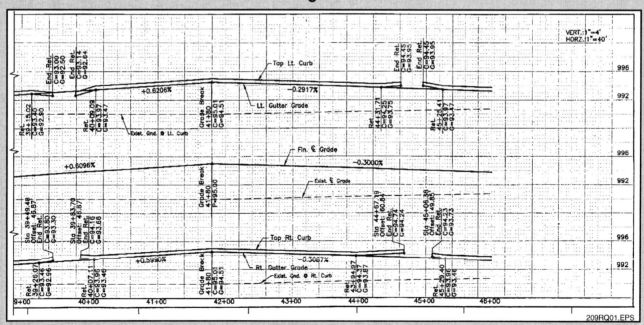

7. You have been assigned to a new job site and need to acquaint yourself with the project. What is the best plan to study to identify the location of existing roads, easements, and utility information, as well as proposed construction?
   a. Structural drawings
   b. Floor plan
   c. Foundation drawings
   d. Site plan

8. Elevation drawings show _____.
   a. a straight ahead view of a building
   b. the natural grade of a project
   c. the proposed grade of a project
   d. the height of the roof line

9. Section drawings show details about how particular features look in a different view, so you can use these drawings to get information about _____.
   a. floor levels in relation to grades
   b. details about footer dimensions
   c. the type of fill needed in an area
   d. property boundary monuments

10. You are working on an addition to a school building and need to find the location of existing underground utilities, so you may need to check the _____.
    a. electrical plans
    b. as-built drawings
    c. plumbing plans
    d. structural plans

11. Contour lines show elevation changes on diagrams. Usually existing contours are shown as _____ lines and proposed contours are shown as _____ lines.
    a. heavy; light
    b. dashed; solid
    c. solid; dashed
    d. light; heavy

12. Contour lines that make closed loops on a plan represent _____.
    a. property boundaries
    b. depressions or hills
    c. building foundations
    d. roadway direction

13. Widely spaced contour lines represent _____ terrain.
    a. steep
    b. level
    c. hilly
    d. rocky

14. You are working on a job and need to know where the property line is located, so you look on the site drawing for the symbol _____.
    a. ++++++++++++++++++
    b. ___ __ __ ___ __ __
    c. __ __ __ __ __ __ __
    d. __ __ __ __ __ __ __

15. You are looking on the site drawings for a bench mark. You know that a benchmark can be one of the following symbols *except* _____.
    a. ✕
    b. △
    c. ✳
    d. ◉

16. You need to find the measurements of the foundation on the plans, so you carefully read the numbers after you have located the foundation _____ lines.
    a. dimension
    b. surface
    c. object
    d. footer plan

17. You are working on a school construction job and need to roughly grade a driveway for the buses. The driveway dimensions are not on the drawing. What do you do?

    a. Use a scale to estimate the driveway dimension based on one that is shown.
    b. Since it is rough grading, use your judgment. It can always be corrected later.
    c. Ask your supervisor to help you to determine what the correct dimensions are.
    d. Grade whatever is convenient. When dimensions are not shown, they do not matter.

18. Specifications are used to expand on the drawing set. When there is a difference between a drawing and a specification, you need to _____.

    a. notify your supervisor
    b. use the drawing set
    c. use the specification
    d. notify the owner

19. When you are new to a job and need to review the plans, you need to look at only the site plan, since all of the grading information can be found on it.

    a. True
    b. False

20. Property boundaries can be found on the _____.

    a. elevation drawing
    b. site plan
    c. specifications
    d. foundation plan

Summary

Construction plans (blueprints) show where a project will be located and how it will be built. All members of the construction team use these drawings, so they are often crowded with information and can be confusing to use, but part of your job is to be able to find the information you need to complete heavy equipment work. Your main concern will be to interpret existing and proposed elevation readings so that you can ensure that the site is prepared for construction according to the specification. Proper leveling and grading are vital to the successful completion and durability of any construction project. Highways and buildings are only as stable as the ground they are built on, so you need to be sure that grading, cut and fill, and compacting tasks are completed as called for on the plans.

Notes

Construction Abbreviations

Common Construction Abbreviations

Above mean sea level	ABMSL	Driveway	drwy.
Abutment	abt.	Drop inlet	DI
Approximate	approx.	Edge of gutter	EG
At	@	Edge of pavement	EP
Avenue	Ave.	Edge of shoulder	ES
Average	avg.	Elevation	el.
Back of sidewalk	BSW	End wall	EW
Back of walk	BW	Equation	eqn.
Backsight	BS	Existing	exist.
Begin curb return	BCR	Expressway	Exwy.
Bench mark	BM	Fahrenheit	F
Between	betw.	Fence	fe.
Bottom	bot.	Fence post	FP
Boulevard	Blvd.	Feet	ft.
Boundary	bndry.	Field book	FB
Bridge	br.	Fill	f
Calculated	calc.	Finished grade	FG
Cast-iron pipe	CIP	Fire hydrant	FH
Catch basin	CB	Flow line	FL
Catch point	CP	Foot	ft.
Cement-treated base	CTB	Footing	ftg.
Concrete block wall	CBW	Foresight	FS
Construction	const.	Found	fd.
Control point	CP	Foundation	fdn.
County	Co.	Freeway	Fwy.
Court	Ct.	Galvanized	galv.
Creek	cr.	Galvanized steel pipe	GSP
Curb	cb.	Gas line	GL
Curb and gutter	C&G	Gas valve	GV
Cut	C	Geodetic	geod.
Description	desc.	Grid	grd.
Destroyed	dest.	Ground	grnd.
Detour	det.	Gutter	gtr.
Direct	D	Head wall	hdwl.
Distance	dist.	Height	ht.
Distance	D	Height of instrument	HI
Distance, horizontal	Dh	Highway	Hwy.
District	Dist.	Hub & tack	H&T
Ditch	dit.	Inch	in.
Drive	Dr.	Inside diameter	ID

Term	Abbr.	Term	Abbr.
Instrument	inst.	Route	Rte.
Intersection	int	Section	S
Iron pipe	IP	Sewer line (sanitary)	SS
Irrigation pipe	irr.P	Shoulder	shldr.
Junction	jct.	Sidewalk	SW
Kilometer	km	Slope stake	SS
Lane	ln.	South	S
Left	lt.	Spike	spk.
Manhole	MH	Stake	stk.
Marker	mkr.	Stand pipe	SP
Maximum	max.	Station	sta.
Measured	meas.	Steel	stl.
Median	med.	Storm drain	SDr.
Mile	mi.	Street	St.
Millimeter	mm	Structure	str.
Minimum	min.	Subdivision	subd.
Minute	min.	Subgrade	SG
Monument	mon.	Tack	tk.
Nail	N	Telephone cable	tel.C.
North	N	Telephone pole	tel.P.
Number	# or no.	Temperature	temp.
Offset	O/S	Temporary bench mark	TBM
Original ground	OG	Top back of curb	TBC
Outside diameter	OD	Top of bank	TB
Overhead	OH	Top of curb	TC
Page	p.	Township	T
Pages	pp.	Tract	tr.
Party chief	PC	Transmission tower	TT
Pavement	pvmt.	Turning point	TP
Perforated metal pipe	PMP	Water line	WL
Pipe	P	Water valve	WV
Place	pl.	Wing wall	WW
Plastic	plas.		
Point	pt.		
Point of intersection	PI		
Portland cement concrete	PCC		
Power pole	PP		
Pressure	press.		
Private	pvt.		
Project control survey	PCS		
Property line	PL		
Punch mark	PM		
Railroad	RR		
Railroad spike	RRspk.		
Read head nail	RH		
Record	rec.		
Reference	ref.		
Reference monument	RM		
Reference point	RP		
Reinforced concrete pipe	RCP		
Retaining wall	ret.W		
Right	rt.		
Right of way	R/W		
River	Riv.		
Road	rd.		
Roadway	rdwy.		
Rock	rk.		

Common Abbreviations Used on Elevations

Term	Abbr.
Aluminum	AL
Asbestos	ASB
Asphalt	ASPH
Basement	BSMT
Beveled	BEV
Brick	BRK
Building	BLDG
Cast iron	CI
Ceiling	CLG
Cement	CEM
Center	CTR
Center line	C or CL
Clear	CLR
Column	COL
Concrete	CONC
Concrete block	CONC B
Copper	COP
Corner	COR
Detail	DET
Diameter	DIA
Dimension	DIM.

Ditto	DO.	Revision	REV
Divided	DIV	Riser	R
Door	DR	Roof	RF
Double-hung window	DHW	Roof drain	RD
Down	DN or D	Roofing	RFG
Downspout	DS	Rough	RGH
Drawing	DWG	Saddle	SDL or S
Drip cap	DC	Scale	SC
Each	EA	Schedule	SCH
East	E	Section	SECT
Elevation	EL	Sheathing	SHTHG
Entrance	ENT	Sheet	SH
Excavate	EXC	Shiplap	SHLP
Exterior	EXT	Siding	SDG
Finish	FIN.	South	S
Flashing	FL	Specifications	SPEC
Floor	FL	Square	SQ
Foot or feet	' or FT	Square inch	SQ. IN.
Foundation	FND	Stainless steel	SST
Full size	FS	Steel	STL
Galvanized	GALV	Stone	STN
Galvanized iron	GI	Terra-cotta	TC
Gauge	GA	Thick or thickness	THK or T
Glass	GL	Typical	TYP
Glass block	GL BL	Vertical	VERT
Grade	GR	Waterproofing	WP
Grade line	GL	West	W
Height	HGT, H, or HT	Width	W or WTH
High point	H PT	Window	WDW
Horizontal	HOR	Wire glass	W GL
Hose bibb	HB	Wood	WD
Inch or inches	" or IN.	Wrought iron	WI
Insulating (insulated)	INS		
Length	LGTH, LG, or L		
Length overall	LOA		
Level	LEV		

Common Abbreviations Used on Plan Views

Light	LT	Access panel	AP
Line	L	Acoustic	ACST
Lining	LN	Acoustical tile	AT
Long	LG	Aggregate	AGGR
Louver	LV	Air conditioning	AIR COND
Low point	LP	Aluminum	AL
Masonry opening	MO	Anchor bolt	AB
Metal	MET. or M	Angle	AN
Molding	MLDG	Apartment	APT
Mullion	MULL	Approximate	APPROX
North	N	Architectural	ARCH
Number	NO. or #	Area	A
Opening	OPNG	Area drain	AD
Outlet	OUT	Asbestos	ASB
Outside diameter	OD	Asbestos board	AB
Overhead	OVHD	Asphalt	ASPH
Panel	PNL	Asphalt tile	AT
Perpendicular	PERP	Basement	BSMT
Plate glass	PL GL	Bathroom	B
Plate height	PL HT	Bathtub	BT
Radius	R	Beam	BM

Bearing plate	BRG PL	Downspout	DS
Bedroom	BR	Drain	D or DR
Blocking	BLKG	Drawing	DWG
Blueprint	BP	Dressed and matched	D & M
Boiler	BLR	Dryer	D
Bookshelves	BK SH	Electric panel	EP
Brass	BRS	End to end	E to E
Brick	BRK	Excavate	EXC
Bronze	BRZ	Expansion joint	EXP JT
Broom closet	BC	Exterior	EXT
Building	BLDG	Finish	FIN.
Building line	BL	Finished floor	FIN. FL
Cabinet	CAB.	Firebrick	FBRK
Caulking	CLKG	Fireplace	FP
Casing	CSG	Fireproof	FPRF
Cast iron	CI	Fixture	FIX.
Cast stone	CS	Flashing	FL
Catch basin	CB	Floor	FL
Cellar	CEL	Floor drain	FD
Cement	CEM	Flooring	FLG
Cement asbestos board	CEM AB	Fluorescent	FLUOR
Cement floor	CEM FL	Flush	FL
Cement mortar	CEM MORT	Footing	FTG
Center	CTR	Foundation	FND
Center to center	C to C	Frame	FR
Center line	C or CL	Full size	FS
Center matched	CM	Furring	FUR
Ceramic	CER	Galvanized iron	GI
Channel	CHAN	Garage	GAR
Cinder block	CIN BL	Gas	G
Circuit breaker	CIR BKR	Glass	GL
Cleanout	CO	Glass block	GL BL
Cleanout door	COD	Grille	G
Clear glass	CL GL	Gypsum	GYP
Closet	C, CL, or CLO	Hardware	HDW
Cold air	CA	Hollow metal door	HMD
Cold water	CW	Hose bibb	HB
Collar beam	COL B	Hot air	HA
Concrete	CONC	Hot water	HW
Concrete block	CONC B	Hot-water heater	HWH
Concrete floor	CONC FL	I beam	I
Conduit	CND	Inside diameter	ID
Construction	CONST	Insulation	INS
Contract	CONT	Interior	INT
Copper	COP	Iron	I
Counter	CTR	Jamb	JB
Cubic feet	CU FT	Kitchen	K
Cutout	CO	Landing	LDG
Detail	DET	Lath	LTH
Diagram	DIAG	Laundry	LAU
Dimension	DIM.	Laundry tray	LT
Dining room	DR	Lavatory	LAV
Dishwasher	DW	Leader	L
Ditto	DO.	Length	L, LG, or LNG
Double acting	DA	Library	LIB
Double-strength glass	DSG	Light	LT
Down	DN	Limestone	LS

Linen closet	L CL	Select	SEL
Lining	LN	Service	SERV
Linoleum	LINO	Sewer	SEW.
Living room	LR	Sheathing	SHTHG
Louver	LV	Sheet	SH
Main	MN	Shelf and rod	SH & RD
Marble	MR	Shelving	SHELV
Masonry opening	MO	Shower	SH
Material	MATL	Sill cock	SC
Maximum	MAX	Single-strength glass	SSG
Medicine cabinet	MC	Sink	SK or S
Minimum	MIN	Soil pipe	SP
Miscellaneous	MISC	Specifications	SPEC
Mixture	MIX	Square feet	SQ FT
Modular	MOD	Stained	STN
Mortar	MOR	Stairs	ST
Molding	MLDG	Stairway	STWY
Nosing	NOS	Standard	STD
Obscure glass	OBSC GL	Steel	ST or STL
On center	OC	Steel sash	SS
Opening	OPNG	Storage	STG
Outlet	OUT	Switch	SW or S
Overall	OA	Telephone	TEL
Overhead	OVHD	Terra cotta	TC
Pantry	PAN	Terrazzo	TER
Partition	PTN	Thermostat	THERMO
Plaster	PL or PLAS	Threshold	TH
Plastered opening	PO	Toilet	T
Plate	PL	Tongue-and-groove	T & G
Plate glass	PL GL	Tread	TR or T
Platform	PLAT	Typical	TYP
Plumbing	PLBG	Unexcavated	UNEXC
Porch	P	Unfinished	UNF
Precast	PRCST	Utility room	URM
Prefabricated	PREFAB	Vent	V
Pull switch	PS	Vent stock	VS
Quarry tile floor	QTF	Vinyl tile	V TILE
Radiator	RAD	Warm air	WA
Random	RDM	Washing machine	WM
Range	R	Water	W
Recessed	REC	Water closet	WC
Refrigerator	REF	Water heater	WH
Register	REG	Waterproof	WP
Reinforce or reinforcing	REINF	Weatherstripping	WS
Revision	REV	Weep hole	WH
Riser	R	White pine	WP
Roof	RF	Wide flange	WF
Roof drain	RD	Wood	WD
Room	RM or R	Wood frame	WF
Rough	RGH	Yellow pine	YP
Rough opening	RGH OPNG		
Rubber tile	R TILE		
Scale	SC		
Schedule	SCH		
Screen	SCR		
Scuttle	S		
Section	SECT.		

Trade Terms Introduced in This Module

Change order: A formal instruction describing and authorizing a project change.

Contour lines: Imaginary lines on a site/plot plan that connect points of the same elevation. Contour lines never cross each other.

Easement: A legal right-of-way provision on another person's property (for example, the right of a neighbor to build a road or a public utility to install water and gas lines on the property). A property owner cannot build on an area where an easement has been identified.

Elevation view: A drawing giving a view from the front or side of a structure.

Front setback: The distance from the property line to the front of the building.

Monuments: Physical structures that mark the locations of survey points.

Plan view: A drawing that represents a view looking down on an object.

Property lines: The recorded legal boundaries of a piece of property.

Request for information (RFI): A form used to question discrepancies on the drawings or to ask for clarification.

Additional Resources

This module is intended to be a thorough resource for task training. The following reference works are suggested for further study. These are optional materials for continued education rather than for task training.

Architectural Graphic Standards, The American Institute of Architects. New York, NY: John Wiley & Sons, Inc.

Reading Architectural Plans for Residential and Commercial Construction, 1998. Ernest R. Weidhaas. Upper Saddle River, NJ: Prentice Hall.

Figure Credits

Kittelson & Associates, 209F03

Sundt Construction, Inc., 209F04, 209RQ01

John Hoerlein, 209F07

Ivey Mechanical Company LLC, 209F12

Ritterbush-Ellig-Hulsing P.C., 209F15

The Numbers and Titles used in this figure , 209F30 are from MasterFormat™ 2004, published by The Construction Specifications Institute (CSI) and Construction Specifications Canada (CSC), and are used with permission from CSI. For those interested in a more in-depth explanation of MasterFormat™ 2004 and its use in the construction industry, visit www.csinet.org/masterformat or contact:

The Construction Specification Institute (CSI)
99 Canal Center Plaza, Suite 300
Alexandria, VA 22314
800-689-2900; 703-684-0300
http://www.csinet.org

CONTREN® LEARNING SERIES — USER UPDATE

The NCCER makes every effort to keep these textbooks up-to-date and free of technical errors. We appreciate your help in this process. If you have an idea for improving this textbook, or if you find an error, a typographical mistake, or an inaccuracy in NCCER's Contren® textbooks, please write us, using this form or a photocopy. Be sure to include the exact module number, page number, a detailed description, and the correction, if applicable. Your input will be brought to the attention of the Technical Review Committee. Thank you for your assistance.

Instructors – If you found that additional materials were necessary in order to teach this module effectively, please let us know so that we may include them in the Equipment/Materials list in the Annotated Instructor's Guide.

Write: Product Development and Revision
          National Center for Construction Education and Research
          P.O. Box 141104, Gainesville, FL 32614-1104

Fax: 352-334-0932

E-mail: curriculum@nccer.org

Craft _____   Module Name _____

Copyright Date _____   Module Number _____   Page Number(s) _____

Description _____

_____

_____

_____

_____

(Optional) Correction _____

_____

_____

(Optional) Your Name and Address _____

_____

_____

_____

Glossary of Trade Terms

Accessories: Attachments used to expand the use of a loader.

Apron: A movable metal plate in front of the scraper bowl that can be raised and lowered to control the flow of material out of the bowl.

Articulated steering: A steering mode where the front and rear wheels may move in opposite directions, allowing for very tight turns, also known as four-wheel steering or circle steering.

Articulating frame: A frame that is hinged in the middle and is able to pivot for better traction and handling.

Articulating: Movement between two parts by means of a joint.

Auxiliary axle: An additional axle that is mounted behind or in front of the truck's drive axles and is used to increase the safe weight capacity of the truck.

Average: The middle point between two numbers or the mean of two or more numbers. It is calculated by adding all numbers together, and then dividing the sum by the quantity of numbers added. For example, the average (or mean) of 3, 7, 11 is 7 ($3 + 7 + 11 = 21$; $21 \div 3 = 7$).

Backhaul: The return trip of a piece of equipment after it has completed dumping its load.

Bail hitch: Used to attach the front end of a tractor to the rear of a scraper.

Balance point: The location on the ground that marks the change from a cut to a fill. On large excavation projects you may have several balance points.

Batter board: A wooden board erected horizontally over trenches and at corners of excavations to provide an attachment and support for a string line.

Bedding material: Select material that is used on the floor of a trench to support the weight of pipe. Bedding material serves as a base for the pipe.

Bedrock: The solid layer of rock under the earth's surface. Solid rock, as distinguished from boulders.

Blade: An attachment on the front end of a loader for scraping and pushing material.

Boots: A special name for laths that are placed by a grade setter to help control the grading operation. The boot can also be the mark on the lath, usually 3, 4, or 5 feet above the finish grade elevation, which can be easily sighted. This allows the grade setter to check the grade alone instead of having to use another person to hold a level rod on the top of the grade stake.

Bowl: The main component at the back of a scraper where the material is loaded and hauled.

Bucket: A U-shaped closed-end scoop that is attached to the front of the loader.

Cab guard: Protects the truck cab from falling rocks and load shift.

Center line: Line that marks the center of a roadway. This is marked on the plans by a line and on the ground by stakes.

Change order: A formal instruction describing and authorizing a project change.

Circle steering: A steering mode where the front and rear wheels may move in opposite directions, allowing for very tight turns. Also known as four-wheel steering or articulated steering.

Clutch: Device used to connect or disconnect two parts of the transmission.

Cohesive: The ability to bond together in a permanent or semi-permanent state. To stick together.

Compaction: Using an engineered process, such as rolling, tamping, or soaking, to reduce the bulk and increase the density of soil.

Consolidation: To become firm by compacting the particles so they will be closer together.

Contour lines: Imaginary lines on a site/plot plan that connect points of the same elevation. Contour lines never cross each other.

Core sample: A sample of earth taken from a test boring.

Crab steering: A steering mode where all wheels may move in the same direction, allowing the machine to move sideways on a diagonal, also known as oblique steering.

Cycle time: The time it takes for a piece of equipment to complete an operation. This normally would include loading, hauling, dumping, and then returning to the starting point.

Debris: Rough broken bits of material such as stone, wood, glass, rubbish, and litter after demolition.

Density: The ratio of the weight of a substance to its volume.

Glossary of Trade Terms

Dewatering: Removing water from an area using a drain or pump.

Discing: The mechanical process of using sharp steel discs to cut through and turn over a top layer of soil, usually 4 to 12 inches deep.

Dozing: Using a blade to scrape or excavate material and move it to another place.

Dragline: An excavating machine having a bucket that is dropped from a boom and dragged toward the machine by a cable.

Easement: A legal right-of-way provision on another person's property (for example, the right of a neighbor to build a road or a public utility to install water and gas lines on the property). A property owner cannot build on an area where an easement has been identified.

Ejector: A large metal plate inside the bowl of a scraper that can be activated to push the material forward, causing it to fall out of the bowl.

Elevation view: A drawing giving a view from the front or side of a structure.

Embankment: Material piled in a uniform manner so as to build up the elevation of an area. Usually, the material is in long narrow strips.

End shoes: Flat pieces of steel on each side of the scraper bowl that keep the material confined to the front of the cutting edge until the paddles can scoop it up. Sometimes referred to as slobber bits.

Expansive soil: A clayey soil that swells with an increase in moisture and shrinks with a decrease in moisture.

Fixed time: Time given to loading, dumping, turning, accelerating, and decelerating.

Foot: In tamping rollers, one of a number of projections from a cylindrical drum that contact the ground.

Four-wheel steering: A steering mode where the front and rear wheels may move in opposite directions, allowing for very tight turns, also known as articulated steering or circle steering.

Front setback: The distance from the property line to the front of the building.

Governor: Device for automatic control of speed, pressure, or temperature.

Gradation: The classification of soils into different particle sizes.

Grade checker: Person who checks grades and gives signals to the equipment operator.

Ground contact pressure (GCP): The weight of the machine divided by the area in square inches of the ground directly supporting it.

Groundwater: Water beneath the surface of the ground.

Grouser: A ridge or cleat across a track that improves the track's grip on the ground.

Grubbing: Digging out roots and other buried material.

Hoist: Hydraulic or mechanical lifting device.

Hydraulic: Powered or moved by liquid under pressure.

Impervious: Not allowing entrance or passage through; for example, soil that will not allow water to pass through it.

Inorganic: Derived from other than living organisms.

Laser: A beam of sharply focused light. Lasers are incorporated into many surveying instruments and replace many functions that levels used to perform. A laser level is commonly known as a laser.

Lift: A layer of material that is a specific thickness; the depth of material that is being rolled by the roller.

Lug down: A slowdown in engine speed (rpm) due to increasing the load beyond capacity. This usually occurs when heavy machinery is crossing soft or unstable soil or is pushing or pulling beyond its capability.

Lug: Effect produced when engine is operating in too high a transmission gear. Engine rotation is jerky, and the engine sounds heavy and labored.

Mat: Asphalt as it comes out of a spreader box or paving machine in a smooth flat form.

Monuments: Physical structures that mark the locations of survey points.

Oblique steering: A steering mode where all wheels may move in the same direction, allowing the machine to move sideways on a diagonal. Also known as crab steering.

Organic: Derived from living organisms such as plants and animals.

Pad: On a segmented or sheepsfoot roller, the part of the roller that contacts the ground; also called the foot.

Glossary of Trade Terms

Parallel: Two lines that are always the same distance apart even if they go on into infinity (forever is called infinity in mathematics).

Parallelogram: A two-dimensional shape that has two sets of parallel lines.

Pay item: A defined piece of material or work that the contractor is paid for. Pay items are usually expressed as unit costs.

Pay material: Deposit of soil valuable enough to stockpile for future use.

Pit: An open excavation that usually does not require vertical shoring or bracing.

Plan view: A drawing that represents a view looking down on an object.

Powered industrial trucks: An OSHA term for several types of light equipment that includes forklifts.

Property lines: The recorded legal boundaries of a piece of property.

psi: Pressure in pounds per square inch.

Puddling: A process in which water is added to the soil until is it semi-liquid; the soil is then allowed to dry before being vibrated.

Quadrilateral: A four-sided closed shape with four angles whose sum is 360 degrees.

Request for information (RFI): A form used to question discrepancies on the drawings or to ask for clarification.

Reservoir: Storage tank.

Ripping: Loosening hard soil, concrete, asphalt, or soft rock with a ripping attachment.

Riprap: Loose pieces of rock that are placed on the slope of an embankment in order to stabilize the soil.

rpm: Revolutions per minute.

Rubble: Fragments of stone, brick, or rock that have broken apart from larger pieces.

Select material: Soil or manufactured material that meets a predetermined specification as to some physical property such as size, shape, or hardness.

Settling: The natural wetting and drying process whereby soil particles become more compact and denser.

Sheepsfoot: A tamping roller with feet expanded at the outer tips.

Shoring: Material used to brace the side of a trench or the vertical face of any excavator.

Soil test: A mechanical or electronic test used to determine the density and moisture of the soil, and therefore the amount of compaction required.

Split load: Load that results when material being dumped does not flow evenly from the dump bed. A split load is more likely to occur with wet sand, clay, or asphaltic materials.

Spoils: Material that has been excavated and stockpiled for future use.

Spot: To line up the haul unit so that it is in the proper position.

Stations: Designated points along a line or a network of points used to survey and lay out construction work. The distance between two stations is normally 100 feet or 100 meters depending on the measurement system used.

Stockpile: Material put into a pile and saved for future use.

Stormwater: Water from rain or snow.

String line: A tough cord or small diameter wire stretched between posts or pins to designate the line and elevation of a grade. String lines take the place of hubs and stakes for some operations.

Stripping: Removal of overburden or thin layers of pay material.

Sump: A small excavation dug below grade for the purpose of draining or retaining subsurface water. The water is then usually pumped out of the sump by mechanical means.

Tag axle: Auxiliary axle that is mounted behind the truck's drive wheels.

Tamping roller: One or more steel drums fitted with projecting feet and towed with a box frame.

Tandem-axle: Usually a double-axle drive unit, but some states call all multiple axle units tandem-axle.

Telehandler: A type of powered industrial truck characterized by a boom with several extendable sections known as a telescoping boom. Another name for a shooting boom forklift.

Test boring: To drill or excavate a hole in the earth in order to take a sample of the material that rests in different layers beneath the surface.

Test pit: See test boring.

Topographic survey: The process of surveying a geographic area to collect data indicating the shape of the terrain and the location of natural and man-made objects.

Trailer: Towed vehicle that rests on its own wheels.

Trench: A temporary long, narrow excavation that will be covered over when work is completed.

Vibratory roller: A compacting device that mechanically vibrates the soils while it rolls. It can be self-propelled or towed.

Water table: The depth below the ground's surface at which the soil is saturated with water.

Winch: A power control unit that is used with wire rope to pull equipment and remove stumps.

Windrow: A long, straight pile of placed material for the purpose of mixing or scraping up.

Index

Index

Concrete-placement attachment, 6.13, 6.14
Conditions, special and general, 9.29–9.30
Confined areas, 1.17–1.18, 5.31, 8.12
Consolidation, 1.23, 1.35
Construction Specifications Canada (CSC), 9.30
Construction Specifications Institute (CSI), 9.30, 9.33–9.34
Contaminants. *See* Pollutants
Contouring procedure, 1.12, 1.20, 1.21
Contour lines, 9.17, 9.20, 9.21, 9.44
Contractor, earthmoving, 1.2, 1.7, 1.9
Contractual requirements, 3.24–3.25, 9.30
Control points. *See* Stakes, control
Controls. *See* Instruments and controls
Cooling system
    dump truck, 2.11
    forklift, 6.5, 6.17, 6.18
    loader, 5.6, 5.16, 5.17
    roller, 3.8, 3.14, 3.17
    scraper, 4.11, 4.16, 4.17, 4.19
Costs
    compaction, 3.5
    delays and rework, 1.7, 9.2, 9.9
    dumping, 1.21
    hauling with a scraper, 4.25
    hourly equipment rates, 1.18
    mass haul strategy, 1.11, 1.18
    rock *vs.* soil excavation, 1.2
Coupler system, 6.27
Cover, roll-out, 2.3
Cross section plan, 1.4, 1.5–1.7, 8.20, 9.4, 9.6
Cryptococcosis, 1.29
CSC. *See* Construction Specifications Canada
CSI. *See* Construction Specifications Institute
Cubes and rectangular objects, 7.15–7.16, 7.19–7.22
Culvert, 1.17–1.18, 1.24, 1.30, 2.21, 3.27
Cut and fill
    balanced job, 1.15, 8.4, 8.7
    calculations, 7.25–7.26
    cross section, 8.9, 8.20
    grading, 4.24, 8.6–8.9
    in mass haul diagram, 1.11
    procedure, 4.21
    in rough terrain, 1.4
    skills, 8.4–8.5
Cutter drum, 5.13
Cycle time, 1.11, 1.14–1.15, 1.21, 1.35, 5.19, 7.2, 8.3
Cylinder, 7.18

D
Dam, 1.16, 1.23, 4.3
Debris, 5.38. *See also* Waste disposal
Demolition work, 1.10, 5.28
Density, 3.2, 3.24, 3.25–3.26, 3.28, 3.32
Designer, 1.9
Dewatering, 1.4, 1.12, 1.35
Differential, 2.7, 5.17
Differential lock, 2.6–2.7, 4.9–4.10, 6.11
Dig Safely card, 1.30, 1.31
Dikes, 4.3
Dimensioning, 9.23, 9.25, 9.29
Dimensions. *See* Scale
Disc harrowing, 1.10, 1.27, 1.35
Ditch, 1.18, 5.12, 7.12, 8.13–8.14, 8.15, 8.20
Doors, symbols, 9.25
DOT. *See* U.S. Department of Transportation
Dozer blades, 5.12, 5.27, 5.32, 6.13
Dragline, 1.2, 1.35

Drainage
    channels, 1.17, 8.14, 8.15–8.16
    excavation requirements, 1.11–1.12, 1.13, 8.14–8.16, 8.20
    french drains, 8.15
    pit, 1.2
    road, 1.4, 8.14–8.15
    setting pipe, 8.16–8.17
    stockpile, 1.12
    sumps, 1.12, 8.15–8.16, 8.20, 8.21
Drawings and plans
    as-built, 1.4, 9.16
    building, 9.5–9.10, 9.11, 9.12
    construction. *See* Blueprint
    cross section, 1.4, 1.5–1.7, 8.20, 9.4, 9.6
    detail, 9.11, 9.15
    electrical, 9.15–9.16
    elevation view, 9.4, 9.5, 9.7, 9.10, 9.13, 9.44
    finish schedule, 9.15, 9.16
    floor, 9.9–9.10, 9.12
    foundation, 9.9, 9.10, 9.11
    grading, 1.3, 8.7, 8.8, 8.9, 8.20
    highway, 9.4–9.5, 9.6
    lines used on, 9.21, 9.22, 9.23
    mass haul diagram, 1.11
    mechanical, 9.15
    plan view and profile sheets, 1.6–1.7, 8.7, 8.8, 8.20, 9.4, 9.5, 9.44
    plumbing, 9.15, 9.18–9.19, 9.27
    reading and interpretation
        abbreviations, 9.17, 9.25, 9.27, 9.39–9.43
        dimensioning, 9.23, 9.25, 9.29
        guidelines, 9.30, 9.32
        lines used on drawings, 9.21, 9.22, 9.23
        site and plot plans, 9.17, 9.20–9.21
        symbols, 9.20, 9.21–9.29
    review and specifications, 1.4–1.7, 3.24–3.25, 9.2, 9.27–9.30, 9.31, 9.33–9.34
    road construction, 1.4, 1.5–1.7
    roof, 9.10
    scale, 9.2, 9.7, 9.23
    section view, 9.10–9.11, 9.14
    shop, 9.16
    site and plot
        for buildings, 8.9, 8.11
        importance of following, 1.30, 8.5, 8.10
        interpretation, 9.17, 9.20–9.21
        overview, 8.11, 9.5, 9.7–9.9
        and soil testing, 1.3–1.4
    soil reports, 9.16–9.17
    structural, 9.15, 9.28
    title sheets, title blocks, and revision blocks, 9.2–9.4, 9.32
    work plan and schedule of operation, 1.9
Drawing set, 9.2–9.17, 9.30, 9.32
Drive, 4.9, 6.2
Drive belt, 2.11, 4.16
Driver's license, 2.1, 2.2, 2.7
Drive wheel, 2.7, 4.2
Driving skills. *See also* Backing up; Turning procedure
    during backhaul, 1.21–1.22
    dump truck, 2.14, 2.17–2.19
    loader, 5.15, 5.22–5.24
    scraper, 4.14, 4.18, 4.20–4.21
Drug use, 2.13, 4.14, 6.15
Dump body, 2.3, 2.6
Dumping and unloading
    illegal, 1.29
    procedure, 1.21–1.22, 2.16–2.17, 5.19, 5.24–5.25, 6.25–6.26

Trencher, 1.18
Trenches
  backfilling, 3.27
  calculation of excavation volumes, 7.25, 7.26
  compaction, 3.24
  definition, 1.18, 1.35
  excavation, 1.17–1.18, 5.31, 8.16–8.17, 8.20
  rock-filled, 8.14
  vs. ditch, 1.18
Triangles, 7.8–7.11
Trimming, fine, 1.27, 4.24, 4.28
Trucks
  articulated, 1.20
  dump. *See* Dump trucks
  effective haul distances, 1.20
  flatbed, unloading procedure, 6.26
  loading procedures, 1.14–1.15, 1.16
  powered industrial, 6.2, 6.39. *See also* Forklifts
  tanker, 1.27
  water, 1.11
Tunnels, 1.18
Turbocharger, 5.22
Turning procedure
  dump truck, 2.16, 2.19, 2.22
  forklift, 6.9, 6.15, 6.21, 6.22
  loader, 5.9, 5.23
  roller, 3.22
  scraper, 4.21

U

Unloading procedure. *See* Dumping and unloading
U.S. Department of Transportation (DOT), 1.31
U.S. Environmental Protection Agency (EPA), 1.11
Utilities
  overhead power lines, 3.19
  underground
    color codes, 1.13, 1.30, 1.31
    locations, 1.9, 1.12–1.13, 1.30
    One-Call system, 1.30, 1.31
    safety, 1.30–1.31, 9.9
    on site plan, 8.11, 9.9
Utility company, 1.12

V

Vandal guards, 3.10, 3.21, 3.22
Vegetation, 1.10, 1.29, 1.30, 4.3
Vibromax VM106 D/PD roller, 3.6
Video camera, 2.5
Visibility while driving
  dump truck, 2.5, 2.11, 2.12, 2.13, 2.19
  forklift, 6.11, 6.17, 6.26
  loader, 5.4, 5.31
  roller, 3.10
  scraper, 4.14
Voltmeter, 4.18
Volume, 7.14–7.18, 7.19–7.26, 7.30, 8.2

W

Walking out (walk-around), 1.28, 3.5, 3.16, 5.15, 6.16
Warmup procedure, 3.14, 3.20, 4.19, 5.20, 5.21, 6.20
Waste disposal, 1.9, 1.10, 1.12, 1.23, 3.14, 5.3, 6.12
Water crossing, 1.31
Water spray unit, 3.12, 3.15, 3.24, 5.12, 6.14
Water supply, 1.27
Water table, 1.12, 1.13, 1.35
Weather considerations
  coolant temperature, 4.19
  drain air brake reservoir, 2.8, 2.11
  engine fluids, 3.14
  equipment warmup times by temperature, 5.21
  fog and laser levels, 8.16
  freeze and thaw cycle, 1.25
  load frozen to the ground, 6.27
  lubrication, 5.16, 6.17
  rain, 1.12. *See also* Ground stability; Runoff
  skidding, 2.18–2.19
  snow. *See* Snow removal attachment
  for a soil stabilization job, 1.27
  tire pressure and temperature, 2.11, 2.12
  wind, 6.15
Weld symbols, 9.29
Wicking, 8.15
Wig-wag, 2.8
Winch, 5.31, 5.38
Windows, symbols, 9.25
Windrow, 1.18, 1.20, 1.21, 1.35, 2.4, 4.24
Windshield and wipers, 2.10, 2.13, 3.18, 4.14, 5.4, 5.13, 6.15
Work site
  access, 1.9, 1.14, 1.15
  cones, tape, and barriers, 6.15
  contamination, 1.29, 1.30
  excavation, 1.12–1.18
  lighting, 6.11
  property adjacent to, 1.30, 8.10, 8.11, 9.4, 9.7, 9.9
  safeguarding property, 1.29–1.30
  safety inspection, 1.28–1.29, 3.19, 5.14, 5.37
  setup of the project, 1.9–1.10
  state bench mark within, 8.5

Y

Y-pattern for loading, 5.26–5.27, 5.37, 6.29